Generalized Locally Toeplitz Sequences: Theory and Applications

Carlo Garoni · Stefano Serra-Capizzano

Generalized Locally Toeplitz Sequences: Theory and Applications

Volume II

Springer

Carlo Garoni
Department of Science
and High Technology
University of Insubria
Como, Italy

Stefano Serra-Capizzano
Department of Science
and High Technology
University of Insubria
Como, Italy

ISBN 978-3-030-02232-7 ISBN 978-3-030-02233-4 (eBook)
https://doi.org/10.1007/978-3-030-02233-4

This book has been realized with the financial support of the Italian INdAM (Istituto Nazionale di Alta Matematica) and the European "Marie-Curie Actions" Programme through the Grant PCOFUND-GA-2012-600198.

Library of Congress Control Number: 2018958367

Mathematics Subject Classification (2010): 15B05, 65N06, 65N25, 65N30, 65N35, 47B06, 35P20, 15A18, 15A60, 15A69

This Springer imprint is published by the registered company Springer Nature Switzerland AG
The registered company address is: Gewerbestrasse 11, 6330 Cham, Switzerland

Preface

Sequences of matrices with increasing size naturally arise in several contexts and especially in the discretization of continuous problems, such as integral and differential equations. The theory of generalized locally Toeplitz (GLT) sequences was developed in order to compute/analyze the asymptotic spectral distribution of these sequences of matrices, which in many cases turn out to be GLT sequences.

In the first volume [22], we presented the theory of *univariate/unilevel* GLT sequences, which arise in the discretization of *unidimensional* integral and differential equations; this is the reason why the first volume addressed only unidimensional applications. In this second volume, we present the theory of *multivariate/multilevel* GLT sequences, which arise in the discretization of *multidimensional* integral and differential equations. The focus here is accordingly on multidimensional applications, especially partial differential equations (PDEs).

It is important to emphasize that the extension from the univariate case addressed in [22] to the multivariate case addressed here, despite being *fundamental for the applications* as it allows one to face concrete PDEs, is essentially a *technical matter* whose purpose is to illustrate the appropriate generalization of ideas already presented in [22]. The fact that all the main "GLT ideas" have been covered in [22] makes it an *essential prerequisite* to this book. In particular, apart from (almost) obvious adaptations, several "multivariate proofs" are the same as their corresponding "univariate versions" from [22]. We have therefore been tempted to omit them here so as to shorten the book, but ultimately we did not opt for this solution in order to help the reader gain familiarity with the multivariate language (especially the multi-index notation).

The book is conceptually divided into two parts. The first part (Chaps. 1–5) covers the theory of multilevel GLT sequences, which is finally summarized in Chap. 6. The second part (Chap. 7) is devoted to PDE applications.

The book is intended for use as a text for graduate or advanced undergraduate courses. It should also be useful as a reference for researchers working in the fields of linear and multilinear algebra, numerical analysis, and matrix analysis. Given its analytic spirit, it could also be of interest to analysts, particularly those working in the fields of measure and operator theory.

As already pointed out, the first volume [22] is an essential prerequisite to this second volume. It also provides detailed motivations to the theory of GLT sequences [22, pp. 1–3] which will not be repeated here for the sake of conciseness. In addition to [22], a basic knowledge of multidimensional integro-differential calculus (partial derivatives, multiple integrals, etc.) is required.

Assuming the reader possesses the necessary prerequisites, most of which, if not already addressed in [22], will be tackled in Chap. 2, there exists a way of reading this book that allows one to omit essentially all the mathematical details/ technicalities without losing the core. This is probably "the best way of reading" for those who love practice more than theory, but it is also advisable for theorists, who can recover the missing details afterward. It consists in reading carefully the summary of the theory in Chap. 6 and the applications in Chap. 7.

To conclude, we wish to express our gratitude to Bruno Iannazzo, Carla Manni, and Hendrik Speleers, who awakened the interest in the theory of GLT sequences and ultimately inspired the writing of this book. We also wish to thank all of our colleagues who have worked in the field of "Toeplitz matrices and spectral distributions" and contributed to laying the foundations of the theory of GLT sequences. We mention in particular Bernhard Beckermann, Albrecht Böttcher, Fabio Di Benedetto, Marco Donatelli, Leonid Golinskii, Sergei Grudsky, Arno Kuijlaars, Maya Neytcheva, Debora Sesana, Bernd Silbermann, Paolo Tilli, Eugene Tyrtyshnikov, and Nickolai Zamarashkin. Finally, special thanks go to those researchers who, possibly attracted by the first volume [22], decided to enter the research field of GLT sequences. We mention in particular Giovanni Barbarino from Scuola Normale Superiore (Pisa, Italy), Davide Bianchi and Isabella Furci from University of Insubria (Como, Italy), Ali Dorostkar and Sven-Erik Ekström from Uppsala University (Uppsala, Sweden), Mariarosa Mazza and Ahmed Ratnani from the Max Planck Institute for Plasma Physics (Munich, Germany). Several of their contributions will certainly appear in a future edition of both volumes I and II.

Based on their research experience, the authors propose a reference textbook in two volumes on the theory of generalized locally Toeplitz sequences and their applications. The first volume focuses on the univariate version of the theory and the related applications in the unidimensional setting, while this second volume, which addresses the multivariate case, is mainly devoted to concrete PDE applications.

Como, Italy
August 2018

Carlo Garoni
Stefano Serra-Capizzano

Contents

About the Authors

Dr. Carlo Garoni graduated in Mathematics at the University of Insubria (Italy) in 2011 and received his Ph.D. in Mathematics at the same university in 2015. He has pursued research at the Universities of Insubria and Rome "Tor Vergata," and he now has a Marie-Curie postdoctoral position at the USI University of Lugano (Switzerland). He has published about 25 research papers in different areas of Mathematics, most of which are connected with the theory of GLT sequences and its applications.

Prof. Stefano Serra-Capizzano is a Full Professor of Numerical Analysis, Deputy Rector of the University of Insubria (Italy), and a long-term Visiting Professor at Uppsala University (Sweden). He has authored over 200 research papers in different areas of Mathematics, with more than 90 collaborators all over the world, and he has recently won a Prodi Chair Professorship in Nonlinear Analysis at Würzburg University (Germany). He is the founder of the Ph.D. program "Mathematics of Computation" at the University of Insubria's Department of Science and High Technology.

Chapter 1
Notes to the Reader

The present book covers the multivariate version of the theory of Generalized Locally Toeplitz (GLT) sequences, also known as the theory of multilevel GLT sequences. In addition, the book presents some emblematic (multidimensional) applications of this theory in the context of the numerical discretization of Partial Differential Equations (PDEs).

The generalization of the theory of GLT sequences from the univariate case addressed in [22] to the multivariate case addressed here is essentially a matter of technicalities, which results in the technical nature of the present volume. *We therefore recommend that, before going into this book, the reader give a reading to* [22, pp. 1–3] *in order to call to mind the motivations behind the theory of (unilevel and multilevel) GLT sequences,* which will not be repeated here for the sake of conciseness. When reading [22, pp. 1–3] in a multidimensional perspective, the GLT sequences and the Differential Equations (DEs) mentioned therein should be understood as multilevel GLT sequences and PDEs, respectively.

After going through [22, pp. 1–3], we encourage the reader to try reading this book according to the scheme suggested in the preface, which consists in reading Chaps. 6 and 7 first, and then coming back to fill the gaps (if necessary or wanted).

When reading the present book, it is advisable that the reader have at hand the first volume [22], for at least two reasons. First, [22] is cited many times throughout the book. Secondly, several "multivariate proofs" from Chaps. 2–5 are essentially the same as their corresponding "univariate versions" from [22], and we recommend that the reader compare them with each other so as to learn the way in which the multilevel language (especially, the multi-index notation) allows one to transfer many results from the univariate to the multivariate case. Roughly speaking, this transfer process is carried out through a sort of "automatic procedure" consisting in turning some letters (n, i, j, x, θ, etc.) in boldface ($\boldsymbol{n}, \boldsymbol{i}, \boldsymbol{j}, \mathbf{x}, \boldsymbol{\theta}$, etc.). Finally, we remark that, as highlighted in the preface, the first volume [22] is an *essential prerequisite* to this second volume. In addition to [22], the other necessary prerequisite for reading this book is a basic knowledge of multidimensional integro-differential calculus (partial derivatives, multiple integrals, etc.).

C. Garoni and S. Serra-Capizzano, *Generalized Locally Toeplitz Sequences: Theory and Applications*, https://doi.org/10.1007/978-3-030-02233-4_1

Chapter 2
Mathematical Background

This chapter collects the necessary preliminaries to develop the multivariate version of the theory of GLT sequences. The reader is supposed to be familiar with the univariate version of the theory [22] and to possess a basic knowledge of multidimensional integro-differential calculus (partial derivatives, multiple integrals, etc.).

2.1 Notation and Terminology

For the reader's convenience, we report in this section some of the most common notations and terminologies that will be used throughout the book. Special attention is devoted to the multi-index notation and the notion of multilevel matrix-sequences. Together with the index at the end, this section can be used as a reference whenever an unknown notation/terminology is encountered.

2.1.1 General Notation and Terminology

- The cardinality of a set S is denoted by $\#S$.
- If S is a subset of a topological space, the closure of S is denoted by $\overline{S}$.
- A permutation σ of the set $\{1, 2, \dots, n\}$ is denoted by $[\sigma(1), \sigma(2), \dots, \sigma(n)]$.
- $\mathbb{R}^{m\times n}$ (resp., $\mathbb{C}^{m\times n}$) is the space of real (resp., complex) $m \times n$ matrices.
- O_m and I_m denote, respectively, the $m \times m$ zero matrix and the $m \times m$ identity matrix. Sometimes, when the size m can be inferred from the context, O and I are used instead of O_m and I_m. The symbol O is also used to indicate rectangular zero matrices whose sizes are clear from the context.
- If $\mathbf{x}$ is a vector and X is a matrix, $\mathbf{x}^T$ and $\mathbf{x}^*$ (resp., X^T and X^*) are the transpose and the conjugate transpose of $\mathbf{x}$ (resp., X).
- If $\mathbf{x}$ is a vector with m components $x_1, \dots, x_m$, $\mathrm{diag}(\mathbf{x})$ is the diagonal matrix whose diagonal entries are $x_1, \dots, x_m$.

C. Garoni and S. Serra-Capizzano, *Generalized Locally Toeplitz Sequences: Theory and Applications*, https://doi.org/10.1007/978-3-030-02233-4_2

- If $\mathbf{x}$, $\mathbf{y}$ are vectors with m components, $\mathbf{x} \cdot \mathbf{y}$ denotes their scalar product.
- We use the abbreviations HPD, HPSD, SPD, SPSD for "Hermitian Positive Definite", "Hermitian Positive SemiDefinite", "Symmetric Positive Definite", "Symmetric Positive SemiDefinite".
- If $X, Y \in \mathbb{C}^{m\times m}$, the notation $X \geq Y$ (resp., $X > Y$) means that X, Y are Hermitian and $X - Y$ is HPSD (resp., HPD).
- If $X, Y \in \mathbb{C}^{m\times m}$, we denote by $X \circ Y$ the componentwise (or Hadamard) product of X and Y: $(X \circ Y)_{ij} = x_{ij} y_{ij},\ i, j = 1, \ldots, m.$
- If $X \in \mathbb{C}^{m\times m}$, we denote by $X^\dagger$ the Moore–Penrose pseudoinverse of X. For more on the Moore–Penrose pseudoinverse, see [22, Sect. 2.4.2].
- If $X \in \mathbb{C}^{m\times m}$, we denote by $\Lambda(X)$ the spectrum of X.
- If $X \in \mathbb{C}^{m\times m}$, the singular values and eigenvalues of X are denoted by $\sigma_j(X)$, $j = 1, \ldots, m$, and $\lambda_j(X)$, $j = 1, \ldots, m$, respectively. The maximum and minimum singular values are also denoted by $\sigma_{\max}(X)$ and $\sigma_{\min}(X)$. If the eigenvalues are real, their maximum and minimum are also denoted by $\lambda_{\max}(X)$ and $\lambda_{\min}(X)$.
- If $1 \leq p \leq \infty$, the symbol $|\cdot|_p$ denotes both the p-norm of vectors and the associated operator norm for matrices:

$$|\mathbf{x}|_p = \begin{cases} \left(\sum_{i=1}^m |x_i|^p\right)^{1/p}, & \text{if } 1 \leq p < \infty, \\ \max_{i=1,\ldots,m} |x_i|, & \text{if } p = \infty, \end{cases} \qquad \mathbf{x} \in \mathbb{C}^m,$$

$$|X|_p = \max_{\substack{\mathbf{x}\in\mathbb{C}^m \\ \mathbf{x}\neq\mathbf{0}}} \frac{|X\mathbf{x}|_p}{|\mathbf{x}|_p}, \qquad X \in \mathbb{C}^{m\times m}.$$

  The 2-norm $|\cdot|_2$ is also known as the spectral (or Euclidean) norm and it will be preferably denoted by $\|\cdot\|$. For more on p-norms, see [22, Sect. 2.4.1].
- Given $X \in \mathbb{C}^{m\times m}$ and $1 \leq p \leq \infty$, $\|X\|_p$ denotes the Schatten p-norm of X, which is defined as the p-norm of the vector $(\sigma_1(X), \ldots, \sigma_m(X))$ formed by the singular values of X. The Schatten 1-norm is also known under the names of trace-norm and nuclear norm. For more on Schatten p-norms, see [22, Sect. 2.4.3].
- $\Re(X)$ and $\Im(X)$ are, respectively, the real and the imaginary part of the square matrix X:

$$\Re(X) = \frac{X + X^*}{2}, \qquad \Im(X) = \frac{X - X^*}{2\mathrm{i}},$$

  where i is the imaginary unit ($\mathrm{i}^2 = -1$). Note that $\Re(X)$, $\Im(X)$ are Hermitian and $X = \Re(X) + \mathrm{i}\,\Im(X)$ for all square matrices X.
- If $X \in \mathbb{C}^{m\times m}$ is diagonalizable and $f : \Lambda(X) \to \mathbb{C}$, we denote by $f(X)$ the matrix obtained by applying the function f to X. For more on matrix functions, see [22, Sect. 2.4.6].
- We use the abbreviations FDs, FEs, IgA for "Finite Differences", "Finite Elements", "Isogeometric Analysis".
- Given two sequences $\{\zeta_n\}_n$ and $\{\xi_n\}_n$, with $\zeta_n \geq 0$ and $\xi_n > 0$ for all n, the notation $\zeta_n = O(\xi_n)$ means that there exists a constant C, independent of n, such that $\zeta_n \leq C\xi_n$ for all n; and the notation $\zeta_n = o(\xi_n)$ means that $\zeta_n/\xi_n \to 0$ as $n \to \infty$.

- $C_c(\mathbb{C})$ (resp.,$C_c(\mathbb{R})$) is the space of complex-valued continuous functions defined on $\mathbb{C}$ (resp., $\mathbb{R}$) with bounded support. Moreover, for $m \in \mathbb{N} \cup \{\infty\}$, $C_c^m(\mathbb{R}) = C_c(\mathbb{R}) \cap C^m(\mathbb{R})$, where $C^m(\mathbb{R})$ is the space of functions $F : \mathbb{R} \to \mathbb{C}$ such that the real and imaginary parts $\Re(F)$, $\Im(F)$ are of class C^m over $\mathbb{R}$ in the classical sense.
- If $w_i : D_i \to \mathbb{C}$, $i = 1, \ldots, d$, we define the tensor-product function $w_1 \otimes \cdots \otimes w_d : D_1 \times \cdots \times D_d \to \mathbb{C}$ as follows: for every $(\xi_1, \ldots, \xi_d) \in D_1 \times \cdots \times D_d$,

$$(w_1 \otimes \cdots \otimes w_d)(\xi_1, \ldots, \xi_d) = w_1(\xi_1) \cdots w_d(\xi_d).$$

- If $f : D \to E$ and $g : E \to F$ are arbitrary functions, the composite function $g \circ f$ is preferably denoted by $g(f)$.
- If $g : D \to \mathbb{C}$, we set $\|g\|_\infty = \sup_{\xi \in D} |g(\xi)|$. If we need/want to specify the domain D, we write $\|g\|_{\infty,D}$ instead of $\|g\|_\infty$. Clearly, $\|g\|_\infty < \infty$ if and only if g is bounded over its domain.
- If $g : D \to \mathbb{C}$ is continuous over D, with $D \subseteq \mathbb{C}^k$ for some k, we denote by $\omega_g(\cdot)$ the modulus of continuity of g,

$$\omega_g(\delta) = \sup_{\substack{\mathbf{x},\mathbf{y} \in D \\ |\mathbf{x}-\mathbf{y}|_\infty \le \delta}} |g(\mathbf{x}) - g(\mathbf{y})|, \qquad \delta > 0.$$

If we need/want to specify D, we will say that $\omega_g(\cdot)$ is the modulus of continuity of g over D.
- χ_E is the characteristic (or indicator) function of the set E,

$$\chi_E(\xi) = \begin{cases} 1, & \text{if } \xi \in E, \\ 0, & \text{otherwise.} \end{cases}$$

- μ_k denotes the Lebesgue measure in $\mathbb{R}^k$. Throughout this book, unless otherwise stated, all the terminology coming from measure theory (such as "measurable set", "measurable function", "almost everywhere (a.e.)", etc.) is always referred to the Lebesgue measure.
- If $E_1, \ldots, E_d \subseteq \mathbb{R}$ are measurable sets and $f : E_1 \times \cdots \times E_d \to \mathbb{C}$, we say that f is d-separable if there exist d measurable functions $f_i : E_i \to \mathbb{C}$, $i = 1, \ldots, d$, such that $f = f_1 \otimes \cdots \otimes f_d$. In this case, $f_1 \otimes \cdots \otimes f_d$ is called a factorization of f. Note that any d-separable function is measurable. Throughout this book, "separable function" is an abbreviation of "d-separable function".
- If $f : D \subseteq \mathbb{R}^k \to \mathbb{C}$ is measurable, we denote by $\mathcal{ER}(f)$ its essential range. For more on the essential range, see [22, Sect. 2.2.1].
- If D is any measurable subset of some $\mathbb{R}^k$, we set

$$\mathfrak{M}_D = \{f : D \to \mathbb{C} : \ f \text{ is measurable}\},$$

$$L^p(D) = \left\{f \in \mathfrak{M}_D : \int_D |f|^p < \infty\right\}, \qquad 1 \le p < \infty,$$

$$L^\infty(D) = \left\{f \in \mathfrak{M}_D : \ \operatorname{ess\,sup}_D |f| < \infty\right\}.$$

If D is the special domain $[0,1]^d \times [-\pi,\pi]^d$, we preferably use the notation $\mathfrak{M}_d$ instead of $\mathfrak{M}_D$:

$$\mathfrak{M}_d = \{\kappa : [0,1]^d \times [-\pi,\pi]^d \to \mathbb{C} : \ \kappa \text{ is measurable}\}.$$

If $f \in L^p(D)$ and the domain D is clear from the context, we write $\|f\|_{L^p}$ instead of $\|f\|_{L^p(D)}$ to indicate the L^p-norm of f, which is defined as

$$\|f\|_{L^p} = \begin{cases} (\int_D |f|^p)^{1/p}, & \text{if } 1 \le p < \infty, \\ \operatorname{ess\,sup}_D |f|, & \text{if } p = \infty. \end{cases}$$

For more on L^p spaces, see [22, Sect. 2.2.2].
- If $D \subseteq \mathbb{R}^k$ is a measurable set with $0 < \mu_k(D) < \infty$, we denote by d_{measure} the pseudometric on $\mathfrak{M}_D$ defined in [22, Eq. (2.14)], which induces on $\mathfrak{M}_D$ the topology τ_{measure} of convergence in measure. For more details on this topic, see [22, Sect. 2.3.2].
- If $f \in L^1([-\pi,\pi]^d)$, the Fourier coefficients of f are denoted by $f_{\mathbf{k}}$ and are defined as follows:

$$f_{\mathbf{k}} = \frac{1}{(2\pi)^d} \int_{[-\pi,\pi]^d} f(\boldsymbol{\theta})\, \mathrm{e}^{-\mathrm{i}\mathbf{k}\cdot\boldsymbol{\theta}} \mathrm{d}\boldsymbol{\theta}, \qquad \mathbf{k} \in \mathbb{Z}^d. \tag{2.1}$$

- We use a notation borrowed from probability theory to indicate sets. For example, if $f, g : D \subseteq \mathbb{R}^k \to \mathbb{C}$, then

$$\{f \ne 1\} = \{\mathbf{x} \in D : \ f(\mathbf{x}) \ne 1\},$$
$$\{0 \le f \le 1, \ g > 2\} = \{\mathbf{x} \in D : \ 0 \le f(\mathbf{x}) \le 1, \ g(\mathbf{x}) > 2\},$$
$\mu_k\{f > 0, \ g < 0\}$ is the measure of the set $\{\mathbf{x} \in D : \ f(\mathbf{x}) > 0, \ g(\mathbf{x}) < 0\}$,
$\chi_{\{f=0\}}$ is the characteristic function of the set where f vanishes,
...

- A functional ϕ is any function defined on some vector space (such as, for example, $C_c(\mathbb{C})$ or $C_c(\mathbb{R})$) and taking values in $\mathbb{C}$.
- If $\mathbb{K}$ is either $\mathbb{R}$ or $\mathbb{C}$ and $g : D \subset \mathbb{R}^k \to \mathbb{K}$ is a measurable function defined on a set D with $0 < \mu_k(D) < \infty$, we denote by ϕ_g the functional

$$\phi_g : C_c(\mathbb{K}) \to \mathbb{C}, \qquad \phi_g(F) = \frac{1}{\mu_k(D)} \int_D F(g(\mathbf{x}))\mathrm{d}\mathbf{x}. \tag{2.2}$$

- A sequence of matrices is a sequence of the form $\{A_n\}_n$, where n varies in some infinite subset of $\mathbb{N}$ and A_n is a square matrix of size d_n such that $d_n \to \infty$ as $n \to \infty$. Throughout this book, unless otherwise specified, the size of the nth matrix of a sequence of matrices is always assumed to be d_n.

2.1.2 Multi-index Notation

Throughout this book, we will systematically use the multi-index notation. When discretizing a linear PDE over a d-dimensional domain $\Omega \subset \mathbb{R}^d$ by means of a linear numerical method, the actual computation of the numerical solution reduces to solving a linear system whose coefficient matrix usually possesses a d-level structure (see Example 2.5 below). As we shall see later on, especially in Chap. 7, the multi-index notation is a powerful tool that allows one to give a compact expression of this matrix by treating the dimensionality parameter d as any other parameter involved in the considered discretization. In this way, the dependence of the matrix structure on d is highlighted and a compact presentation is made possible.

A multi-index $\boldsymbol{i}$ of size d, also called a d-index, is simply a (row) vector in $\mathbb{Z}^d$; its components are denoted by $i_1, \ldots, i_d$.

- $\mathbf{0}, \mathbf{1}, \mathbf{2}, \ldots$ are the vectors of all zeros, all ones, all twos, ... (their size will be clear from the context).
- For any d-index $\boldsymbol{m}$, we set $N(\boldsymbol{m}) = \prod_{j=1}^{d} m_j$ and we write $\boldsymbol{m} \to \infty$ to indicate that $\min(\boldsymbol{m}) \to \infty$. The notation $N(\boldsymbol{\alpha}) = \prod_{j=1}^{d} \alpha_j$ will be actually used for any vector $\boldsymbol{\alpha}$ with d components and not only for d-indices.
- Let $\{a_{\boldsymbol{n}}\}_{\boldsymbol{n}\in\mathbb{N}^d}$ be a family of numbers parameterized by a d-index $\boldsymbol{n}$. The limit of $a_{\boldsymbol{n}}$ as $\boldsymbol{n} \to \infty$ is defined, as in the case of a traditional sequence $\{a_n\}_{n\in\mathbb{N}}$, in the following way: $\lim_{\boldsymbol{n}\to\infty} a_{\boldsymbol{n}} = a$ if and only if for every neighborhood U of a there exists $\boldsymbol{N}$ such that $a_{\boldsymbol{n}} \in U$ for $\boldsymbol{n} \geq \boldsymbol{N}$. Moreover, we define

$$\limsup_{\boldsymbol{n}\to\infty} a_{\boldsymbol{n}} = \lim_{\boldsymbol{n}\to\infty}\Big(\sup_{\boldsymbol{m}\geq\boldsymbol{n}} a_{\boldsymbol{m}}\Big), \qquad \liminf_{\boldsymbol{n}\to\infty} a_{\boldsymbol{n}} = \lim_{\boldsymbol{n}\to\infty}\Big(\inf_{\boldsymbol{m}\geq\boldsymbol{n}} a_{\boldsymbol{m}}\Big).$$

- If $\boldsymbol{h}, \boldsymbol{k}$ are d-indices, $\boldsymbol{h} \leq \boldsymbol{k}$ means that $h_r \leq k_r$ for all $r = 1, \ldots, d$, while $\boldsymbol{h} \not\leq \boldsymbol{k}$ means that $h_r > k_r$ for at least one $r \in \{1, \ldots, d\}$.
- If $\boldsymbol{h}, \boldsymbol{k}$ are d-indices such that $\boldsymbol{h} \leq \boldsymbol{k}$, the multi-index range $\boldsymbol{h}, \ldots, \boldsymbol{k}$ (or, more precisely, the d-index range $\boldsymbol{h}, \ldots, \boldsymbol{k}$) is the set of cardinality $N(\boldsymbol{k} - \boldsymbol{h} + \mathbf{1})$ given by $\{\boldsymbol{j} \in \mathbb{Z}^d : \boldsymbol{h} \leq \boldsymbol{j} \leq \boldsymbol{k}\}$. We assume for this set the standard lexicographic ordering:

$$\Big[\ldots \big[\, [\, (j_1, \ldots, j_d) \,]_{j_d=h_d,\ldots,k_d} \, \big]_{j_{d-1}=h_{d-1},\ldots,k_{d-1}} \ldots \Big]_{j_1=h_1,\ldots,k_1}. \tag{2.3}$$

  For instance, in the case $d = 2$ the ordering is

$$\begin{aligned} &(h_1, h_2),\ (h_1, h_2+1),\ \ldots,\ (h_1, k_2),\\ &(h_1+1, h_2),\ (h_1+1, h_2+1),\ \ldots,\ (h_1+1, k_2),\\ &\ldots\ldots\ldots,\ (k_1, h_2),\ (k_1, h_2+1),\ \ldots,\ (k_1, k_2). \end{aligned}$$

- When a d-index $\boldsymbol{j}$ varies over a d-index range $\boldsymbol{h}, \ldots, \boldsymbol{k}$ (this is often written as $\boldsymbol{j} = \boldsymbol{h}, \ldots, \boldsymbol{k}$), it is understood that $\boldsymbol{j}$ varies from $\boldsymbol{h}$ to $\boldsymbol{k}$ following the specific ordering (2.3). For instance, if $\boldsymbol{m} \in \mathbb{N}^d$ and we write $\mathbf{x} = [x_{\boldsymbol{i}}]_{\boldsymbol{i}=\mathbf{1}}^{\boldsymbol{m}}$, then $\mathbf{x}$ is a vector of size $N(\boldsymbol{m})$ whose components $x_{\boldsymbol{i}}$, $\boldsymbol{i} = \mathbf{1}, \dots, \boldsymbol{m}$, are ordered in accordance with (2.3): the first component is $x_{\mathbf{1}} = x_{(1,\dots,1,1)}$, the second component is $x_{(1,\dots,1,2)}$, and so on until the last component, which is $x_{\boldsymbol{m}} = x_{(m_1,\dots,m_d)}$. Similarly, if

$$X = [x_{\boldsymbol{ij}}]_{\boldsymbol{i},\boldsymbol{j}=\mathbf{1}}^{\boldsymbol{m}}, \tag{2.4}$$

then X is an $N(\boldsymbol{m}) \times N(\boldsymbol{m})$ matrix whose components are indexed by a pair of d-indices $\boldsymbol{i}, \boldsymbol{j}$, both varying from $\mathbf{1}$ to $\boldsymbol{m}$ according to the lexicographic ordering (2.3).

- If $\boldsymbol{h}, \boldsymbol{k}$ are d-indices such that $\boldsymbol{h} \le \boldsymbol{k}$, the notation $\sum_{\boldsymbol{j}=\boldsymbol{h}}^{\boldsymbol{k}}$ indicates the summation over all $\boldsymbol{j}$ in $\boldsymbol{h}, \dots, \boldsymbol{k}$.
- If $\boldsymbol{i}, \boldsymbol{j}$ are d-indices, $\boldsymbol{i} \preceq \boldsymbol{j}$ means that $\boldsymbol{i}$ precedes (or equals) $\boldsymbol{j}$ in the lexicographic ordering (which is a total ordering on $\mathbb{Z}^d$). Moreover, we define

$$\boldsymbol{i} \wedge \boldsymbol{j} = \begin{cases} \boldsymbol{i}, & \text{if } \boldsymbol{i} \preceq \boldsymbol{j}, \\ \boldsymbol{j}, & \text{if } \boldsymbol{i} \succ \boldsymbol{j}. \end{cases} \tag{2.5}$$

  Note that $\boldsymbol{i} \wedge \boldsymbol{j}$ is the minimum among $\boldsymbol{i}$ and $\boldsymbol{j}$ with respect to the lexicographic ordering. In the case where i and j are 1-indices (i.e., normal scalar indices), it is clear that $i \wedge j = \min(i, j)$.
- Operations involving d-indices that have no meaning in the vector space $\mathbb{Z}^d$ must always be interpreted in the componentwise sense. For instance, $\boldsymbol{np} = (n_1p_1, \dots, n_dp_d)$, $\alpha\boldsymbol{i}/\boldsymbol{j} = (\alpha i_1/j_1, \dots, \alpha i_d/j_d)$ for all $\alpha \in \mathbb{C}$, $\boldsymbol{i}^2 = (i_1^2, \dots, i_d^2)$, $\max(\boldsymbol{i}, \boldsymbol{j}) = (\max(i_1, j_1), \dots, \max(i_d, j_d))$, $\boldsymbol{i} \bmod \boldsymbol{m} = (i_1 \bmod m_1, \dots, i_d \bmod m_d)$, etc.
- When a multi-index appears as subscript or superscript, we sometimes suppress the brackets to simplify the notation. For instance, the component of the vector $\mathbf{x} = [x_{\boldsymbol{i}}]_{\boldsymbol{i}=\mathbf{1}}^{\boldsymbol{m}}$ corresponding to the d-index $\boldsymbol{i}$ is denoted by $x_{\boldsymbol{i}}$ or $x_{i_1,\dots,i_d}$, and we often avoid the heavy notation $x_{(i_1,\dots,i_d)}$.

We provide below a few examples to help the reader become familiar with the multi-index notation.

Example 2.1 Let $\boldsymbol{h} = \mathbf{1} = (1, 1)$ and $\boldsymbol{k} = (4, 2)$. The multi-index range $\boldsymbol{h}, \dots, \boldsymbol{k}$ consists of $N(\boldsymbol{k} - \boldsymbol{h} + \mathbf{1}) = N(\boldsymbol{k}) = 8$ elements which are sorted according to the lexicographic ordering (2.3) as follows:

$$(1, 1),\ (1, 2),\ (2, 1),\ (2, 2),\ (3, 1),\ (3, 2),\ (4, 1),\ (4, 2).$$

Note that $\sum_{\boldsymbol{j}=\boldsymbol{h}}^{\boldsymbol{k}} \boldsymbol{j}^2 = (60, 20)$.

Example 2.2 Let $a : [0, 1]^2 \to \mathbb{C}$ and $\boldsymbol{n} \in \mathbb{N}^2$. Set

$$\mathbf{x} = \left[a\left(\frac{\boldsymbol{i}}{\boldsymbol{n}}\right)\right]_{\boldsymbol{i}=\mathbf{1}}^{\boldsymbol{n}}.$$

Then, $\mathbf{x}$ is the vector of size $N(\boldsymbol{n})$ given by

$$
\mathbf{x} = \begin{bmatrix} a(\frac{1}{n_1}, \frac{1}{n_2}) \\ a(\frac{1}{n_1}, \frac{2}{n_2}) \\ \vdots \\ a(\frac{1}{n_1}, 1) \\ \hline a(\frac{2}{n_1}, \frac{1}{n_2}) \\ a(\frac{2}{n_1}, \frac{2}{n_2}) \\ \vdots \\ a(\frac{2}{n_1}, 1) \\ \hline \vdots \\ \vdots \\ \vdots \\ \hline a(1, \frac{1}{n_2}) \\ a(1, \frac{2}{n_2}) \\ \vdots \\ a(1, 1) \end{bmatrix} = \begin{bmatrix} \mathbf{x}_1 \\ \hline \mathbf{x}_2 \\ \hline \vdots \\ \hline \mathbf{x}_{n_1} \end{bmatrix}, \qquad \mathbf{x}_{i_1} = \begin{bmatrix} a(\frac{i_1}{n_1}, \frac{1}{n_2}) \\ a(\frac{i_1}{n_1}, \frac{2}{n_2}) \\ \vdots \\ a(\frac{i_1}{n_1}, 1) \end{bmatrix}, \qquad i_1 = 1, \dots, n_1.
$$

Moreover, $\|\mathbf{x}\|^2 = \sum_{\boldsymbol{i}=\mathbf{1}}^{\boldsymbol{n}} |x_{\boldsymbol{i}}|^2 = \sum_{\boldsymbol{i}=\mathbf{1}}^{\boldsymbol{n}} |a(\boldsymbol{i}/\boldsymbol{n})|^2$.

Example 2.3 Consider the matrix

$$
A = \left[\begin{array}{cc|cc} 4 & 4 & 0 & 0 \\ 4 & 4 & 0 & 0 \\ \hline 0 & 0 & 1 & 1 \\ 0 & 0 & 1 & 1 \end{array}\right]. \tag{2.6}
$$

Instead of indexing the entries of A in the standard way, i.e., by means of two traditional scalar indices $i, j = 1, \dots, 4$, we can decide to index the entries of A by means of two 2-indices (or bi-indices) $\boldsymbol{i}, \boldsymbol{j} = \mathbf{1}, \dots, \mathbf{2} = (1, 1), (1, 2), (2, 1), (2, 2)$. This is possible because both the ranges $1, \dots, 4$ and $\mathbf{1}, \dots, \mathbf{2}$ have 4 elements, and 4 is the size of A. The two writings $A = [a_{ij}]_{i,j=1}^4$ and $A = [a_{\boldsymbol{i}\boldsymbol{j}}]_{\boldsymbol{i},\boldsymbol{j}=\mathbf{1}}^{\mathbf{2}}$ correspond to the two different indicizations. We have

$$
\begin{aligned}
&a_{(1,1),(1,1)} = 4, && a_{(1,1),(1,2)} = 4, && a_{(1,1),(2,1)} = 0, && a_{(1,1),(2,2)} = 0, \\
&a_{(1,2),(1,1)} = 4, && a_{(1,2),(1,2)} = 4, && a_{(1,2),(2,1)} = 0, && a_{(1,2),(2,2)} = 0, \\
&a_{(2,1),(1,1)} = 0, && a_{(2,1),(1,2)} = 0, && a_{(2,1),(2,1)} = 1, && a_{(2,1),(2,2)} = 1, \\
&a_{(2,2),(1,1)} = 0, && a_{(2,2),(1,2)} = 0, && a_{(2,2),(2,1)} = 1, && a_{(2,2),(2,2)} = 1.
\end{aligned}
$$

The indicization of the entries of A with the bi-indices $\boldsymbol{i}, \boldsymbol{j}$ reflects the fact that we are thinking of A as a block matrix partitioned into 4 blocks as indicated in (2.6): for all $\boldsymbol{i}, \boldsymbol{j} = \mathbf{1}, \ldots, \mathbf{2}$, the entry $a_{\boldsymbol{i}\boldsymbol{j}}$ is the (i_2, j_2) entry of the (i_1, j_1) block of A. For example,

$$a_{(2,1),(2,2)} = \text{entry } (1,2) \text{ in the } (2,2) \text{ block of } A = 1.$$

We can therefore write

$$a_{\boldsymbol{i}\boldsymbol{j}} = \begin{cases} 4, & \text{if } i_1 = j_1 = 1, \\ 0, & \text{if } i_1 = 1 \text{ and } j_1 = 2, \\ 0, & \text{if } i_1 = 2 \text{ and } j_1 = 1, \\ 1, & \text{if } i_1 = j_1 = 2. \end{cases}$$

Note that $a_{\boldsymbol{i}\boldsymbol{j}}$ depends only on the first components of the bi-indices $\boldsymbol{i}$ and $\boldsymbol{j}$.

Example 2.4 (*tensor products*) Let $X \in \mathbb{C}^{m_1 \times m_2}$ and $Y \in \mathbb{C}^{\ell_1 \times \ell_2}$, and define the block matrix

$$Z = [x_{ij}Y]_{\substack{i=1,\ldots,m_1 \\ j=1,\ldots,m_2}} = \begin{bmatrix} x_{11}Y & x_{12}Y & \cdots & x_{1m_2}Y \\ x_{21}Y & x_{22}Y & \cdots & x_{2m_2}Y \\ \vdots & \vdots & & \vdots \\ x_{m_1 1}Y & x_{m_1 2}Y & \cdots & x_{m_1 m_2}Y \end{bmatrix} \in \mathbb{C}^{m_1\ell_1 \times m_2\ell_2}. \qquad (2.7)$$

Using the identities

$$r = \lfloor r/s \rfloor s + r \bmod s, \qquad \lfloor r/s \rfloor = \lceil (r+1)/s \rceil - 1,$$

which are satisfied for all integers $r \geq 0$ and $s \geq 1$, for every $i = 1, \ldots, m_1\ell_1$ and $j = 1, \ldots, m_2\ell_2$ we can write

$$i = (\lceil i/\ell_1 \rceil - 1)\ell_1 + ((i-1) \bmod \ell_1) + 1,$$
$$j = (\lceil j/\ell_2 \rceil - 1)\ell_2 + ((j-1) \bmod \ell_2) + 1.$$

The (i, j) entry of the matrix Z is then given by

$$z_{ij} = x_{\lceil i/\ell_1 \rceil, \lceil j/\ell_2 \rceil}\, y_{((i-1) \bmod \ell_1)+1, ((j-1) \bmod \ell_2)+1}.$$

It is clear that this expression is rather complicated. Now, suppose we decide to index the entries of Z by two bi-indices $\boldsymbol{i}, \boldsymbol{j}$ such that $\boldsymbol{i} = \mathbf{1}, \ldots, \boldsymbol{n}$ and $\boldsymbol{j} = \mathbf{1}, \ldots, \boldsymbol{k}$, where $\boldsymbol{n} = (m_1, \ell_1)$ and $\boldsymbol{k} = (m_2, \ell_2)$. This indicization, which is possible because

$$\#\{\mathbf{1}, \ldots, \boldsymbol{n}\} = N(\boldsymbol{n}) = m_1\ell_1 = \text{number of rows of } Z,$$
$$\#\{\mathbf{1}, \ldots, \boldsymbol{k}\} = N(\boldsymbol{k}) = m_2\ell_2 = \text{number of columns of } Z,$$

reflects the fact that we are thinking of Z as an $m_1 \times m_2$ block matrix in which each of the $m_1 m_2$ blocks is of size $\ell_1 \times \ell_2$. Actually, this is the natural way of thinking in view of the block structure of Z; see (2.7). With such indicization, for all $\boldsymbol{i} = \boldsymbol{1}, \ldots, \boldsymbol{n}$ and $\boldsymbol{j} = \boldsymbol{1}, \ldots, \boldsymbol{k}$ we have

$$z_{\boldsymbol{i}\boldsymbol{j}} = \text{entry } (i_2, j_2) \text{ in the } (i_1, j_1) \text{ block of } Z = x_{i_1 j_1} y_{i_2 j_2}.$$

We then see that $z_{\boldsymbol{i}\boldsymbol{j}}$ has a much simpler expression than z_{ij}. To conclude, we remark that the matrix Z is the so-called tensor (Kronecker) product of X and Y, and it is usually denoted by $X \otimes Y$; we shall come back to tensor products in Sect. 2.5.

Example 2.5 (*multilevel matrices*) In many cases, it is convenient to partition matrices into blocks, which are partitioned into smaller blocks, which are partitioned into smaller blocks, and so on until a certain nesting level d is reached. Such matrices are called multilevel matrices. More precisely, following Tyrtyshnikov [41, Sect. 6], we say that a square matrix A of size N is a d-level matrix with level orders $n_1, n_2, \ldots, n_d$ if $N = n_1 n_2 \cdots n_d$ and A is partitioned into n_1^2 square blocks of size N/n_1, each of which is partitioned into n_2^2 square blocks of size $N/(n_1 n_2)$, each of which is partitioned into n_3^2 square blocks of size $N/(n_1 n_2 n_3)$, and so on until the last n_d^2 square blocks of size $N/(n_1 n_2 \cdots n_d) = 1$, which are scalars. In formulas,

$$\begin{aligned}
&A = [A_{i_1 j_1}]_{i_1, j_1 = 1}^{n_1}, && A_{i_1 j_1} \in \mathbb{C}^{\tilde{n}_1 \times \tilde{n}_1}, && \tilde{n}_1 = \tfrac{N}{n_1};\\
&A_{i_1 j_1} = [A_{i_1 j_1;\, i_2 j_2}]_{i_2, j_2 = 1}^{n_2}, && A_{i_1 j_1;\, i_2 j_2} \in \mathbb{C}^{\tilde{n}_2 \times \tilde{n}_2}, && \tilde{n}_2 = \tfrac{N}{n_1 n_2};\\
&\vdots\\
&A_{i_1 j_1; \ldots; i_{d-1} j_{d-1}} = [A_{i_1 j_1; \ldots; i_d j_d}]_{i_d, j_d = 1}^{n_d}, && A_{i_1 j_1; \ldots; i_d j_d} \in \mathbb{C}^{\tilde{n}_d \times \tilde{n}_d}, && \tilde{n}_d = \tfrac{N}{n_1 \cdots n_d} = 1.
\end{aligned}$$

The level orders $n_1, n_2, \ldots, n_d$ and the order N are also referred to as partial orders and total order, respectively. Indexing the entries of a d-level matrix A by two traditional scalar indices $i, j = 1, \ldots, N$ is a nightmare. On the contrary, A admits a natural indicization by means of two d-indices $\boldsymbol{i}, \boldsymbol{j} = \boldsymbol{1}, \ldots, \boldsymbol{n}$, where $\boldsymbol{n} = (n_1, \ldots, n_d)$. Indeed, A can be written in the form (2.4) as follows:

$$A = [A_{\boldsymbol{i}\boldsymbol{j}}]_{\boldsymbol{i}, \boldsymbol{j} = \boldsymbol{1}}^{\boldsymbol{n}},$$

where $A_{\boldsymbol{i}\boldsymbol{j}} = A_{i_1 j_1; \ldots; i_d j_d}$ for $\boldsymbol{i}, \boldsymbol{j} = \boldsymbol{1}, \ldots, \boldsymbol{n}$. We remark that, as we shall see in Chap. 7, when a linear PDE over a d-dimensional hyperrectangle is discretized by a linear numerical method, the resulting discretization matrix is normally a d-level matrix with level orders $n_1, n_2, \ldots, n_d$ and total order $N(\boldsymbol{n}) = n_1 n_2 \cdots n_d$, where n_i is the discretization parameter in the ith direction.

Example 2.6 (*matrix computations with multi-indices*) Let $\boldsymbol{n} \in \mathbb{N}^d$ and let

$$A = [a_{\boldsymbol{i}\boldsymbol{j}}]_{\boldsymbol{i}, \boldsymbol{j} = \boldsymbol{1}}^{\boldsymbol{n}}, \qquad B = [b_{\boldsymbol{i}\boldsymbol{j}}]_{\boldsymbol{i}, \boldsymbol{j} = \boldsymbol{1}}^{\boldsymbol{n}}, \qquad \mathbf{x} = [x_{\boldsymbol{j}}]_{\boldsymbol{j} = \boldsymbol{1}}^{\boldsymbol{n}}, \qquad \mathbf{y} = [y_{\boldsymbol{j}}]_{\boldsymbol{j} = \boldsymbol{1}}^{\boldsymbol{n}}.$$

It is not difficult to see that the following properties hold.

- $A^* = [\overline{a_{ji}}]_{i,j=1}^{n}$.
- $\alpha A + \beta B = [\alpha a_{ij} + \beta b_{ij}]_{i,j=1}^{n}$ for all $\alpha, \beta \in \mathbb{C}$.
- $AB = [\sum_{k=1}^{n} a_{ik} b_{kj}]_{i,j=1}^{n}$.
- $A\mathbf{x} = [\sum_{j=1}^{n} a_{ij} x_j]_{i=1}^{n}$.
- $\mathbf{x}^* A\mathbf{y} = \sum_{i,j=1}^{n} a_{ij} \overline{x_i} y_j$.

These properties show that the usual matrix computation rules remain formally the same when passing from standard scalar indices to multi-indices. We invite the reader to prove them as an exercise.

2.1.3 Multilevel Matrix-Sequences

We recall from Sect. 2.1.1 that a sequence of matrices is a sequence of the form $\{A_n\}_n$, where n varies in some infinite subset of $\mathbb{N}$ and A_n is a square matrix of size $d_n \to \infty$. A d-level matrix-sequence is a special sequence of matrices of the form $\{A_{\boldsymbol{n}}\}_n$, where:

- n varies in some infinite subset of $\mathbb{N}$;
- $\boldsymbol{n} = \boldsymbol{n}(n)$ is a d-index with positive components which depends on n and satisfies $\boldsymbol{n} \to \infty$ as $n \to \infty$;
- $A_{\boldsymbol{n}}$ is a square matrix of size $N(\boldsymbol{n})$.

Recall from Sect. 2.1.2 that $\boldsymbol{n} \to \infty$ means $\min(\boldsymbol{n}) \to \infty$.

The name "d-level matrix-sequence" is due to the fact that, in practical applications, especially in the context of PDE discretizations, each matrix $A_{\boldsymbol{n}}$ of a d-level matrix-sequence $\{A_{\boldsymbol{n}}\}_n$ is normally a d-level matrix with level orders $(n_1, \ldots, n_d) = \boldsymbol{n}$; see Example 2.5. Throughout this book, we often use the abbreviation "matrix-sequence" for both "d-level matrix-sequence" and "multilevel matrix-sequence".

2.2 Multivariate Trigonometric Polynomials

A d-variate trigonometric polynomial is a finite linear combination of the d-variate Fourier frequencies

$$\mathrm{e}^{\mathrm{i}\boldsymbol{k}\cdot\boldsymbol{\theta}} = \mathrm{e}^{\mathrm{i}(k_1\theta_1 + \ldots + k_d\theta_d)}, \qquad \boldsymbol{k} \in \mathbb{Z}^d,$$

that is, a function of the form

$$f(\boldsymbol{\theta}) = \sum_{\boldsymbol{k}=-\boldsymbol{N}}^{\boldsymbol{N}} f_{\boldsymbol{k}}\, \mathrm{e}^{\mathrm{i}\boldsymbol{k}\cdot\boldsymbol{\theta}}, \qquad f_{-\boldsymbol{N}}, \ldots, f_{\boldsymbol{N}} \in \mathbb{C}, \qquad \boldsymbol{N} \in \mathbb{N}^d. \tag{2.8}$$

Note that, as a consequence of the orthogonality relations

$$\int_{[-\pi,\pi]^d} e^{i\boldsymbol{\ell}\cdot\boldsymbol{\theta}} e^{-i\boldsymbol{k}\cdot\boldsymbol{\theta}} d\boldsymbol{\theta} = \begin{cases} (2\pi)^d, & \text{if } \boldsymbol{k} = \boldsymbol{\ell}, \\ 0, & \text{if } \boldsymbol{k} \neq \boldsymbol{\ell}, \end{cases} \tag{2.9}$$

the numbers $f_{-N}, \ldots, f_N$ appearing in (2.8) are the (only possible nonzero) Fourier coefficients of f according to the definition (2.1). Note also that a 1-variate (or univariate) trigonometric polynomial is just a trigonometric polynomial in the classical sense. In what follows, we say that a d-variate trigonometric polynomial is d-separable (or simply separable) if it is a d-separable function from $\mathbb{R}^d$ to $\mathbb{C}$ according to the definition in Sect. 2.1.1.

Lemma 2.1 *Let $f : \mathbb{R}^d \to \mathbb{C}$ be a separable d-variate trigonometric polynomial and let $f = f_1 \otimes \cdots \otimes f_d$ be a factorization of f. If f is not identically 0, then $f_1, \ldots, f_d$ are (univariate) trigonometric polynomials.*

Proof Since $f_2, \ldots, f_d$ are not identically 0, there exists $(\vartheta_2, \ldots, \vartheta_d)$ such that $f_2(\vartheta_2) \cdots f_d(\vartheta_d) \neq 0$. The definition of d-variate trigonometric polynomials implies that $\theta_1 \mapsto f(\theta_1, \vartheta_2, \ldots, \vartheta_d) = f_1(\theta_1) f_2(\vartheta_2) \cdots f_d(\vartheta_d)$ is a (univariate) trigonometric polynomial, and this means that f_1 is a trigonometric polynomial. With the same argument, one can show that $f_2, \ldots, f_d$ are trigonometric polynomials as well. □

Corollary 2.1 *Let $f : \mathbb{R}^d \to \mathbb{C}$ be a separable d-variate trigonometric polynomial. Then, there exist (univariate) trigonometric polynomials $f_1, \ldots, f_d : \mathbb{R} \to \mathbb{C}$ such that $f = f_1 \otimes \cdots \otimes f_d$.*

The next lemma shows that the set of zeros of every non-trivial d-variate trigonometric polynomial has zero measure.

Lemma 2.2 *Let $f : \mathbb{R}^d \to \mathbb{C}$ be a d-variate trigonometric polynomial with at least one nonzero Fourier coefficient. Then $\mu_d\{f = 0\} = 0$.*

Proof The proof proceeds by induction on d. For $d = 1$, the result has already been proved in [22, solution of Exercise 6.2, pp. 286–287]. Suppose $d > 1$ and assume that the lemma holds for dimensions up to $d - 1$. Let $f(\boldsymbol{\theta}) = \sum_{\boldsymbol{k}=-\boldsymbol{N}}^{\boldsymbol{N}} f_{\boldsymbol{k}}\, e^{i\boldsymbol{k}\cdot\boldsymbol{\theta}}$ and set $Z = \{f = 0\}$. By Fubini's theorem,

$$\begin{aligned} \mu_d(Z) &= \int_{\mathbb{R}^d} \chi_Z(\theta_1, \ldots, \theta_d) d\theta_1 \ldots d\theta_d = \int_{\mathbb{R}^{d-1}} d\theta_2 \ldots d\theta_d \int_{\mathbb{R}} \chi_Z(\theta_1, \ldots, \theta_d) d\theta_1 \\ &= \int_{\mathbb{R}^{d-1}} d\theta_2 \ldots d\theta_d \int_{Z_{\theta_2,\ldots,\theta_d}} d\theta_1, \end{aligned} \tag{2.10}$$

where, for each fixed $(\theta_2, \ldots, \theta_d) \in \mathbb{R}^{d-1}$, the set $Z_{\theta_2,\ldots,\theta_d}$ is defined by

$$Z_{\theta_2,\ldots,\theta_d} = \{\theta_1 \in \mathbb{R} : f(\theta_1, \theta_2, \ldots, \theta_d) = 0\}.$$

Write

$$f(\boldsymbol{\theta}) = \sum_{k_1=-N_1}^{N_1} p_{k_1}(\theta_2, \ldots, \theta_d)\, \mathrm{e}^{\mathrm{i}k_1\theta_1}, \tag{2.11}$$

where $p_{-N_1}, \ldots, p_{N_1}$ are $(d-1)$-variate trigonometric polynomials given by

$$p_{k_1}(\theta_2, \ldots, \theta_d) = \sum_{(k_2,\ldots,k_d)=-(N_2,\ldots,N_d)}^{(N_2,\ldots,N_d)} f_{\boldsymbol{k}}\, \mathrm{e}^{\mathrm{i}(k_2,\ldots,k_d)\cdot(\theta_2,\ldots,\theta_d)}, \qquad k_1 = -N_1, \ldots, N_1.$$

Let

$$A = \{(\theta_2, \ldots, \theta_d) \in \mathbb{R}^{d-1} : \ p_{k_1}(\theta_2, \ldots, \theta_d) = 0 \text{ for all } k_1 = -N_1, \ldots, N_1\}.$$

Since not all the Fourier coefficients $f_{\boldsymbol{k}}$ are equal to 0, at least one of the polynomials p_{k_1} has at least a nonzero Fourier coefficient. Thus, by induction hypothesis, $\mu_{d-1}(A) = 0$. Moreover, by (2.11),

- if $(\theta_2, \ldots, \theta_d) \in A$ then $Z_{\theta_2,\ldots,\theta_d} = \mathbb{R}$;
- if $(\theta_2, \ldots, \theta_d) \in A^c = \mathbb{R}^{d-1}\backslash A$ then there exists an index $k_1 \in \{-N_1, \ldots, N_1\}$ such that $p_{k_1}(\theta_2, \ldots, \theta_d) \neq 0$, and so $\mu_1(Z_{\theta_2,\ldots,\theta_d}) = 0$ by induction hypothesis.

Going back to (2.10), we see that

$$\mu_d(Z) = \int_A \mathrm{d}\theta_2 \ldots \mathrm{d}\theta_d \int_{Z_{\theta_2,\ldots,\theta_d}} \mathrm{d}\theta_1 + \int_{A^c} \mathrm{d}\theta_2 \ldots \mathrm{d}\theta_d \int_{Z_{\theta_2,\ldots,\theta_d}} \mathrm{d}\theta_1 = 0,$$

and the proof is complete. □

The next two lemmas are the multivariate versions of [22, Lemmas 2.7 and 2.8]. They show how it is possible to approximate an L^1 (resp., a measurable) function by means of standard (resp., weighted) multivariate trigonometric polynomials. In what follows, for any $D \subseteq \mathbb{R}^k$ we denote by $C_c(D)$ the space of continuous functions $f : D \to \mathbb{C}$ such that support the $\mathrm{supp}(f) = \overline{\{\mathbf{x} \in D : \ f(\mathbf{x}) = 0\}} = \overline{\{f = 0\}}$ is compact. We recall that, if D is measurable (so that it makes sense to talk about $L^p(D)$), then the space $C_c(D)$ is dense in $L^p(D)$ for all $1 \le p < \infty$ [22, p. 13].

Lemma 2.3 *Let $f \in L^1([-\pi, \pi]^d)$. Then, there exists a sequence of d-variate trigonometric polynomials $\{p_m\}_m$ such that $\|p_m\|_\infty \le \mathrm{ess\,sup}_{[-\pi,\pi]^d}|f|$ for all m and $p_m \to f$ a.e. and in $L^1([-\pi, \pi]^d)$.*

Proof The proof follows the same pattern as [22, proof of Lemma 2.7]. It suffices to show that, for each $\varepsilon > 0$, there exists a d-variate trigonometric polynomial p_ε such that

$$\|p_\varepsilon\|_\infty \le \mathrm{ess\,sup}_{[-\pi,\pi]^d}|f|, \qquad \|f - p_\varepsilon\|_{L^1} \le \varepsilon.$$

Indeed, this shows the existence of a sequence of d-variate trigonometric polynomials $\{p_m\}_m$ such that $\|p_m\|_\infty \le \mathrm{ess\,sup}_{[-\pi,\pi]^d}|f|$ for all m and $p_m \to f$ in $L^1([-\pi, \pi]^d)$.

Recalling that the L^1 convergence of a sequence implies the a.e. convergence of an appropriate subsequence, passing to an appropriate subsequence of $\{p_m\}_m$ (if necessary), we may assume that $p_m \to f$ a.e.

Let $\varepsilon > 0$. By [22, Theorem 2.2] in combination with the dominated convergence theorem and the density of $C_c((-\pi,\pi)^d)$ in $L^1((-\pi,\pi)^d)$, there exists $f_\varepsilon \in C_c((-\pi,\pi)^d)$ such that

$$\|f_\varepsilon\|_\infty \le \operatorname{ess\,sup}_{[-\pi,\pi]^d}|f|, \qquad \|f - f_\varepsilon\|_{L^1} < \varepsilon. \tag{2.12}$$

The function f_ε is continuous on $[-\pi,\pi]^d$ and satisfies $f_\varepsilon(\boldsymbol{\theta}) = 0$ for every $\boldsymbol{\theta} \in \partial([-\pi,\pi]^d)$. Thus, by the multivariate version of Fejér's theorem, which is proved essentially in the same way as the classical (univariate) Fejér theorem [28, Theorem 3.1], there exists a d-variate trigonometric polynomial p_ε such that

$$\|p_\varepsilon\|_\infty \le \|f_\varepsilon\|_\infty, \qquad \|f_\varepsilon - p_\varepsilon\|_\infty < \varepsilon. \tag{2.13}$$

By combining (2.12) and (2.13), we arrive at

$$\|p_\varepsilon\|_\infty \le \operatorname{ess\,sup}_{[-\pi,\pi]^d}|f|, \qquad \|f - p_\varepsilon\|_{L^1} \le \varepsilon(1 + (2\pi)^d),$$

which proves the thesis. □

Lemma 2.4 *Let $\kappa : [0,1]^d \times [-\pi,\pi]^d \to \mathbb{C}$ be a measurable function. Then, there exists a sequence $\{\kappa_m\}_m$ such that $\kappa_m : [0,1]^d \times [-\pi,\pi]^d \to \mathbb{C}$ is a function of the form*

$$\kappa_m(\mathbf{x},\boldsymbol{\theta}) = \sum_{\boldsymbol{j}=-\boldsymbol{N}_m}^{\boldsymbol{N}_m} a_{\boldsymbol{j}}^{(m)}(\mathbf{x})\,\mathrm{e}^{\mathrm{i}\boldsymbol{j}\cdot\boldsymbol{\theta}}, \qquad a_{\boldsymbol{j}}^{(m)} \in C^\infty([0,1]^d), \qquad \boldsymbol{N}_m \in \mathbb{N}^d, \tag{2.14}$$

and $\kappa_m \to \kappa$ a.e.

Proof The proof is essentially the same as [22, proof of Lemma 2.8]. The function $\tilde{\kappa}_m = \kappa\chi_{\{|\kappa|\le 1/m\}}$ belongs to $L^\infty([0,1]^d \times [-\pi,\pi]^d)$ and converges to κ in measure. Indeed, $\tilde{\kappa}_m \to \kappa$ pointwise over $[0,1]^d \times [-\pi,\pi]^d$, and the pointwise (a.e.) convergence on a set of finite measure implies the convergence in measure [22, Lemma 2.4]. By [22, Lemma 2.2], the space generated by the trigonometric monomials

$$\left\{\mathrm{e}^{2\pi\mathrm{i}\boldsymbol{\ell}\cdot\mathbf{x}}\mathrm{e}^{\mathrm{i}\boldsymbol{j}\cdot\boldsymbol{\theta}} = \mathrm{e}^{\mathrm{i}(2\pi\ell_1x_1+\ldots+2\pi\ell_dx_d+j_1\theta_1+\ldots+j_d\theta_d)} \;:\; \boldsymbol{\ell},\boldsymbol{j} \in \mathbb{Z}^d\right\}$$

is dense in $L^1([0,1]^d \times [-\pi,\pi]^d)$, so we can choose a function κ_m belonging to this space such that $\|\kappa_m - \tilde{\kappa}_m\|_{L^1} \le 1/m$. Note that κ_m is a function of the form (2.14). Moreover, for each $\varepsilon > 0$, using Chebyshev's inequality [22, Eq. (2.4)] we obtain

$$\begin{aligned}\mu_{2d}\{|\kappa_m - \kappa| > \varepsilon\} &\le \mu_{2d}(\{|\kappa_m - \tilde{\kappa}_m| > \varepsilon/2\} \cup \{|\tilde{\kappa}_m - \kappa| > \varepsilon/2\}) \\ &\le \mu_{2d}\{|\kappa_m - \tilde{\kappa}_m| > \varepsilon/2\} + \mu_{2d}\{|\tilde{\kappa}_m - \kappa| > \varepsilon/2\} \\ &\le \frac{\|\kappa_m - \tilde{\kappa}_m\|_{L^1}}{(\varepsilon/2)} + \mu_{2d}\{|\tilde{\kappa}_m - \kappa| > \varepsilon/2\},\end{aligned}$$

which converges to 0 as $m \to \infty$. Hence, $\kappa_m \to \kappa$ in measure. Since the convergence in measure on a set of finite measure implies the existence of a subsequence that converges a.e. [22, Lemma 2.4], passing to a subsequence of $\{\kappa_m\}_m$ (if necessary) we may assume that $\kappa_m \to \kappa$ a.e. □

2.3 Multivariate Riemann-Integrable Functions

A function $a : [0, 1]^d \to \mathbb{C}$ is said to be Riemann-integrable if its real and imaginary parts $\Re(a), \Im(a) : [0, 1]^d \to \mathbb{R}$ are Riemann-integrable in the classical sense. Recall that any Riemann-integrable function is *bounded* by definition. We report below a list of properties possessed by Riemann-integrable functions that will be used in this book, either explicitly or implicitly.

- If $\alpha, \beta \in \mathbb{C}$ and $a, b : [0, 1]^d \to \mathbb{C}$ are Riemann-integrable, then $\alpha a + \beta b$ is Riemann-integrable.
- If $a, b : [0, 1]^d \to \mathbb{C}$ are Riemann-integrable, then ab is Riemann-integrable.
- If $a : [0, 1]^d \to \mathbb{C}$ is Riemann-integrable and $F : \mathbb{C} \to \mathbb{C}$ is continuous, then $F(a) : [0, 1]^d \to \mathbb{C}$ is Riemann-integrable.
- If $a : [0, 1]^d \to \mathbb{C}$ is Riemann-integrable, then a belongs to $L^\infty([0, 1]^d)$ and its Lebesgue and Riemann integrals over $[0, 1]^d$ coincide.
- If $a : [0, 1]^d \to \mathbb{C}$ is bounded, then a is Riemann-integrable if and only if a is continuous a.e.

Note that the last two properties imply the first three. The proof of the second-to-last property can be found in [30, pp. 73–74], while the last property is Lebesgue's characterization theorem of Riemann-integrable functions [30, p. 104]. Note that the proofs in [30] are made for the case $d = 1$ only, but the generalization to the case $d > 1$ is straightforward. A further property of Riemann-integrable functions that will be used in this book is stated and proved in the next lemma, which is the multivariate version of [22, Lemma 2.9].

Lemma 2.5 *Let* $a : [0, 1]^d \to \mathbb{R}$ *be Riemann-integrable. For each* $\boldsymbol{n} \in \mathbb{N}^d$*, consider the partition of* $(0, 1]^d$ *given by the* d*-dimensional hyperrectangles*

$$I_{\boldsymbol{i},\boldsymbol{n}} = \left(\frac{\boldsymbol{i}-\boldsymbol{1}}{\boldsymbol{n}}, \frac{\boldsymbol{i}}{\boldsymbol{n}}\right] = \left(\frac{i_1 - 1}{n_1}, \frac{i_1}{n_1}\right] \times \cdots \times \left(\frac{i_d - 1}{n_d}, \frac{i_d}{n_d}\right], \qquad \boldsymbol{i} = \boldsymbol{1}, \ldots, \boldsymbol{n},$$

and let

$$a_{i,n} \in \left[\inf_{\mathbf{x} \in I_{i,n}} a(\mathbf{x}), \ \sup_{\mathbf{x} \in I_{i,n}} a(\mathbf{x}) \right], \qquad \boldsymbol{i} = \mathbf{1}, \dots, \boldsymbol{n}.$$

Then

$$\sum_{\boldsymbol{i}=\mathbf{1}}^{\boldsymbol{n}} a_{\boldsymbol{i},\boldsymbol{n}} \chi_{I_{\boldsymbol{i},\boldsymbol{n}}} \to a \ \ a.e. \ in \ [0,1]^d \tag{2.15}$$

and

$$\lim_{\boldsymbol{n}\to\infty} \frac{1}{N(\boldsymbol{n})} \sum_{\boldsymbol{i}=\mathbf{1}}^{\boldsymbol{n}} a_{\boldsymbol{i},\boldsymbol{n}} = \int_{[0,1]^d} a(\mathbf{x}) \mathrm{d}\mathbf{x}. \tag{2.16}$$

Proof The proof is essentially the same as [22, proof of Lemma 2.9]. Fix $\varepsilon > 0$ and let $\mathbf{x} \in (0,1]^d$ be a continuity point of a. Then, there is a $\delta > 0$ such that $|a(\mathbf{y}) - a(\mathbf{x})| \le \varepsilon$ whenever $\mathbf{y} \in [0,1]^d$ and $|\mathbf{y} - \mathbf{x}|_\infty \le \delta$. Take $\boldsymbol{n} \ge (1/\delta)\mathbf{1}$ and call $I_{\boldsymbol{k},\boldsymbol{n}}$ the unique hyperrectangle of the partition $(0,1]^d = \bigcup_{\boldsymbol{i}=\mathbf{1}}^{\boldsymbol{n}} I_{\boldsymbol{i},\boldsymbol{n}}$ containing $\mathbf{x}$. For $\mathbf{y} \in I_{\boldsymbol{k},\boldsymbol{n}}$, we have $\mathbf{y} \in [0,1]^d$ and $|\mathbf{y} - \mathbf{x}|_\infty \le \delta$, hence $|a(\mathbf{y}) - a(\mathbf{x})| \le \varepsilon$. It follows that

$$\begin{aligned} \left| \sum_{\boldsymbol{i}=\mathbf{1}}^{\boldsymbol{n}} a_{\boldsymbol{i},\boldsymbol{n}} \chi_{I_{\boldsymbol{i},\boldsymbol{n}}}(\mathbf{x}) - a(\mathbf{x}) \right| &= |a_{\boldsymbol{k},\boldsymbol{n}} - a(\mathbf{x})| \\ &\le \max\Big(a(\mathbf{x}) - \inf_{\mathbf{y} \in I_{\boldsymbol{k},\boldsymbol{n}}} a(\mathbf{y}), \ \sup_{\mathbf{y} \in I_{\boldsymbol{k},\boldsymbol{n}}} a(\mathbf{y}) - a(\mathbf{x}) \Big) \le \varepsilon. \end{aligned}$$

As a consequence, $\sum_{\boldsymbol{i}=\mathbf{1}}^{\boldsymbol{n}} a_{\boldsymbol{i},\boldsymbol{n}} \chi_{I_{\boldsymbol{i},\boldsymbol{n}}}(\mathbf{x}) \to a(\mathbf{x})$ whenever $\mathbf{x}$ is a continuity point of a in $(0,1]^d$. This implies (2.15), because a is Riemann-integrable and hence continuous a.e. in $[0,1]^d$. Since

$$\left| \sum_{\boldsymbol{i}=\mathbf{1}}^{\boldsymbol{n}} a_{\boldsymbol{i},\boldsymbol{n}} \chi_{I_{\boldsymbol{i},\boldsymbol{n}}} \right| \le \|a\|_\infty < \infty, \qquad \frac{1}{N(\boldsymbol{n})} \sum_{\boldsymbol{i}=\mathbf{1}}^{\boldsymbol{n}} a_{\boldsymbol{i},\boldsymbol{n}} = \int_{[0,1]^d} \Big(\sum_{\boldsymbol{i}=\mathbf{1}}^{\boldsymbol{n}} a_{\boldsymbol{i},\boldsymbol{n}} \chi_{I_{\boldsymbol{i},\boldsymbol{n}}} \Big),$$

(2.16) follows from (2.15) and from the dominated convergence theorem. □

2.4 Matrix Norms

For the reader's convenience, we report from [22, Sects. 2.4.1 and 2.4.3] several matrix-norm inequalities that we shall use throughout the book. First, we recall the expressions of the p-norms for $p = 1, \infty$:

$$|X|_1 = \max_{j=1,\dots,m} \sum_{i=1}^m |x_{ij}|, \qquad |X|_\infty = \max_{i=1,\dots,m} \sum_{j=1}^m |x_{ij}|, \qquad X \in \mathbb{C}^{m\times m}.$$

An important bound for $\|X\|$ in terms of $|X|_1$ and $|X|_\infty$ (and hence in terms of the components of X) is the following:

$$\|X\| \le \sqrt{|X|_1|X|_\infty} \le \max(|X|_1, |X|_\infty), \qquad X \in \mathbb{C}^{m\times m}. \tag{2.17}$$

Given $1 \le p, q \le \infty$, we say that p, q are conjugate exponents if $\frac{1}{p} + \frac{1}{q} = 1$ (it is understood that $\frac{1}{\infty} = 0$). The following Hölder-type inequality holds for the Schatten norms:

$$\|XY\|_1 \le \|X\|_p\|Y\|_q, \qquad X, Y \in \mathbb{C}^{m\times m}. \tag{2.18}$$

We will also need the following trace-norm inequalities:

$$\|X\|_1 \le \text{rank}(X)\|X\| \le m\|X\|, \qquad X \in \mathbb{C}^{m\times m}, \tag{2.19}$$

$$\|X\|_1 \le \sum_{i,j=1}^{m} |x_{ij}|, \qquad X \in \mathbb{C}^{m\times m}. \tag{2.20}$$

2.5 Tensor Products and Direct Sums

If X, Y are matrices of any dimension, say $X \in \mathbb{C}^{m_1\times m_2}$ and $Y \in \mathbb{C}^{\ell_1\times \ell_2}$, the tensor (Kronecker) product of X and Y is the $m_1\ell_1 \times m_2\ell_2$ matrix defined by

$$X \otimes Y = \left[x_{ij}Y\right]_{\substack{i=1,\dots,m_1\\ j=1,\dots,m_2}} = \begin{bmatrix} x_{11}Y & \cdots & x_{1m_2}Y \\ \vdots & & \vdots \\ x_{m_1 1}Y & \cdots & x_{m_1 m_2}Y \end{bmatrix},$$

and the direct sum of X and Y is the $(m_1 + \ell_1) \times (m_2 + \ell_2)$ matrix defined by

$$X \oplus Y = \text{diag}(X, Y) = \begin{bmatrix} X & O \\ O & Y \end{bmatrix}.$$

Tensor products and direct sums possess a lot of nice algebraic properties.

(i) Associativity: for all matrices X, Y, Z,

$$(X \otimes Y) \otimes Z = X \otimes (Y \otimes Z),$$
$$(X \oplus Y) \oplus Z = X \oplus (Y \oplus Z).$$

We can therefore omit parentheses in expressions like $X_1 \otimes X_2 \otimes \cdots \otimes X_d$ or $X_1 \oplus X_2 \oplus \cdots \oplus X_d$.

(ii) If X_1, X_2 can be multiplied and Y_1, Y_2 can be multiplied, then

$$(X_1 \otimes Y_1)(X_2 \otimes Y_2) = (X_1X_2) \otimes (Y_1Y_2),$$
$$(X_1 \oplus Y_1)(X_2 \oplus Y_2) = (X_1X_2) \oplus (Y_1Y_2).$$

(iii) For all matrices X, Y,

$$(X \otimes Y)^* = X^* \otimes Y^*, \qquad (X \otimes Y)^T = X^T \otimes Y^T$$
$$(X \oplus Y)^* = X^* \oplus Y^*, \qquad (X \oplus Y)^T = X^T \oplus Y^T.$$

(iv) Bilinearity (of tensor products): for each fixed matrix X, the application

$$Y \mapsto X \otimes Y$$

is linear on $\mathbb{C}^{\ell_1 \times \ell_2}$ for all $\ell_1, \ell_2 \in \mathbb{N}$; for each fixed matrix Y, the application

$$X \mapsto X \otimes Y$$

is linear on $\mathbb{C}^{m_1 \times m_2}$ for all $m_1, m_2 \in \mathbb{N}$.

From (i)–(iv), a lot of other interesting properties follow. For example, if X, Y are invertible, then $X \otimes Y$ is invertible, with inverse $X^{-1} \otimes Y^{-1}$. If X, Y are normal (resp., Hermitian, symmetric, unitary) then $X \otimes Y$ is also normal (resp., Hermitian, symmetric, unitary). If $X \in \mathbb{C}^{m \times m}$ and $Y \in \mathbb{C}^{\ell \times \ell}$, the eigenvalues and singular values of $X \otimes Y$ are

$$\{\lambda_i(X)\lambda_j(Y) : i = 1, \ldots, m, \; j = 1, \ldots, \ell\}, \tag{2.21}$$
$$\{\sigma_i(X)\sigma_j(Y) : i = 1, \ldots, m, \; j = 1, \ldots, \ell\}; \tag{2.22}$$

and the eigenvalues and singular values of $X \oplus Y$ are

$$\{\lambda_i(X) : i = 1, \ldots, m\} \cup \{\lambda_j(Y) : \; j = 1, \ldots, \ell\}, \tag{2.23}$$
$$\{\sigma_i(X) : i = 1, \ldots, m\} \cup \{\sigma_j(Y) : \; j = 1, \ldots, \ell\}; \tag{2.24}$$

see [22, Exercise 2.5]. In particular, for all $X \in \mathbb{C}^{m \times m}$, $Y \in \mathbb{C}^{\ell \times \ell}$, and $1 \le p \le \infty$, we have

$$\|X \otimes Y\|_p = \|X\|_p \, \|Y\|_p, \tag{2.25}$$
$$\|X \oplus Y\|_p = \big|(\|X\|_p, \|Y\|_p)\big|_p = \begin{cases} (\|X\|_p^p + \|Y\|_p^p)^{1/p}, & \text{if } 1 \le p < \infty, \\ \max(\|X\|_\infty, \|Y\|_\infty), & \text{if } p = \infty, \end{cases} \tag{2.26}$$

and

$$\operatorname{rank}(X \otimes Y) = \operatorname{rank}(X)\operatorname{rank}(Y), \tag{2.27}$$
$$\operatorname{rank}(X \oplus Y) = \operatorname{rank}(X) + \operatorname{rank}(Y). \tag{2.28}$$

In addition to the properties considered so far, we need to highlight two further properties of tensor products, which are very important for the "multidimensional purposes" of this book. The first one is the *multi-index formula for tensor products*: if we have d matrices $X_k \in \mathbb{C}^{m_k \times m_k}$, $k = 1, \ldots, d$, then

$$(X_1 \otimes X_2 \otimes \cdots \otimes X_d)_{\boldsymbol{i}\boldsymbol{j}} = (X_1)_{i_1 j_1}(X_2)_{i_2 j_2} \cdots (X_d)_{i_d j_d}, \quad \boldsymbol{i}, \boldsymbol{j} = \mathbf{1}, \ldots, \boldsymbol{m}, \tag{2.29}$$

where $\boldsymbol{m} = (m_1, m_2, \ldots, m_d)$. Note that (2.29) can be rewritten in the form (2.4) as follows:

$$X_1 \otimes X_2 \otimes \cdots \otimes X_d = \left[(X_1)_{i_1 j_1}(X_2)_{i_2 j_2} \cdots (X_d)_{i_d j_d}\right]_{\boldsymbol{i},\boldsymbol{j}=\mathbf{1}}^{\boldsymbol{m}}. \tag{2.30}$$

Note also that $X_1 \otimes X_2 \otimes \cdots \otimes X_d$ is one of the most eminent example of a d-level matrix with level orders $m_1, m_2, \ldots, m_d$ and total order $N(\boldsymbol{m})$; see Example 2.5 for the corresponding definitions. Equation (2.29) is of fundamental importance and, indeed, it motivates the introduction of multi-indices to index the entries of a matrix formed by a sum of tensor products. To better understand the importance of (2.29), try to write the (i, j) entry of $X_1 \otimes X_2 \otimes \cdots \otimes X_d$ as a function of two scalar indices $i, j = 1, \ldots, N(\boldsymbol{m})$; see also Example 2.4.

The second property is a natural *upper bound for the rank of the difference of two tensor products* formed by d factors. More precisely, suppose we have $2d$ matrices $X_1, \ldots, X_d, Y_1, \ldots, Y_d$, with $X_i, Y_i \in \mathbb{C}^{m_i \times m_i}$ for all $i = 1, \ldots, d$; then,

$$\operatorname{rank}(X_1 \otimes \cdots \otimes X_d - Y_1 \otimes \cdots \otimes Y_d) \le N(\boldsymbol{m}) \sum_{i=1}^{d} \frac{\operatorname{rank}(X_i - Y_i)}{m_i}, \tag{2.31}$$

where $\boldsymbol{m} = (m_1, \ldots, m_d)$. This is true because

$$\begin{aligned}
&\operatorname{rank}(X_1 \otimes \cdots \otimes X_d - Y_1 \otimes \cdots \otimes Y_d)\\
&= \operatorname{rank}\left(\sum_{i=1}^{d} Y_1 \otimes \cdots \otimes Y_{i-1} \otimes (X_i - Y_i) \otimes X_{i+1} \otimes \cdots \otimes X_d\right)\\
&\le \sum_{i=1}^{d} \operatorname{rank}(Y_1 \otimes \cdots \otimes Y_{i-1} \otimes (X_i - Y_i) \otimes X_{i+1} \otimes \cdots \otimes X_d)\\
&= \sum_{i=1}^{d} \operatorname{rank}(Y_1 \otimes \cdots \otimes Y_{i-1})\operatorname{rank}(X_i - Y_i)\operatorname{rank}(X_{i+1} \otimes \cdots \otimes X_d)
\end{aligned}$$

$$\leq \sum_{i=1}^{d} m_1 \cdots m_{i-1} \operatorname{rank}(X_i - Y_i) m_{i+1} \cdots m_d.$$

We conclude this section with a few results concerning the commutative properties of tensor products and direct sums. We also discuss the distributive properties of tensor products with respect to direct sums.

Lemma 2.6 *For every $\boldsymbol{m} \in \mathbb{N}^d$ and every permutation σ of the set $\{1, \ldots, d\}$, there exists a permutation matrix $\Pi_{\boldsymbol{m};\sigma}$ of size $N(\boldsymbol{m})$ such that*

$$X_{\sigma(1)} \otimes X_{\sigma(2)} \otimes \cdots \otimes X_{\sigma(d)} = \Pi_{\boldsymbol{m};\sigma}(X_1 \otimes X_2 \otimes \cdots \otimes X_d)\Pi_{\boldsymbol{m};\sigma}^T$$

for all matrices $X_1 \in \mathbb{C}^{m_1 \times m_1}$, $X_2 \in \mathbb{C}^{m_2 \times m_2}$, $\ldots$, $X_d \in \mathbb{C}^{m_d \times m_d}$.

Proof The proof proceeds by induction on d. The case $d = 1$ is trivial. For $d = 2$, the result is clear if σ is the identity $[1, 2]$, so we only have to prove it for $\sigma = [2, 1]$. In other words, we have to show that for every $\boldsymbol{m} \in \mathbb{N}^2$ there exists a permutation matrix $\Pi_{\boldsymbol{m};[2,1]}$ such that

$$X_2 \otimes X_1 = \Pi_{\boldsymbol{m};[2,1]}(X_1 \otimes X_2)\Pi_{\boldsymbol{m};[2,1]}^T \tag{2.32}$$

for all $X_1 \in \mathbb{C}^{m_1 \times m_1}$ and $X_2 \in \mathbb{C}^{m_2 \times m_2}$. Let $\Pi_{\boldsymbol{m};[2,1]}$ be the permutation matrix associated with the permutation ς of $\{1, \ldots, m_1m_2\}$ given by

$$\begin{aligned}\varsigma = [&1, m_2 + 1, 2m_2 + 1, \ldots, (m_1 - 1)m_2 + 1,\\ &2, m_2 + 2, 2m_2 + 2, \ldots, (m_1 - 1)m_2 + 2,\\ &\ldots\ldots\ldots,\\ &m_2, 2m_2, 3m_2 \ldots, m_1m_2],\end{aligned}$$

i.e.,

$$\varsigma(i) = ((i - 1) \bmod m_1)m_2 + \left\lfloor \frac{i-1}{m_1} \right\rfloor + 1, \qquad i = 1, \ldots, m_1m_2.$$

In other words, $\Pi_{\boldsymbol{m};[2,1]}$ is the matrix whose rows are, in this order, $\mathbf{e}_{\varsigma(1)}^T, \ldots, \mathbf{e}_{\varsigma(m_1m_2)}^T$, where $\mathbf{e}_1, \ldots, \mathbf{e}_{m_1m_2}$ are the vectors of the canonical basis of $\mathbb{C}^{m_1m_2}$. It can be verified that $\Pi_{\boldsymbol{m};[2,1]}$ satisfies (2.32) for all $X_1 \in \mathbb{C}^{m_1 \times m_1}$ and $X_2 \in \mathbb{C}^{m_2 \times m_2}$. The verification can be done componentwise, by showing that the (i, j) entry of the first matrix in (2.32) is equal to the (i, j) entry of the second matrix for all $i, j = 1, \ldots, m_1m_2$. This completes the proof of the lemma for $d = 2$.

For $d \geq 3$, we assume that the lemma holds for indices up to $d - 1$ and we prove that it holds for d. Let $\boldsymbol{m} \in \mathbb{N}^d$ and let σ be a permutation of $\{1, \ldots, d\}$. Let $1 \leq i \leq d$ be the index such that $\sigma(i) = d$, and let τ be the permutation of $\{1, \ldots, d - 1\}$ defined by $\tau(j) = \sigma(j)$ for $j = 1, \ldots, i - 1$ and $\tau(j) = \sigma(j + 1)$ for $j = i, \ldots, d - 1$. If $i = d$, then, by induction hypothesis and the properties of tensor products,

$$\begin{aligned}
X_{\sigma(1)} \otimes \cdots \otimes X_{\sigma(d)} &= X_{\tau(1)} \otimes \cdots \otimes X_{\tau(d-1)} \otimes X_d \\
&= \left[\Pi_{(m_1,\ldots,m_{d-1});\tau}(X_1 \otimes \cdots \otimes X_{d-1})\Pi^T_{(m_1,\ldots,m_{d-1});\tau}\right] \otimes X_d \\
&= (\Pi_{(m_1,\ldots,m_{d-1});\tau} \otimes I_{m_d})(X_1 \otimes \cdots \otimes X_{d-1} \otimes X_d)(\Pi^T_{(m_1,\ldots,m_{d-1});\tau} \otimes I_{m_d}),
\end{aligned}$$

and the thesis holds with $\Pi_{\boldsymbol{m};\sigma} = \Pi_{(m_1,\ldots,m_{d-1});\tau} \otimes I_{m_d}$. If $i < d$, then

$$\begin{aligned}
X_{\sigma(1)} \otimes \cdots \otimes X_{\sigma(d)} &= X_{\sigma(1)} \otimes \cdots \otimes X_{\sigma(i-1)} \otimes X_d \otimes X_{\sigma(i+1)} \otimes \cdots \otimes X_{\sigma(d)} \\
&= X_{\sigma(1)} \otimes \cdots \otimes X_{\sigma(i-1)} \otimes \\
&\quad \left[\Pi_{(m_{\sigma(i+1)}\cdots m_{\sigma(d)},m_d);[2,1]}(X_{\sigma(i+1)} \otimes \cdots \otimes X_{\sigma(d)} \otimes X_d)\Pi^T_{(m_{\sigma(i+1)}\cdots m_{\sigma(d)},m_d);[2,1]}\right] \\
&= (I_{m_{\sigma(1)}\cdots m_{\sigma(i-1)}} \otimes \Pi_{(m_{\sigma(i+1)}\cdots m_{\sigma(d)},m_d);[2,1]}) \\
&\quad \cdot (X_{\sigma(1)} \otimes \cdots \otimes X_{\sigma(i-1)} \otimes X_{\sigma(i+1)} \otimes \cdots \otimes X_{\sigma(d)} \otimes X_d) \\
&\quad \cdot (I_{m_{\sigma(1)}\cdots m_{\sigma(i-1)}} \otimes \Pi^T_{(m_{\sigma(i+1)}\cdots m_{\sigma(d)},m_d);[2,1]}) \\
&= P_{\boldsymbol{m};\sigma}(X_{\tau(1)} \otimes \cdots \otimes X_{\tau(d-1)} \otimes X_d)P^T_{\boldsymbol{m};\sigma},
\end{aligned} \tag{2.33}$$

where $P_{\boldsymbol{m};\sigma} = I_{m_{\sigma(1)}\cdots m_{\sigma(i-1)}} \otimes \Pi_{(m_{\sigma(i+1)}\cdots m_{\sigma(d)},m_d);[2,1]}$. By induction hypothesis,

$$X_{\tau(1)} \otimes \cdots \otimes X_{\tau(d-1)} = \Pi_{(m_1,\ldots,m_{d-1});\tau}(X_1 \otimes \cdots \otimes X_{d-1})\Pi^T_{(m_1,\ldots,m_{d-1});\tau}.$$

Substituting in (2.33) and using the properties of tensor products, we see that the thesis holds with $\Pi_{\boldsymbol{m};\sigma} = P_{\boldsymbol{m};\sigma}(\Pi_{(m_1,\ldots,m_{d-1});\tau} \otimes I_{m_d})$, which is a permutation matrix because it is the product of two permutation matrices. □

Lemma 2.7 *For every $\boldsymbol{m} \in \mathbb{N}^d$ and every permutation σ of the set $\{1,\ldots,d\}$, there exists a permutation matrix $V_{\boldsymbol{m};\sigma}$ of size $m_1 + \ldots + m_d$ such that*

$$X_{\sigma(1)} \oplus X_{\sigma(2)} \oplus \cdots \oplus X_{\sigma(d)} = V_{\boldsymbol{m};\sigma}(X_1 \oplus X_2 \oplus \cdots \oplus X_d)V^T_{\boldsymbol{m};\sigma}$$

for all matrices $X_1 \in \mathbb{C}^{m_1\times m_1}$, $X_2 \in \mathbb{C}^{m_2\times m_2}$, $\ldots$, $X_d \in \mathbb{C}^{m_d\times m_d}$.

Proof The proof proceeds by induction on d. The case $d = 1$ is trivial. For $d = 2$, the only possible permutations are the identity $\sigma = [1, 2]$ and the transposition $\sigma = [2, 1]$, and we can take

$$V_{\boldsymbol{m};[1,2]} = I_{m_1+m_2}, \qquad V_{\boldsymbol{m};[2,1]} = \begin{bmatrix} O & I_{m_2} \\ I_{m_1} & O \end{bmatrix}.$$

For $d \geq 3$, we assume that the lemma holds for indices up to $d - 1$ and we prove that it holds for d. Let $1 \leq i \leq d$ be the index such that $\sigma(i) = d$, and let τ be the permutation of $\{1,\ldots,d-1\}$ defined be $\tau(j) = \sigma(j)$ for $j = 1,\ldots,i-1$ and $\tau(j) = \sigma(j+1)$ for $j = i,\ldots,d-1$. If $i = d$, then, by induction hypothesis and the properties of direct sums,

$$
\begin{aligned}
X_{\sigma(1)} \oplus \cdots \oplus X_{\sigma(d)} &= X_{\tau(1)} \oplus \cdots \oplus X_{\tau(d-1)} \oplus X_d \\
&= V_{(m_1,\dots,m_{d-1});\tau}(X_1 \oplus \cdots \oplus X_{d-1})V^T_{(m_1,\dots,m_{d-1});\tau} \oplus X_d \\
&= (V_{(m_1,\dots,m_{d-1});\tau} \oplus I_{m_d})(X_1 \oplus \cdots \oplus X_{d-1} \oplus X_d)(V^T_{(m_1,\dots,m_{d-1});\tau} \oplus I_{m_d})
\end{aligned}
$$

and the thesis holds with $V_{\boldsymbol{m};\sigma} = V_{(m_1,\dots,m_{d-1});\tau} \oplus I_{m_d}$. If $i < d$, then

$$
\begin{aligned}
& X_{\sigma(1)} \oplus \cdots \oplus X_{\sigma(d)} = X_{\sigma(1)} \oplus \cdots \oplus X_{\sigma(i-1)} \oplus X_d \oplus X_{\sigma(i+1)} \oplus \cdots \oplus X_{\sigma(d)} \\
&= X_{\sigma(1)} \oplus \cdots \oplus X_{\sigma(i-1)} \oplus \\
&\quad \left[V_{(m_{\sigma(i+1)}+\dots+m_{\sigma(d)},m_d);[2,1]}(X_{\sigma(i+1)} \oplus \cdots \oplus X_{\sigma(d)} \oplus X_d)V^T_{(m_{\sigma(i+1)}+\dots+m_{\sigma(d)},m_d);[2,1]}\right] \\
&= (I_{m_{\sigma(1)}+\dots+m_{\sigma(i-1)}} \oplus V_{(m_{\sigma(i+1)}+\dots+m_{\sigma(d)},m_d);[2,1]}) \\
&\quad \cdot (X_{\sigma(1)} \oplus \cdots \oplus X_{\sigma(i-1)} \oplus X_{\sigma(i+1)} \oplus \cdots \oplus X_{\sigma(d)} \oplus X_d) \\
&\quad \cdot (I_{m_{\sigma(1)}+\dots+m_{\sigma(i-1)}} \oplus V^T_{(m_{\sigma(i+1)}+\dots+m_{\sigma(d)},m_d);[2,1]}) \\
&= U_{\boldsymbol{m};\sigma}\left(X_{\tau(1)} \oplus \cdots \oplus X_{\tau(d-1)} \oplus X_d\right) U^T_{\boldsymbol{m};\sigma},
\end{aligned} \tag{2.34}
$$

where $U_{\boldsymbol{m};\sigma} = I_{m_{\sigma(1)}+\dots+m_{\sigma(i-1)}} \oplus V_{(m_{\sigma(i+1)}+\dots+m_{\sigma(d)},m_d);[2,1]}$. By induction hypothesis,

$$X_{\tau(1)} \oplus \cdots \oplus X_{\tau(d-1)} = V_{(m_1,\dots,m_{d-1});\tau}(X_1 \oplus \cdots \oplus X_{d-1})V^T_{(m_1,\dots,m_{d-1});\tau}.$$

Substituting in (2.34) and using the properties of direct sums, we see that the thesis holds with $V_{\boldsymbol{m};\sigma} = U_{\boldsymbol{m};\sigma}(V_{(m_1,\dots,m_{d-1});\tau} \oplus I_{m_d})$, which is a permutation matrix because it is the product of two permutation matrices. □

Lemmas 2.6 and 2.7 show that tensor products and direct sums are “almost” commutative. More precisely, Lemma 2.6 says that *the tensor product operation is commutative, up to a permutation transformation $\Pi_{\boldsymbol{m};\sigma}$ which depends on $\boldsymbol{m}$ and σ, but not on the specific matrices $X_1, X_2, \dots, X_d$*. Lemma 2.7 says the same thing for the direct sum operation.

Concerning the distributive properties of tensor products with respect to direct sums, it follows directly from the definitions that the distributive law on the right holds without permutation transformations. In other words, for all matrices $X_1, \dots, X_d, Y$ we have

$$(X_1 \oplus X_2 \oplus \cdots \oplus X_d) \otimes Y = (X_1 \otimes Y) \oplus (X_2 \otimes Y) \oplus \cdots \oplus (X_d \otimes Y). \tag{2.35}$$

As for the distributive law on the left, a result analogous to Lemmas 2.6 and 2.7 holds, showing that this property holds modulo permutation transformations which only depend on the dimensions of the involved matrices. More precisely, for every $\ell \in \mathbb{N}$ and $\boldsymbol{m} \in \mathbb{N}^d$, there exists a permutation matrix $Q_{\ell,\boldsymbol{m}}$ of size $\ell(m_1 + \ldots + m_d)$ such that

$$X \otimes (Y_1 \oplus Y_2 \oplus \cdots \oplus Y_d) = Q_{\ell,\boldsymbol{m}}\big[(X \otimes Y_1) \oplus (X \otimes Y_2) \oplus \cdots \oplus (X \otimes Y_d)\big]Q^T_{\ell,\boldsymbol{m}} \tag{2.36}$$

for all matrices $X \in \mathbb{C}^{\ell\times\ell}$, $Y_1 \in \mathbb{C}^{m_1\times m_1}$, $Y_2 \in \mathbb{C}^{m_2\times m_2}$, ..., $Y_d \in \mathbb{C}^{m_d\times m_d}$. In this book, however, we will only need the distributive law on the right (2.35), and so we do not provide the proof of the distributive law on the left (2.36); the interested reader is referred to [24, Lemma 2].

2.6 Singular Value and Eigenvalue Distribution of a Sequence of Matrices

We recall that, throughout this book, a sequence of matrices is a sequence of the form $\{A_n\}_n$, where n varies in some infinite subset of $\mathbb{N}$ and A_n is a square matrix of size $d_n \to \infty$. In [22, Chap. 3] we studied the notion of (asymptotic) singular value and eigenvalue distribution, as well as other related concepts such as clustering and attraction, for sequences of matrices $\{A_n\}_n$ such that $d_n = n$. We also studied the so-called zero-distributed sequences $\{Z_n\}_n$ in the case where the size of the nth matrix Z_n is $d_n = n$. In [22, Chap. 4] we investigated the spectral distribution of sequences of perturbed Hermitian matrices, i.e., sequences of the form $\{A_n = X_n + Y_n\}_n$, where X_n is Hermitian and Y_n is a small perturbation of X_n; once again, the size of X_n and Y_n was supposed to be $d_n = n$. The crucial observation is that the assumption $d_n = n$ was made only for reasons of notational simplicity and elegance. Indeed, apart from obvious adaptations, *nothing changes in* [22, Chaps. 3 and 4] *if the assumption* "$d_n = n$" *is replaced by* "$d_n \to \infty$ *as* $n \to \infty$". In this section, for the reader's convenience, we rewrite definitions, results, etc., from [22, Chaps. 3 and 4] for general sequences of matrices, i.e., without the constraint $d_n = n$. Of course, all the proofs are omitted because, up to obvious adaptations, they are the same as in [22]; it is therefore not instructive to reproduce them here.

2.6.1 *The Notion of Singular Value and Eigenvalue Distribution*

Let $\mathbb{K}$ be either $\mathbb{R}$ or $\mathbb{C}$ and let $g : D \subset \mathbb{R}^k \to \mathbb{K}$ be a measurable function defined on a set D with $0 < \mu_k(D) < \infty$. To any such g we associate the functional ϕ_g defined by Eq. (2.2). This functional will play an important role in what follows.

Definition 2.1 (*singular value and eigenvalue distribution of a sequence of matrices*) Let $\{A_n\}_n$ be a sequence of matrices, with A_n of size d_n.

- We say that $\{A_n\}_n$ has an asymptotic singular value distribution described by a functional $\phi : C_c(\mathbb{R}) \to \mathbb{C}$, and we write $\{A_n\}_n \sim_\sigma \phi$, if

$$\lim_{n\to\infty} \frac{1}{d_n} \sum_{j=1}^{d_n} F(\sigma_j(A_n)) = \phi(F), \qquad \forall\, F \in C_c(\mathbb{R}). \tag{2.37}$$

If $\phi = \phi_{|f|}$ for some measurable $f : D \subset \mathbb{R}^k \to \mathbb{C}$ defined on a set D with $0 < \mu_k(D) < \infty$, we say that $\{A_n\}_n$ has an asymptotic singular value distribution described by f and we write $\{A_n\}_n \sim_\sigma f$. In this case, the function f is referred to as the singular value symbol of $\{A_n\}_n$.

- We say that $\{A_n\}_n$ has an asymptotic eigenvalue (or spectral) distribution described by a functional $\phi : C_c(\mathbb{C}) \to \mathbb{C}$, and we write $\{A_n\}_n \sim_\lambda \phi$, if

$$\lim_{n\to\infty} \frac{1}{d_n} \sum_{j=1}^{d_n} F(\lambda_j(A_n)) = \phi(F), \qquad \forall F \in C_c(\mathbb{C}). \tag{2.38}$$

If $\phi = \phi_f$ for some measurable $f : D \subset \mathbb{R}^k \to \mathbb{C}$ defined on a set D with $0 < \mu_k(D) < \infty$, we say that $\{A_n\}_n$ has an asymptotic eigenvalue (or spectral) distribution described by f and we write $\{A_n\}_n \sim_\lambda f$. In this case, the function f is referred to as the eigenvalue (or spectral) symbol of $\{A_n\}_n$.

The adjective "asymptotic" is often omitted, i.e., one often simply says that $\{A_n\}_n$ has a singular value or eigenvalue/spectral distribution (described by something). When we write a relation such as $\{A_n\}_n \sim_\sigma \phi$ (resp., $\{A_n\}_n \sim_\lambda \phi$), it is understood that ϕ is a functional on $C_c(\mathbb{R})$ (resp., $C_c(\mathbb{C})$), as in Definition 2.1. Similarly, when we write a relation such as $\{A_n\}_n \sim_\sigma f$ or $\{A_n\}_n \sim_\lambda f$, it is understood that f is as in Definition 2.1; that is, f is a measurable function defined on a subset D of some $\mathbb{R}^k$ with $0 < \mu_k(D) < \infty$. Sometimes, for brevity, we will write $\{A_n\}_n \sim_{\sigma,\lambda} f$ to indicate that $\{A_n\}_n \sim_\sigma f$ and $\{A_n\}_n \sim_\lambda f$.

Remark 2.1 (*informal meaning of the singular value and eigenvalue distribution*) By definition of ϕ_f, the spectral distribution $\{A_n\}_n \sim_\lambda f$ means that

$$\lim_{n\to\infty} \frac{1}{d_n} \sum_{j=1}^{d_n} F(\lambda_j(A_n)) = \frac{1}{\mu_k(D)} \int_D F(f(\mathbf{x}))\mathrm{d}\mathbf{x}, \qquad \forall F \in C_c(\mathbb{C}). \tag{2.39}$$

Intuitively, if the function f were continuous a.e. and the eigenvalues of A_n were exact samples of f over an equispaced grid in D, then (2.39) would be satisfied as it would simply say that a special Riemann sum of a Riemann-integrable function converges to the corresponding integral. The informal meaning behind (2.39) is in fact the following. Assuming that f is continuous a.e., if n is large enough and $\{\mathbf{x}_{j,n} : j = 1, \ldots, d_n\}$ is an equispaced grid in D, then a suitable ordering of the eigenvalues of A_n is such that the pairs $\{(\mathbf{x}_{j,n}, \lambda_j(A_n)) : j = 1, \ldots, d_n\}$ reconstruct approximately the hypersurface $\{(\mathbf{x}, f(\mathbf{x})) : \mathbf{x} \in D\}$. In other words, the eigenvalues of A_n, except possibly for $o(d_n)$ outliers, are approximately equal to the samples of f over a uniform grid in D (for n large enough). For instance, if $k = 1$ and $D = [a, b]$, then, assuming we have no outliers, the eigenvalues of A_n are approximately equal to

$$f\Big(a + i\,\frac{b-a}{d_n}\Big), \qquad i = 1, \ldots, d_n,$$

for n large enough. Similarly, if $k = 2$, d_n is a perfect square and $D = [a_1, b_1] \times [a_2, b_2]$, then, assuming we have no outliers, the eigenvalues of A_n are approximately equal to

$$f\Big(a_1 + i_1\,\frac{b_1 - a_1}{\sqrt{d_n}},\; a_2 + i_2\,\frac{b_2 - a_2}{\sqrt{d_n}}\Big), \qquad i_1, i_2 = 1, \ldots, \sqrt{d_n},$$

for n large enough. A completely analogous meaning can also be given for the singular value distribution $\{A_n\}_n \sim_\sigma f$, which is equivalent to

$$\lim_{n\to\infty} \frac{1}{d_n}\sum_{j=1}^{d_n} F(\sigma_j(A_n)) = \frac{1}{\mu_k(D)}\int_D F(|f(\mathbf{x})|)\mathrm{d}\mathbf{x}, \qquad \forall\, F \in C_c(\mathbb{R}). \tag{2.40}$$

Remark 2.2 It is clear that $\{A_n\}_n \sim_\sigma f$ is equivalent to $\{A_n\}_n \sim_\sigma |f|$. Moreover, if every A_n is normal, then $\{A_n\}_n \sim_\lambda f$ implies $\{A_n\}_n \sim_\sigma f$. Indeed, the singular values of a normal matrix coincide with the moduli of the eigenvalues [22, p. 30]. Therefore, for any fixed $F \in C_c(\mathbb{R})$, by applying the eigenvalue distribution (2.39) with the test function $F(|\cdot|) \in C_c(\mathbb{C})$, we get

$$\lim_{n\to\infty} \frac{1}{d_n}\sum_{j=1}^{d_n} F(\sigma_j(A_n)) = \lim_{n\to\infty} \frac{1}{d_n}\sum_{j=1}^{d_n} F(|\lambda_j(A_n)|) = \frac{1}{\mu_k(D)}\int_D F(|f(\mathbf{x})|)\mathrm{d}\mathbf{x}.$$

2.6.2 Clustering and Attraction

This subsection introduces the notions of clustering and attraction for general sequences of matrices. Throughout the book, if $z \in \mathbb{C}$ and $\varepsilon > 0$, we denote by $D(z, \varepsilon)$ the disk with center z and radius ε, i.e., $D(z, \varepsilon) = \{w \in \mathbb{C} : \; |w - z| < \varepsilon\}$. If $S \subseteq \mathbb{C}$ and $\varepsilon > 0$, we denote by $D(S, \varepsilon)$ the ε-expansion of S, which is defined as $D(S, \varepsilon) = \bigcup_{z\in S} D(z, \varepsilon)$.

Definition 2.2 (*clustering of a sequence of matrices*) Let $\{A_n\}_n$ be a sequence of matrices, with A_n of size d_n, and let $S \subseteq \mathbb{C}$ be a nonempty subset of $\mathbb{C}$.

- We say that $\{A_n\}_n$ is strongly clustered at S (in the sense of the eigenvalues), or equivalently that the eigenvalues of $\{A_n\}_n$ are strongly clustered at S, if, for every $\varepsilon > 0$, the number of eigenvalues of A_n lying outside $D(S, \varepsilon)$ is bounded by a constant C_ε independent of n. In other words, for every $\varepsilon > 0$ we have

$$\#\{j \in \{1, \ldots, d_n\} : \; \lambda_j(A_n) \notin D(S, \varepsilon)\} = O(1). \tag{2.41}$$

- We say that $\{A_n\}$ is weakly clustered at S (in the sense of the eigenvalues), or equivalently that the eigenvalues of $\{A_n\}_n$ are weakly clustered at S, if, for every $\varepsilon > 0$,

$$\#\{j \in \{1, \ldots, d_n\} : \lambda_j(A_n) \notin D(S, \varepsilon)\} = o(d_n). \tag{2.42}$$

By replacing "eigenvalues" with "singular values" and $\lambda_j(A_n)$ with $\sigma_j(A_n)$ in (2.41)–(2.42), we obtain the definitions of a sequence of matrices strongly or weakly clustered at a nonempty subset of $\mathbb{C}$ in the sense of the singular values.

Throughout this book, when we speak of strong/weak cluster, sequence of matrices strongly/weakly clustered, etc., without further specifications, it is understood "in the sense of the eigenvalues". When the clustering is intended in the sense of the singular values, this is specified every time.

Definition 2.3 (*spectral attraction*) Let $\{A_n\}_n$ be a sequence of matrices, with A_n of size d_n, and let $z \in \mathbb{C}$. We say that z strongly attracts the spectrum $\Lambda(A_n)$ with infinite order if, once we have ordered the eigenvalues of A_n according to their distance from z,

$$|\lambda_1(A_n) - z| \le |\lambda_2(A_n) - z| \le \ldots \le |\lambda_{d_n}(A_n) - z|,$$

the following limit relation holds for each fixed $j \ge 1$:

$$\lim_{n\to\infty} |\lambda_j(A_n) - z| = 0.$$

The next theorem and its corollary are the versions of [22, Theorem 3.1 and Corollary 3.1] for general sequences of matrices.

Theorem 2.1 *If $\{A_n\}_n \sim_\lambda f$, then $\{A_n\}_n$ is weakly clustered at the essential range $\mathcal{ER}(f)$ and every point of $\mathcal{ER}(f)$ strongly attracts the spectrum $\Lambda(A_n)$ with infinite order.*

Corollary 2.2 *If $\{A_n\}_n \sim_\lambda f$ and $\Lambda(A_n)$ is contained in $S \subseteq \mathbb{C}$ for all n, then $\mathcal{ER}(f)$ is contained in the closure $\overline{S}$. In particular, if $\{A_n\}_n \sim_\lambda f$ and the matrices A_n are Hermitian (resp., HPSD), then $\mathcal{ER}(f) \subseteq \mathbb{R}$ (resp., $\mathcal{ER}(f) \subseteq [0, \infty)$).*

2.6.3 Zero-Distributed Sequences

A sequence of matrices $\{Z_n\}_n$ is said to be a zero-distributed sequence if we have $\{Z_n\}_n \sim_\sigma 0$, where 0 is the identically zero function (defined on some unspecified measurable subset D of some $\mathbb{R}^k$ with $0 < \mu_k(D) < \infty$). In other words, $\{Z_n\}_n$ is zero-distributed if and only if

$$\lim_{n\to\infty} \frac{1}{d_n} \sum_{j=1}^{d_n} F(\sigma_j(Z_n)) = F(0), \qquad \forall F \in C_c(\mathbb{R}),$$

where, of course, d_n is the size of Z_n. Theorems 2.2 and 2.3 are the versions of [22, Theorems 3.2 and 3.3] for general sequences of matrices. In the statement of Theorem 2.3 and throughout the book, we use the natural convention $1/\infty = 0$.

Theorem 2.2 *Let $\{Z_n\}_n$ be a sequence of matrices, with Z_n of size d_n. The following conditions are equivalent.*

1. $\{Z_n\}_n \sim_\sigma 0$.
2. *For every* $\varepsilon > 0$, $\displaystyle\lim_{n\to\infty} \frac{\#\{j \in \{1,\dots,d_n\} : \sigma_j(Z_n) > \varepsilon\}}{d_n} = 0$.
3. *For every n we have* $Z_n = R_n + N_n$, *where* $\displaystyle\lim_{n\to\infty} \frac{\mathrm{rank}(R_n)}{d_n} = \lim_{n\to\infty} \|N_n\| = 0$.

With the terminology of clustering introduced in Sect. 2.6.2, condition 2 in Theorem 2.2 can be reformulated by saying that $\{Z_n\}_n$ is weakly clustered at $\{0\}$ in the sense of the singular values.

Theorem 2.3 *Let $\{Z_n\}_n$ be a sequence of matrices, with Z_n of size d_n, and suppose that* $\lim_{n\to\infty} \|Z_n\|_p/(d_n)^{1/p} = 0$ *for some* $p \in [1,\infty]$. *Then* $\{Z_n\}_n \sim_\sigma 0$.

Remark 2.3 (*algebra of zero-distributed sequences*) It follows from the equivalence $1 \iff 3$ in Theorem 2.2 that, for each fixed sequence of positive integers d_n such that $d_n \to \infty$ as $n \to \infty$, the set of zero-distributed sequences

$$\mathscr{Z} = \Big\{\{Z_n\}_n : Z_n \in \mathbb{C}^{d_n\times d_n},\ \{Z_n\}_n \sim_\sigma 0\Big\} \tag{2.43}$$

is a *-algebra over the complex field $\mathbb{C}$ with respect to the natural operations of conjugate transposition, addition, scalar-multiplication and product:

$$\begin{aligned} \{A_n\}_n^* &= \{A_n^*\}_n, \\ \{A_n\}_n + \{B_n\}_n &= \{A_n + B_n\}_n, \\ \alpha\{A_n\}_n &= \{\alpha A_n\}_n, \\ \{A_n\}_n\{B_n\}_n &= \{A_n B_n\}_n. \end{aligned} \tag{2.44}$$

2.6.4 Sparsely Unbounded and Sparsely Vanishing Sequences of Matrices

We introduce in this subsection the notions of sparsely unbounded and sparsely vanishing sequences of matrices, along with a number of related results, which are the versions of [22, Propositions 5.3 and 5.4, Remark 8.6 and Proposition 8.4] for general sequences of matrices.

Definition 2.4 (*sparsely unbounded sequence of matrices*) A sequence of matrices $\{A_n\}_n$, with A_n of size d_n, is said to be sparsely unbounded (s.u.) if for every $M > 0$ there exists n_M such that, for $n \ge n_M$,

$$\frac{\#\{i \in \{1, \ldots, d_n\} : \sigma_i(A_n) > M\}}{d_n} \le r(M),$$

where $\lim_{M\to\infty} r(M) = 0$.

Definition 2.5 (*sparsely vanishing sequence of matrices*) A sequence of matrices $\{A_n\}_n$, with A_n of size d_n, is said to be sparsely vanishing (s.v.) if for every $M > 0$ there exists n_M such that, for $n \ge n_M$,

$$\frac{\#\{i \in \{1, \ldots, d_n\} : \sigma_i(A_n) < 1/M\}}{d_n} \le r(M),$$

where $\lim_{M\to\infty} r(M) = 0$.

It is clear that if $\{A_n\}_n$ is s.v. then $\{A_n^\dagger\}_n$ is s.u.; it suffices to recall that the singular values of $A^\dagger$ are $1/\sigma_1(A), \ldots, 1/\sigma_r(A), 0, \ldots, 0$, where $\sigma_1(A), \ldots, \sigma_r(A)$ are the nonzero singular values of A ($r = \text{rank}(A)$).

Proposition 2.1 *Let $\{A_n\}_n$ be a sequence of matrices, with A_n of size d_n. The following conditions are equivalent.*

1. *$\{A_n\}_n$ is s.u.*
2. $\displaystyle\lim_{M\to\infty} \limsup_{n\to\infty} \frac{\#\{i \in \{1, \ldots, d_n\} : \sigma_i(A_n) > M\}}{d_n} = 0.$
3. *For every $M > 0$ there exists n_M such that, for $n \ge n_M$,*

$$A_n = \hat{A}_{n,M} + \tilde{A}_{n,M}, \qquad \text{rank}(\hat{A}_{n,M}) \le r(M)d_n, \qquad \|\tilde{A}_{n,M}\| \le M,$$

where $\lim_{M\to\infty} r(M) = 0$.

Note that condition 2 can be rewritten as $\displaystyle\lim_{M\to\infty} \limsup_{n\to\infty} \frac{1}{d_n} \sum_{i=1}^{d_n} \chi_{(M,\infty)}(\sigma_i(A_n)) = 0.$

Proposition 2.2 *If $\{A_n\}_n \sim_\sigma f$ then $\{A_n\}_n$ is s.u.*

Proposition 2.3 *Let $\{A_n\}_n$ be a sequence of matrices, with A_n of size d_n. The following conditions are equivalent.*

1. *$\{A_n\}_n$ is s.v.*
2. $\displaystyle\lim_{M\to\infty} \limsup_{n\to\infty} \frac{\#\{i \in \{1, \ldots, d_n\} : \sigma_i(A_n) < 1/M\}}{d_n} = 0.$

Note that condition 2 can be rewritten as $\displaystyle\lim_{M\to\infty} \limsup_{n\to\infty} \frac{1}{d_n} \sum_{i=1}^{d_n} \chi_{[0,1/M)}(\sigma_i(A_n)) = 0.$

Proposition 2.4 *If $\{A_n\}_n \sim_\sigma f$ then $\{A_n\}_n$ is s.v. if and only if $f \ne 0$ a.e.*

Remark 2.4 Let $\{A_n\}_n$ be an s.u. sequence of Hermitian matrices, with A_n of size d_n. Then, the following stronger version of condition 3 in Proposition 2.1 is satisfied: for every $M > 0$ there exists n_M such that, for $n \geq n_M$,

$$A_n = \hat{A}_{n,M} + \tilde{A}_{n,M}, \qquad \text{rank}(\hat{A}_{n,M}) \leq r(M)d_n, \qquad \|\tilde{A}_{n,M}\| \leq M,$$

where $\lim_{M\to\infty} r(M) = 0$, the matrices $\hat{A}_{n,M}$ and $\tilde{A}_{n,M}$ are Hermitian, and for all functions $g : \mathbb{R} \to \mathbb{R}$ we have

$$g(\hat{A}_{n,M} + \tilde{A}_{n,M}) = g(\hat{A}_{n,M}) + g(\tilde{A}_{n,M}).$$

This stronger version of condition 3 has been proved in [22, p. 157 (lines 21–34) and p. 158 (lines 1–8)] for the case "$d_n = n$", but the extension to the case "$d_n \to \infty$ as $n \to \infty$" is obvious.

2.6.5 Spectral Distribution of Sequences of Perturbed Hermitian Matrices

The spectral distribution of sequences of perturbed Hermitian matrices has been investigated in [22, Chap. 4]. Here, we report the version of the main result obtained therein [22, Theorem 4.3] for general sequences of matrices.

Theorem 2.4 *Let $\{X_n\}_n$, $\{Y_n\}_n$ be sequences of matrices, with X_n, Y_n of size d_n, and set $A_n = X_n + Y_n$. Assume that the following conditions are met.*

1. $\|X_n\|, \|Y_n\| \leq C$ *for all n, where C is a constant independent of n.*
2. *Every X_n is Hermitian and $\{X_n\}_n \sim_\lambda \phi$.*
3. $\|Y_n\|_1 = o(d_n)$.

Then $\{A_n\}_n \sim_\lambda \phi$.

Corollary 2.3 *Let $\{X_n\}_n$, $\{Y_n\}_n$ be sequences of matrices, with X_n, Y_n of size d_n, and set $A_n = X_n + Y_n$. Assume that the following conditions are met.*

1. $\|X_n\|, \|Y_n\| \leq C$ *for all n, where C is a constant independent of n.*
2. *Every X_n is Hermitian and $\{X_n\}_n \sim_\lambda f$.*
3. $\|Y_n\|_1 = o(d_n)$.

Then $\{A_n\}_n \sim_\lambda f$.

Proof Recall from Definition 2.1 that $\{A_n\}_n \sim_\lambda f$ means $\{A_n\}_n \sim_\lambda \phi_f$ and apply Theorem 2.4 with $\phi = \phi_f$. □

2.7 Approximating Classes of Sequences

In [22, Chap. 5] we studied the theory of approximating classes of sequences (a.c.s.). For notational simplicity and elegance, the theory presented therein refers to sequences of matrices $\{A_n\}_n$ such that the size of A_n is $d_n = n$. However, apart from obvious adaptations, *nothing changes in* [22, Chap. 5] *if the assumption* "$d_n = n$" *is replaced by* "$d_n \to \infty$ *as* $n \to \infty$". In this section, for the reader's convenience, we rewrite definitions, results, etc., from [22, Chap. 5] for general sequences of matrices, i.e., without the constraint $d_n = n$. Of course, all the proofs are omitted because, up to obvious adaptations, they are the same as in [22]; it is therefore not instructive to reproduce them here. We also avoid repeating again the motivations behind the notion of a.c.s., which can be found in [22, p. 65]. Throughout the book, we use the abbreviation "a.c.s." for both the singular "approximating class of sequences" and the plural "approximating classes of sequences"; it will be clear from the context whether "a.c.s." is singular or plural.

2.7.1 Definition of a.c.s. and a.c.s. Topology

Here is the formal definition of a.c.s., as appeared in the original paper [33].

Definition 2.6 (*approximating class of sequences*) Let $\{A_n\}_n$ be a sequence of matrices, with A_n of size d_n, and let $\{\{B_{n,m}\}_n\}_m$ be a sequence of sequences of matrices, with $B_{n,m}$ of size d_n. We say that $\{\{B_{n,m}\}_n\}_m$ is an approximating class of sequences (a.c.s.) for $\{A_n\}_n$ if the following condition is met: for every m there exists n_m such that, for $n \geq n_m$,

$$A_n = B_{n,m} + R_{n,m} + N_{n,m}, \qquad \text{rank}(R_{n,m}) \leq c(m)d_n, \qquad \|N_{n,m}\| \leq \omega(m), \quad (2.45)$$

where the quantities n_m, $c(m)$, $\omega(m)$ depend only on m, and

$$\lim_{m\to\infty} c(m) = \lim_{m\to\infty} \omega(m) = 0.$$

Roughly speaking, $\{\{B_{n,m}\}_n\}_m$ is an a.c.s. for $\{A_n\}_n$ if, for all sufficiently large m, the sequence $\{B_{n,m}\}_n$ approximates (asymptotically) the sequence $\{A_n\}_n$, in the sense that A_n is eventually equal to $B_{n,m}$ plus a small-rank matrix (with respect to the matrix size d_n) plus a small-norm matrix. Note that an equivalent definition of a.c.s. is obtained by replacing, in Definition 2.6, "for every m" with "for every sufficiently large m". Indeed, suppose that the splitting (2.45) holds for $m \geq M$. For $m < M$, define $n_m = 1$, $c(m) = 1$, $\omega(m) = 0$ and $R_{n,m} = A_{n,m} - B_{n,m}$, $N_{n,m} = O_{d_n}$. Then, we see that (2.45) actually holds for every m.

It turns out that, for each fixed sequence of positive integers d_n such that $d_n \to \infty$, the notion of a.c.s. is a notion of convergence in the space of all sequences of matrices corresponding to $\{d_n\}_n$, i.e.,

$$\mathscr{E} = \{\{A_n\}_n : A_n \in \mathbb{C}^{d_n \times d_n} \text{ for every } n\}. \tag{2.46}$$

To be precise, for every square matrix $A \in \mathbb{C}^{\ell \times \ell}$, let

$$\begin{aligned} p(A) &= \inf\left\{\frac{\operatorname{rank}(R)}{\ell} + \|N\| : \quad R, N \in \mathbb{C}^{\ell \times \ell}, \quad R + N = A\right\} \\ &= \min_{i=0,\ldots,\ell}\left(\frac{i}{\ell} + \sigma_{i+1}(A)\right), \end{aligned} \tag{2.47}$$

where $\sigma_1(A) \geq \ldots \geq \sigma_\ell(A)$ and $\sigma_{\ell+1}(A) = 0$ by convention; note that Eq. (2.47) is proved in [22, Theorem 5.3]. For any $\{A_n\}_n \in \mathscr{E}$, define

$$p_{\text{a.c.s.}}(\{A_n\}_n) = \limsup_{n\to\infty} p(A_n), \qquad \{A_n\}_n \in \mathscr{E}, \tag{2.48}$$

$$d_{\text{a.c.s.}}(\{A_n\}_n, \{B_n\}_n) = p_{\text{a.c.s.}}(\{A_n - B_n\}_n), \qquad \{A_n\}_n, \{B_n\}_n \in \mathscr{E}. \tag{2.49}$$

Then, the following theorem holds [22, Sect. 5.2.1].

Theorem 2.5 *Fix a sequence of positive integers d_n such that $d_n \to \infty$, and let $\mathscr{E}$ be the space* (2.46)*. The following properties hold.*

1. *$d_{\text{a.c.s.}}$ in* (2.49) *is a pseudometric on $\mathscr{E}$ such that $d_{\text{a.c.s.}}(\{A_n\}_n, \{B_n\}_n) = 0$ if and only if $\{A_n - B_n\}_n$ is zero-distributed.*
2. *Suppose $\{A_n\}_n \in \mathscr{E}$ and $\{\{B_{n,m}\}_n\}_m \subset \mathscr{E}$. Then, $\{\{B_{n,m}\}_n\}_m$ is an a.c.s. for $\{A_n\}_n$ if and only if $d_{\text{a.c.s.}}(\{A_n\}_n, \{B_{n,m}\}_n) \to 0$ as $m \to \infty$.*

Theorem 2.5 justifies the convergence notation $\{B_{n,m}\}_n \xrightarrow{\text{a.c.s.}} \{A_n\}_n$, which will be used to indicate that $\{\{B_{n,m}\}_n\}_m$ is an a.c.s. for $\{A_n\}_n$. The so-called a.c.s. topology $\tau_{\text{a.c.s.}}$ is defined as the topology induced on $\mathscr{E}$ by the pseudometric $d_{\text{a.c.s.}}$. As explained in [22, Sect. 5.2.3], the a.c.s. topology is strongly connected with the topology τ_{measure} associated with the convergence in measure of functions. This connection becomes absolutely evident in the light of the two research issues proposed in [16, Sect. 4], which have been positively solved in two recent papers [3, 5]. It is not to be excluded that the deep connections between the a.c.s. convergence and the convergence in measure highlighted in [3, 5, 16] and [22, Sect. 5.2.3] may lead to a "bridge", in the precise mathematical sense established in [11], between measure theory and the asymptotic linear algebra theory underlying the notion of a.c.s.; a bridge that could be exploited to obtain matrix theory results from measure theory results and vice versa. For deeper insights on this topic, we suggest reading [5, Sect. 1].

2.7.2 The a.c.s. Tools for Computing Singular Value and Spectral Distributions

The importance of the a.c.s. resides in Theorems 2.6 and 2.7, which are the versions of [22, Theorems 5.4 and 5.6] for general sequences of matrices.

Theorem 2.6 *Let $\{A_n\}_n$, $\{B_{n,m}\}_n$ be sequences of matrices and let $\phi, \phi_m : C_c(\mathbb{R}) \to \mathbb{C}$ be functionals. Suppose that*

1. $\{B_{n,m}\}_n \sim_\sigma \phi_m$ *for every* m,
2. $\{B_{n,m}\}_n \xrightarrow{\text{a.c.s.}} \{A_n\}_n$,
3. $\phi_m \to \phi$ *pointwise over* $C_c(\mathbb{R})$.

Then $\{A_n\}_n \sim_\sigma \phi$.

Theorem 2.7 *Let $\{A_n\}_n$, $\{B_{n,m}\}_n$ be sequences of Hermitian matrices and let $\phi, \phi_m : C_c(\mathbb{C}) \to \mathbb{C}$ be functionals. Suppose that*

1. $\{B_{n,m}\}_n \sim_\lambda \phi_m$ *for every* m,
2. $\{B_{n,m}\}_n \xrightarrow{\text{a.c.s.}} \{A_n\}_n$,
3. $\phi_m \to \phi$ *pointwise over* $C_c(\mathbb{C})$.

Then $\{A_n\}_n \sim_\lambda \phi$.

Theorem 2.6 admits the following converse, which is the version of [22, Theorem 5.5] for general sequences of matrices.

Theorem 2.8 *Let $\{A_n\}_n$, $\{B_{n,m}\}_n$ be sequences of matrices and let $\phi, \phi_m : C_c(\mathbb{R}) \to \mathbb{C}$ be functionals. Suppose that*

1. $\{A_n\}_n \sim_\sigma \phi$,
2. $\{B_{n,m}\}_n \sim_\sigma \phi_m$ *for every* m,
3. $\{B_{n,m}\}_n \xrightarrow{\text{a.c.s.}} \{A_n\}_n$.

Then $\phi_m \to \phi$ *pointwise over* $C_c(\mathbb{R})$.

We report below two important corollaries of Theorems 2.6 and 2.7.

Corollary 2.4 *Let $\{A_n\}_n$, $\{B_{n,m}\}_n$ be sequences of matrices and let $f, f_m : D \subset \mathbb{R}^k \to \mathbb{C}$ be measurable functions defined on a set D with $0 < \mu_k(D) < \infty$. Suppose that*

1. $\{B_{n,m}\}_n \sim_\sigma f_m$ *for every* m,
2. $\{B_{n,m}\}_n \xrightarrow{\text{a.c.s.}} \{A_n\}_n$,
3. $f_m \to f$ *in measure.*

Then $\{A_n\}_n \sim_\sigma f$.

Proof Apply Theorem 2.6 with $\phi_m = \phi_{|f_m|}$ and $\phi = \phi_{|f|}$. Note that $\phi_m \to \phi$ pointwise over $C_c(\mathbb{R})$ by [22, Lemma 2.5], since $|f_m| \to |f|$ in measure. □

Corollary 2.5 *Let $\{A_n\}_n$, $\{B_{n,m}\}_n$ be sequences of Hermitian matrices and let $f, f_m : D \subset \mathbb{R}^k \to \mathbb{C}$ be measurable functions defined on a set D with $0 < \mu_k(D) < \infty$. Suppose that*

1. $\{B_{n,m}\}_n \sim_\lambda f_m$ *for every* m,
2. $\{B_{n,m}\}_n \xrightarrow{\text{a.c.s.}} \{A_n\}_n$,
3. $f_m \to f$ *in measure.*

Then $\{A_n\}_n \sim_\lambda f$.

Proof Apply Theorem 2.7 with $\phi_m = \phi_{f_m}$ and $\phi = \phi_f$. Note that $\phi_m \to \phi$ pointwise over $C_c(\mathbb{C})$ by [22, Lemma 2.5]. □

Remark 2.5 (*topological interpretation of Corollaries* 2.4 *and* 2.5) It is interesting to give a topological interpretation of Corollaries 2.4 and 2.5. We only focus on Corollary 2.4 as the discussion for Corollary 2.5 is similar. Fix a sequence of positive integers d_n such that $d_n \to \infty$, and let

$$\mathscr{E} = \{\{A_n\}_n : A_n \in \mathbb{C}^{d_n \times d_n} \text{ for every } n\},$$
$$\mathfrak{M}_D = \{f : D \to \mathbb{C} : f \text{ is measurable}\}.$$

We have seen in Sect. 2.7.1 and [22, Sect. 2.3.2] that $\mathscr{E}$ (resp., $\mathfrak{M}_D$) is a pseudometric space with respect to the pseudometric $d_{\text{a.c.s.}}$ (resp., d_{measure}) which induces the a.c.s. topology $\tau_{\text{a.c.s.}}$ (resp., the topology of convergence in measure τ_{measure}). Corollary 2.4 is then equivalent to saying that the set of "σ-pairs"

$$\{(\{A_n\}_n, f) \in \mathscr{E} \times \mathfrak{M}_D : \{A_n\}_n \sim_\sigma f\}$$

is closed in $\mathscr{E} \times \mathfrak{M}_D$ equipped with the product (pseudometrizable) topology $\tau_{\text{a.c.s.}} \times \tau_{\text{measure}}$ induced, for example, by the pseudometric

$$d_{\text{a.c.s.}\times\text{measure}}((\{A_n\}_n, \kappa), (\{B_n\}_n, \xi)) = d_{\text{a.c.s.}}(\{A_n\}_n, \{B_n\}_n) + d_{\text{measure}}(\kappa, \xi);$$

see [22, Exercise 2.2]. Indeed, Corollary 2.4 reads as follows: if a sequence of σ-pairs $(\{B_{n,m}\}_n, f_m)$ converges to a pair $(\{A_n\}_n, f)$ in $\mathscr{E} \times \mathfrak{M}_D$, then $(\{A_n\}_n, f)$ is a σ-pair.

2.7.3 The a.c.s. Algebra

In this subsection, we formulate the algebraic properties of a.c.s. [22, Sect. 5.4]. The next theorem is the version of [22, Propositions 5.1, 5.2 and 5.5] for general sequences of matrices.

Theorem 2.9 *Let $\{A_n\}_n$, $\{A'_n\}_n$ be sequences of matrices, with A_n, A'_n of size d_n, and suppose that $\{B_{n,m}\}_n \xrightarrow{\text{a.c.s.}} \{A_n\}_n$ and $\{B'_{n,m}\}_n \xrightarrow{\text{a.c.s.}} \{A'_n\}_n$. The following properties hold.*

- $\{B_{n,m}^*\}_n \xrightarrow{\text{a.c.s.}} \{A_n^*\}_n$.
- $\{\alpha B_{n,m} + \beta B_{n,m}'\}_n \xrightarrow{\text{a.c.s.}} \{\alpha A_n + \beta A_n'\}_n$ *for all* $\alpha, \beta \in \mathbb{C}$.
- *If* $\{A_n\}_n$ *and* $\{A_n'\}_n$ *are s.u. then* $\{B_{n,m}B_{n,m}'\}_n \xrightarrow{\text{a.c.s.}} \{A_n A_n'\}_n$.

It is worth mentioning two further properties of a.c.s., which will not be used in this book, but are anyway interesting to know. They are the versions for general sequences of matrices of the two properties stated at the end of [22, Sect. 5.4].

- Let $\{A_n\}_n$ be an s.u. sequence of matrices and suppose that $\{B_{n,m}\}_n \xrightarrow{\text{a.c.s.}} \{A_n\}_n$. Assume also that the A_n and $B_{n,m}$ are Hermitian. Then $\{f(B_{n,m})\}_n \xrightarrow{\text{a.c.s.}} \{f(A_n)\}_n$ for all continuous functions $f : \mathbb{C} \to \mathbb{C}$.
- Let $\{A_n\}_n$ be an s.v. sequence of matrices and suppose that $\{B_{n,m}\}_n \xrightarrow{\text{a.c.s.}} \{A_n\}_n$. Then $\{B_{n,m}^\dagger\}_n \xrightarrow{\text{a.c.s.}} \{A_n^\dagger\}_n$.

2.7.4 Some Criteria to Identify a.c.s.

Two useful criteria to identify an a.c.s. without constructing the splitting (2.45) are given below. They are the versions of [22, Corollaries 5.3 and 5.4] for general sequences of matrices.

Theorem 2.10 *Let* $\{A_n\}_n$ *be a sequence of matrices, with* A_n *of size* d_n, *let* $\{\{B_{n,m}\}_n\}_m$ *be a sequence of sequences of matrices, with* $B_{n,m}$ *of size* d_n, *and let* $p \in [1, \infty]$. *Suppose that for every* m *there exists* n_m *such that, for* $n \ge n_m$,

$$\|A_n - B_{n,m}\|_p \le \varepsilon(m,n)(d_n)^{1/p},$$

where $\lim_{m\to\infty} \limsup_{n\to\infty} \varepsilon(m,n) = 0$. *Then* $\{B_{n,m}\}_n \xrightarrow{\text{a.c.s.}} \{A_n\}_n$.

Theorem 2.11 *Let* $\{A_n\}_n$ *be a sequence of matrices, with* A_n *of size* d_n, *and let* $\{\{B_{n,m}\}_n\}_m$ *be a sequence of sequences of matrices, with* $B_{n,m}$ *of size* d_n. *Suppose that* $\{A_n - B_{n,m}\}_n \sim_\sigma g_m$ *for some* $g_m : D \subset \mathbb{R}^k \to \mathbb{C}$ *such that* $g_m \to 0$ *in measure. Then* $\{B_{n,m}\}_n \xrightarrow{\text{a.c.s.}} \{A_n\}_n$.

We provide the following converse of Theorem 2.11 for future use.

Proposition 2.5 *Let* $\{A_n\}_n$ *be a sequence of matrices, with* A_n *of size* d_n, *and let* $\{\{B_{n,m}\}_n\}_m$ *be a sequence of sequences of matrices, with* $B_{n,m}$ *of size* d_n. *Suppose that* $\{A_n - B_{n,m}\}_n \sim_\sigma g_m$ *for some* $g_m : D \subset \mathbb{R}^k \to \mathbb{C}$ *and that* $\{B_{n,m}\}_n \xrightarrow{\text{a.c.s.}} \{A_n\}_n$. *Then* $g_m \to 0$ *in measure.*

Proof Since $\{A_n - B_{n,m}\}_n \xrightarrow{\text{a.c.s.}} \{O_{d_n}\}_n$ and $\{O_{d_n}\}_n \sim_\sigma 0$, Theorem 2.8 implies that $\phi_{g_m} \to \phi_0$ pointwise over $C_c(\mathbb{R})$. It follows that $g_m \to 0$ in measure by [22, Lemma 2.6]. □

2.7.5 An Extension of the Concept of a.c.s.

We now provide a natural extension of the a.c.s. notion. The underlying idea is that, in Definition 2.6, one could choose to approximate $\{A_n\}_n$ by a class of sequences $\{\{B_{n,\alpha}\}_n\}_{\alpha\in\mathcal{A}}$ parameterized by a not necessarily integer parameter α. For example, one may want to use a parameter $\varepsilon > 0$ and claim that a given class of sequences $\{\{B_{n,\varepsilon}\}_n\}_{\varepsilon>0}$ is an a.c.s. for $\{A_n\}_n$ as $\varepsilon \to 0$. Intuitively, this assertion should have the following meaning: for every $\varepsilon > 0$ there exists n_ε such that, for $n \ge n_\varepsilon$,

$$A_n = B_{n,\varepsilon} + R_{n,\varepsilon} + N_{n,\varepsilon}, \qquad \text{rank}(R_{n,\varepsilon}) \le c(\varepsilon)d_n, \qquad \|N_{n,\varepsilon}\| \le \omega(\varepsilon),$$

where n_ε, $c(\varepsilon)$, $\omega(\varepsilon)$ depend only on ε and both $c(\varepsilon)$ and $\omega(\varepsilon)$ tend to 0 as $\varepsilon \to 0$. This is in fact the correct meaning.

Definition 2.7 *(approximating class of sequences as* $\varepsilon \to 0$*)* Let $\{A_n\}_n$ be a sequence of matrices, with A_n of size d_n, and let $\{\{B_{n,\varepsilon}\}_n\}_{\varepsilon>0}$ be a class of sequences of matrices, with $B_{n,\varepsilon}$ of size d_n. We say that $\{\{B_{n,\varepsilon}\}_n\}_{\varepsilon>0}$ is an a.c.s. for $\{A_n\}_n$ as $\varepsilon \to 0$ if the following property holds: for every $\varepsilon > 0$ there exists n_ε such that, for $n \ge n_\varepsilon$,

$$A_n = B_{n,\varepsilon} + R_{n,\varepsilon} + N_{n,\varepsilon}, \qquad \text{rank}(R_{n,\varepsilon}) \le c(\varepsilon)d_n, \qquad \|N_{n,\varepsilon}\| \le \omega(\varepsilon),$$

where the quantities n_ε, $c(\varepsilon)$, $\omega(\varepsilon)$ depend only on ε and

$$\lim_{\varepsilon\to 0} c(\varepsilon) = \lim_{\varepsilon\to 0} \omega(\varepsilon) = 0.$$

Clearly, if $\{\{B_{n,\varepsilon}\}_n\}_{\varepsilon>0}$ is an a.c.s. for $\{A_n\}_n$ as $\varepsilon \to 0$, then $\{\{B_{n,\varepsilon(m)}\}_n\}_m$ is an a.c.s. for $\{A_n\}_n$ (in the sense of the classical Definition 2.6) for all sequences of positive numbers $\{\varepsilon(m)\}_m$ such that $\varepsilon(m) \to 0$ as $m \to \infty$. A.c.s. parameterized by a positive $\varepsilon \to 0$ appear, for example, in the definition of multilevel GLT sequences (Definition 5.1). Thanks to the topological results in Sect. 2.7.1, we can give the following elegant characterization of an a.c.s. of this kind:

> a class of sequences of matrices $\{\{B_{n,\varepsilon}\}_n\}_{\varepsilon>0}$ is an a.c.s. for $\{A_n\}_n$ as $\varepsilon \to 0$ if and only if $d_{\text{a.c.s.}}(\{A_n\}_n, \{B_{n,\varepsilon}\}_n) \to 0$ as $\varepsilon \to 0$.

Throughout this book, to indicate that $\{\{B_{n,\varepsilon}\}_n\}_{\varepsilon>0}$ is an a.c.s. for $\{A_n\}_n$ as $\varepsilon \to 0$, we will write

$$\{B_{n,\varepsilon}\}_n \xrightarrow{\text{a.c.s.}} \{A_n\}_n \text{ as } \varepsilon \to 0.$$

For the definition of multilevel LT sequences (Definition 4.3), we need the concept of a.c.s. parameterized by a multi-index $\boldsymbol{m} \to \infty$. In what follows, a multi-index sequence of sequences of matrices is any class of sequences of the form $\{\{B_{n,\boldsymbol{m}}\}_n\}_{\boldsymbol{m}\in\mathcal{M}}$ which satisfies the following two properties.

1. $\mathcal{M} \subseteq \mathbb{N}^q$ for some $q \geq 1$ and $\mathcal{M} \cap \{\boldsymbol{i} \in \mathbb{N}^q : \boldsymbol{i} \geq \boldsymbol{k}\}$ is nonempty for every $\boldsymbol{k} \in \mathbb{N}^q$. We express the latter condition by saying that ∞ is an *accumulation point* for $\mathcal{M}$. This is required to ensure that $\boldsymbol{m}$ can tend to ∞ inside $\mathcal{M}$.
2. For every $\boldsymbol{m} \in \mathcal{M}$, $\{B_{n,\boldsymbol{m}}\}_n$ is a sequence of matrices as defined at the end of Sect. 2.1.1.

Definition 2.8 (*approximating class of sequences as* $\boldsymbol{m} \to \infty$) Let $\{A_n\}_n$ be a sequence of matrices, with A_n of size d_n, and let $\{\{B_{n,\boldsymbol{m}}\}_n\}_{\boldsymbol{m}\in\mathcal{M}}$ be a multi-index sequence of sequences of matrices, with $B_{n,\boldsymbol{m}}$ of size d_n. We say that $\{\{B_{n,\boldsymbol{m}}\}_n\}_{\boldsymbol{m}\in\mathcal{M}}$ is an a.c.s. for $\{A_n\}_n$ as $\boldsymbol{m} \to \infty$ if the following property holds: for every $\boldsymbol{m} \in \mathcal{M}$ there exists $n_{\boldsymbol{m}}$ such that, for $n \geq n_{\boldsymbol{m}}$,

$$A_n = B_{n,\boldsymbol{m}} + R_{n,\boldsymbol{m}} + N_{n,\boldsymbol{m}}, \qquad \text{rank}(R_{n,\boldsymbol{m}}) \leq c(\boldsymbol{m})d_n, \qquad \|N_{n,\boldsymbol{m}}\| \leq \omega(\boldsymbol{m}), \quad (2.50)$$

where the quantities $n_{\boldsymbol{m}}$, $c(\boldsymbol{m})$, $\omega(\boldsymbol{m})$ depend only on $\boldsymbol{m}$ and

$$\lim_{\boldsymbol{m}\to\infty} c(\boldsymbol{m}) = \lim_{\boldsymbol{m}\to\infty} \omega(\boldsymbol{m}) = 0.$$

Note that an equivalent definition is obtained by replacing, in Definition 2.8, "for all $\boldsymbol{m} \in \mathcal{M}$" with "for all sufficiently large $\boldsymbol{m} \in \mathcal{M}$" (i.e., "for every $\boldsymbol{m} \in \mathcal{M}$ that is greater than or equal to some $\hat{\boldsymbol{m}}$"). Indeed, suppose the splitting (2.50) holds for $\boldsymbol{m} \geq \hat{\boldsymbol{m}}$. For the other values of $\boldsymbol{m}$, define $n_{\boldsymbol{m}} = 1$, $c(\boldsymbol{m}) = 1$, $\omega(\boldsymbol{m}) = 0$ and $R_{n,\boldsymbol{m}} = A_n - B_{n,\boldsymbol{m}}$, $N_{n,\boldsymbol{m}} = O_{d_n}$. Then, we see that (2.50) holds for every $\boldsymbol{m} \in \mathcal{M}$. Definition 2.8 extends the classical definition of a.c.s. (Definition 2.6). Indeed, a classical a.c.s. $\{\{B_{n,m}\}_n\}_m$ for $\{A_n\}_n$ is an a.c.s. also in the sense of Definition 2.8 (take $\mathcal{M}$ as the infinite subset of $\mathbb{N}$ where m varies). In addition, if $\left\{\{B_{n,\boldsymbol{m}}\}_n\right\}_{\boldsymbol{m}\in\mathcal{M}}$ is an a.c.s. for $\{A_n\}_n$ in the sense of Definition 2.8, then $\left\{\{B_{n,\boldsymbol{m}}\}_n\right\}_m$ is an a.c.s. for $\{A_n\}_n$ (in the sense of the classical Definition 2.6) for all sequences of multi-indices $\{\boldsymbol{m} = \boldsymbol{m}(m)\}_m \subseteq \mathcal{M}$ such that $\boldsymbol{m} \to \infty$ as $m \to \infty$.

Remark 2.6 Let $\{B_{n,\boldsymbol{m}}\}_n \xrightarrow{\text{a.c.s.}} \{A_n\}_n$ and $\{B'_{n,\boldsymbol{m}}\}_n \xrightarrow{\text{a.c.s.}} \{A'_n\}_n$ as $\boldsymbol{m} \to \infty$. The following properties hold.

- $\{B^*_{n,\boldsymbol{m}}\}_n \xrightarrow{\text{a.c.s.}} \{A^*_n\}_n$ as $\boldsymbol{m} \to \infty$.
- $\{\alpha B_{n,\boldsymbol{m}} + \beta B'_{n,\boldsymbol{m}}\}_n \xrightarrow{\text{a.c.s.}} \{\alpha A_n + \beta A'_n\}_n$ as $\boldsymbol{m} \to \infty$ for all $\alpha, \beta \in \mathbb{C}$.
- If $\{A_n\}_n$ and $\{A'_n\}_n$ are s.u. then $\{B_{n,\boldsymbol{m}} B'_{n,\boldsymbol{m}}\}_n \xrightarrow{\text{a.c.s.}} \{A_n A'_n\}_n$.

The proof of these results is essentially the same as the proof of the analogous results for standard a.c.s.; see Theorem 2.9.

As in the case of a.c.s. parameterized by a positive $\varepsilon \to 0$, also for a.c.s. parameterized by a multi-index $\boldsymbol{m} \to \infty$ we can give the following elegant characterization based on the topological results of Sect. 2.7.1:

a multi-index sequence of sequences of matrices $\{\{B_{n,\boldsymbol{m}}\}_n\}_{\boldsymbol{m}\in\mathcal{M}}$ is an a.c.s. for $\{A_n\}_n$ as $\boldsymbol{m} \to \infty$ if and only if $d_{\text{a.c.s.}}(\{A_n\}_n, \{B_{n,\boldsymbol{m}}\}_n) \to 0$ as $\boldsymbol{m} \to \infty$.

Throughout this book, to indicate that $\{\{B_{n,\boldsymbol{m}}\}_n\}_{\boldsymbol{m}\in\mathcal{M}}$ is an a.c.s. for $\{A_n\}_n$ as $\boldsymbol{m} \to \infty$, we will write

$$\{B_{n,\boldsymbol{m}}\}_n \xrightarrow{\text{a.c.s.}} \{A_n\}_n \text{ as } \boldsymbol{m} \to \infty.$$

Chapter 3
Multilevel Toeplitz Sequences

This chapter is devoted to multilevel Toeplitz matrices. More precisely, we focus on the sequences of multilevel Toeplitz matrices generated by a multivariate L^1 function. These sequences, together with the sequences of multilevel diagonal sampling matrices (to be introduced afterwards) and the zero-distributed sequences (already studied in Sect. 2.6.3), should be regarded as the "building blocks" of the theory of multilevel GLT sequences. Despite its conciseness, this chapter contains everything we need about multilevel Toeplitz matrices to fully develop the theory of multilevel GLT sequences. In particular, it contains an a.c.s.-based proof of the multivariate L^1 versions of Szegő's first limit theorem and the Avram–Parter theorem about the singular value and eigenvalue distributions of multilevel Toeplitz sequences. We stress that *almost all the results and the proofs contained in this chapter have an exact analog in* [22, Chap. 6], where we dealt with classical (i.e., unilevel) Toeplitz matrices. However, we decided to reproduce here all the proofs without omitting details, in order to help the reader become familiar with the multilevel language (especially, the multi-index notation). In this regard, the reader is invited to compare the "multivariate proofs" presented in this chapter with the corresponding "univariate proofs" in [22, Chap. 6], in order to learn the way in which the multilevel language allows one to transfer many results from the univariate to the multivariate case by simply turning some letters (n, i, j, x, θ, etc.) in boldface ($\boldsymbol{n}, \boldsymbol{i}, \boldsymbol{j}, \mathbf{x}, \boldsymbol{\theta}$, etc.).

3.1 Multilevel Toeplitz Matrices and Multilevel Toeplitz Sequences

Given a d-index $\boldsymbol{n} \in \mathbb{N}^d$, a matrix of the form

$$\left[a_{\boldsymbol{i}-\boldsymbol{j}}\right]_{\boldsymbol{i},\boldsymbol{j}=\mathbf{1}}^{\boldsymbol{n}} \in \mathbb{C}^{N(\boldsymbol{n})\times N(\boldsymbol{n})}, \tag{3.1}$$

C. Garoni and S. Serra-Capizzano, *Generalized Locally Toeplitz Sequences: Theory and Applications*, https://doi.org/10.1007/978-3-030-02233-4_3

whose $(\boldsymbol{i}, \boldsymbol{j})$ entry depends only on the difference $\boldsymbol{i} - \boldsymbol{j}$, is called a multilevel Toeplitz matrix (or, more precisely, a d-level Toeplitz matrix). In the case $d = 1$, the matrix (3.1) becomes

$$[a_{i-j}]_{i,j=1}^{n} = \begin{bmatrix} a_0 & a_{-1} & a_{-2} & \cdots & \cdots & a_{-(n-1)} \\ a_1 & \ddots & \ddots & \ddots & & \vdots \\ a_2 & \ddots & \ddots & \ddots & \ddots & \vdots \\ \vdots & \ddots & \ddots & \ddots & \ddots & a_{-2} \\ \vdots & & \ddots & \ddots & \ddots & a_{-1} \\ a_{n-1} & \cdots & \cdots & a_2 & a_1 & a_0 \end{bmatrix},$$

which means that a 1-level (or unilevel) Toeplitz matrix is just a matrix whose entries are constant along each diagonal. Unilevel Toeplitz matrices are nothing else than classical Toeplitz matrices, which have been the subject of [22, Chap. 6]. In the case $d = 2$, the matrix (3.1) can be written as

$$[a_{\boldsymbol{i}-\boldsymbol{j}}]_{\boldsymbol{i},\boldsymbol{j}=\boldsymbol{1}}^{\boldsymbol{n}} = \Big[[a_{i_1-j_1,i_2-j_2}]_{i_2,j_2=1}^{n_2}\Big]_{i_1,j_1=1}^{n_1} = [\mathbf{a}_{i_1-j_1}]_{i_1,j_1=1}^{n_1}$$
$$= \begin{bmatrix} \mathbf{a}_0 & \mathbf{a}_{-1} & \mathbf{a}_{-2} & \cdots & \cdots & \mathbf{a}_{-(n_1-1)} \\ \mathbf{a}_1 & \ddots & \ddots & \ddots & & \vdots \\ \mathbf{a}_2 & \ddots & \ddots & \ddots & \ddots & \vdots \\ \vdots & \ddots & \ddots & \ddots & \ddots & \mathbf{a}_{-2} \\ \vdots & & \ddots & \ddots & \ddots & \mathbf{a}_{-1} \\ \mathbf{a}_{n_1-1} & \cdots & \cdots & \mathbf{a}_2 & \mathbf{a}_1 & \mathbf{a}_0 \end{bmatrix}, \tag{3.2}$$

where, for all $k = -(n_1 - 1), \ldots, n_1 - 1$,

$$\mathbf{a}_k = [a_{k,i_2-j_2}]_{i_2,j_2=1}^{n_2} = \begin{bmatrix} a_{k,0} & a_{k,-1} & a_{k,-2} & \cdots & \cdots & a_{k,-(n_2-1)} \\ a_{k,1} & \ddots & \ddots & \ddots & & \vdots \\ a_{k,2} & \ddots & \ddots & \ddots & \ddots & \vdots \\ \vdots & \ddots & \ddots & \ddots & \ddots & a_{k,-2} \\ \vdots & & \ddots & \ddots & \ddots & a_{k,-1} \\ a_{k,n_2-1} & \cdots & \cdots & a_{k,2} & a_{k,1} & a_{k,0} \end{bmatrix}. \tag{3.3}$$

A matrix of this form, which we call a 2-level Toeplitz matrix, is also known as a block Toeplitz matrix with Toeplitz blocks, or BTTB matrix. In general, we can say that a d-level Toeplitz matrix is a block Toeplitz matrix with $(d - 1)$-level Toeplitz blocks.

For $n \in \mathbb{N}$ and $k \in \mathbb{Z}$, let $J_n^{(k)}$ be the $n \times n$ matrix whose (i, j) entry equals 1 if $i - j = k$ and 0 otherwise, that is,

$$(J_n^{(k)})_{ij} = \delta_{i-j-k}, \qquad i, j = 1, \ldots, n, \qquad n \in \mathbb{N}, \qquad k \in \mathbb{Z}, \tag{3.4}$$

where $\delta_r = 1$ if $r = 0$ and $\delta_r = 0$ otherwise. For $\boldsymbol{n} \in \mathbb{N}^d$ and $\boldsymbol{k} \in \mathbb{Z}^d$, let

$$J_{\boldsymbol{n}}^{(\boldsymbol{k})} = J_{n_1}^{(k_1)} \otimes J_{n_2}^{(k_2)} \otimes \cdots \otimes J_{n_d}^{(k_d)}. \tag{3.5}$$

Lemma 3.1 *The d-level Toeplitz matrix* (3.1) *admits the following expression:*

$$\left[a_{\boldsymbol{i}-\boldsymbol{j}}\right]_{\boldsymbol{i},\boldsymbol{j}=\boldsymbol{1}}^{\boldsymbol{n}} = \sum_{\boldsymbol{k}=-(\boldsymbol{n}-\boldsymbol{1})}^{\boldsymbol{n}-\boldsymbol{1}} a_{\boldsymbol{k}} J_{\boldsymbol{n}}^{(\boldsymbol{k})}, \tag{3.6}$$

where $J_{\boldsymbol{n}}^{(\boldsymbol{k})}$ is defined in (3.5).

Proof Equation (3.6) is proved componentwise, by showing that the $(\boldsymbol{i}, \boldsymbol{j})$ entry of the matrix in the right-hand side is equal to $a_{\boldsymbol{i}-\boldsymbol{j}}$. By (3.4) and the crucial property (2.29), for all $\boldsymbol{i}, \boldsymbol{j} = \boldsymbol{1}, \ldots, \boldsymbol{n}$ we have

$$\begin{aligned}(J_{\boldsymbol{n}}^{(\boldsymbol{k})})_{\boldsymbol{i}\boldsymbol{j}} &= (J_{n_1}^{(k_1)} \otimes J_{n_2}^{(k_2)} \otimes \cdots \otimes J_{n_d}^{(k_d)})_{\boldsymbol{i}\boldsymbol{j}} = (J_{n_1}^{(k_1)})_{i_1 j_1} (J_{n_2}^{(k_2)})_{i_2 j_2} \cdots (J_{n_d}^{(k_d)})_{i_d j_d} \\ &= \delta_{i_1-j_1-k_1} \delta_{i_2-j_2-k_2} \cdots \delta_{i_d-j_d-k_d} = \delta_{\boldsymbol{i}-\boldsymbol{j}-\boldsymbol{k}},\end{aligned} \tag{3.7}$$

where $\delta_{\boldsymbol{r}} = 1$ if $\boldsymbol{r} = \boldsymbol{0}$ and $\delta_{\boldsymbol{r}} = 0$ otherwise. Therefore,

$$\left(\sum_{\boldsymbol{k}=-(\boldsymbol{n}-\boldsymbol{1})}^{\boldsymbol{n}-\boldsymbol{1}} a_{\boldsymbol{k}} J_{\boldsymbol{n}}^{(\boldsymbol{k})} \right)_{\boldsymbol{i}\boldsymbol{j}} = \sum_{\boldsymbol{k}=-(\boldsymbol{n}-\boldsymbol{1})}^{\boldsymbol{n}-\boldsymbol{1}} a_{\boldsymbol{k}} (J_{\boldsymbol{n}}^{(\boldsymbol{k})})_{\boldsymbol{i}\boldsymbol{j}} = \sum_{\boldsymbol{k}=-(\boldsymbol{n}-\boldsymbol{1})}^{\boldsymbol{n}-\boldsymbol{1}} a_{\boldsymbol{k}} \delta_{\boldsymbol{i}-\boldsymbol{j}-\boldsymbol{k}} = a_{\boldsymbol{i}-\boldsymbol{j}},$$

and (3.6) is proved. □

Given a function $f : [-\pi, \pi]^d \to \mathbb{C}$ belonging to $L^1([-\pi, \pi]^d)$, its Fourier coefficients are denoted by

$$f_{\boldsymbol{k}} = \frac{1}{(2\pi)^d} \int_{[-\pi,\pi]^d} f(\boldsymbol{\theta})\, \mathrm{e}^{-\mathrm{i}\boldsymbol{k}\cdot\boldsymbol{\theta}} \mathrm{d}\boldsymbol{\theta}, \qquad \boldsymbol{k} \in \mathbb{Z}^d. \tag{3.8}$$

The $\boldsymbol{n}$th (d-level) Toeplitz matrix associated with f is defined as

$$T_{\boldsymbol{n}}(f) = \left[f_{\boldsymbol{i}-\boldsymbol{j}}\right]_{\boldsymbol{i},\boldsymbol{j}=\boldsymbol{1}}^{\boldsymbol{n}} = \sum_{\boldsymbol{k}=-(\boldsymbol{n}-\boldsymbol{1})}^{\boldsymbol{n}-\boldsymbol{1}} f_{\boldsymbol{k}} J_{\boldsymbol{n}}^{(\boldsymbol{k})}, \tag{3.9}$$

where the second equality follows from Lemma 3.1. We call $\{T_{\boldsymbol{n}}(f)\}_{\boldsymbol{n}\in\mathbb{N}^d}$ the family of (d-level) Toeplitz matrices (or simply the Toeplitz family) generated by f, which in turn is referred to as the generating function of $\{T_{\boldsymbol{n}}(f)\}_{\boldsymbol{n}\in\mathbb{N}^d}$. Any matrix-sequence extracted from the Toeplitz family $\{T_{\boldsymbol{n}}(f)\}_{\boldsymbol{n}\in\mathbb{N}^d}$, i.e., any matrix-sequence of the form $\{T_{\boldsymbol{n}}(f)\}_n$ with $\boldsymbol{n}=\boldsymbol{n}(n)\to\infty$ as $n\to\infty$, is referred to as a (d-level) Toeplitz sequence generated by f. Note that if $g=f$ a.e. then $T_{\boldsymbol{n}}(g)=T_{\boldsymbol{n}}(f)$ for all $\boldsymbol{n}\in\mathbb{N}^d$, i.e., g generates the same Toeplitz family (and hence also the same Toeplitz sequences) as f. This is due to the fact that the Fourier coefficients of g and f coincide.

If f is a d-variate trigonometric polynomial, say

$$f(\boldsymbol{\theta})=\sum_{\boldsymbol{k}=-\boldsymbol{r}}^{\boldsymbol{r}} f_{\boldsymbol{k}}\,\mathrm{e}^{\mathrm{i}\boldsymbol{k}\cdot\boldsymbol{\theta}}, \tag{3.10}$$

then, as indicated by the notation, the Fourier coefficients of f are given by the coefficients $f_{\boldsymbol{k}}$ in (3.10) for $\boldsymbol{k}\in\{-\boldsymbol{r},\ldots,\boldsymbol{r}\}$ and are equal to 0 for $\boldsymbol{k}\notin\{-\boldsymbol{r},\ldots,\boldsymbol{r}\}$. This is a consequence of the orthogonality relations (2.9). It follows that every Toeplitz matrix $T_{\boldsymbol{n}}(f)$ generated by f has a number of nonzero entries in each row and column which is bounded by $(2|\boldsymbol{r}|_\infty+1)^d$, a constant independent of $\boldsymbol{n}$. Indeed, considering for instance the $\boldsymbol{i}$th row, we have $(T_{\boldsymbol{n}}(f))_{\boldsymbol{i}\boldsymbol{j}}=f_{\boldsymbol{i}-\boldsymbol{j}}=0$ whenever $|\boldsymbol{i}-\boldsymbol{j}|_\infty>|\boldsymbol{r}|_\infty$, which means that the only possible nonzero entries of $T_{\boldsymbol{n}}(f)$ in the $\boldsymbol{i}$th row are those corresponding to the column multi-indices $\boldsymbol{j}\in\{\boldsymbol{1},\ldots,\boldsymbol{n}\}$ such that $|\boldsymbol{i}-\boldsymbol{j}|_\infty\le|\boldsymbol{r}|_\infty$; and the number of all multi-indices $\boldsymbol{j}\in\mathbb{Z}^d$ satisfying $|\boldsymbol{i}-\boldsymbol{j}|_\infty\le|\boldsymbol{r}|_\infty$ is $(2|\boldsymbol{r}|_\infty+1)^d$. We also note that, by (3.7) and the equation

$$(T_{\boldsymbol{n}}(\mathrm{e}^{\mathrm{i}\boldsymbol{k}\cdot\boldsymbol{\theta}}))_{\boldsymbol{i}\boldsymbol{j}}=(\mathrm{e}^{\mathrm{i}\boldsymbol{k}\cdot\boldsymbol{\theta}})_{\boldsymbol{i}-\boldsymbol{j}}=\begin{cases}1, & \text{if } \boldsymbol{i}-\boldsymbol{j}=\boldsymbol{k},\\ 0, & \text{if } \boldsymbol{i}-\boldsymbol{j}\neq\boldsymbol{k},\end{cases}\qquad \boldsymbol{i},\boldsymbol{j}=\boldsymbol{1},\ldots,\boldsymbol{n},$$

which is satisfied for all $\boldsymbol{n}\in\mathbb{N}^d$ and $\boldsymbol{k}\in\mathbb{Z}^d$, we have

$$T_{\boldsymbol{n}}(\mathrm{e}^{\mathrm{i}\boldsymbol{k}\cdot\boldsymbol{\theta}})=J_{\boldsymbol{n}}^{(\boldsymbol{k})},\qquad \boldsymbol{n}\in\mathbb{N}^d,\qquad \boldsymbol{k}\in\mathbb{Z}^d.$$

In other words, $\{J_{\boldsymbol{n}}^{(\boldsymbol{k})}\}_{\boldsymbol{n}\in\mathbb{N}^d}$ is the Toeplitz family generated by the d-variate Fourier frequency $\mathrm{e}^{\mathrm{i}\boldsymbol{k}\cdot\boldsymbol{\theta}}$.

3.2 Basic Properties of Multilevel Toeplitz Matrices

In this section, we study the basic properties of the Toeplitz matrices $T_{\boldsymbol{n}}(f)$ generated by a function $f\in L^1([-\pi,\pi]^d)$. For each fixed $\boldsymbol{n}\in\mathbb{N}^d$, the map

$$T_{\boldsymbol{n}}(\cdot): L^1([-\pi,\pi]^d)\to\mathbb{C}^{N(\boldsymbol{n})\times N(\boldsymbol{n})},\qquad f\mapsto T_{\boldsymbol{n}}(f),$$

is linear, i.e.,

$$T_{\boldsymbol{n}}(\alpha f+\beta g)=\alpha T_{\boldsymbol{n}}(f)+\beta T_{\boldsymbol{n}}(g),\qquad \alpha,\beta\in\mathbb{C},\qquad f,g\in L^1([-\pi,\pi]^d).\quad(3.11)$$

This follows from the relation

$$(\alpha f+\beta g)_{\boldsymbol{k}}=\alpha f_{\boldsymbol{k}}+\beta g_{\boldsymbol{k}},\qquad \boldsymbol{k}\in\mathbb{Z}^d,$$

which is a consequence of the linearity of the integral in (3.8). It is clear from the definition that, if C is a constant (considered as a constant function from $[-\pi,\pi]^d$ to $\mathbb{C}$), then

$$T_{\boldsymbol{n}}(C)=C I_{N(\boldsymbol{n})}.\qquad(3.12)$$

For every $f\in L^1([-\pi,\pi]^d)$, the Fourier coefficients of f are related to the Fourier coefficients of its conjugate $\overline{f}$ by the equation

$$(\overline{f})_{\boldsymbol{j}}=\frac{1}{(2\pi)^d}\int_{[-\pi,\pi]^d}\overline{f(\boldsymbol{\theta})}\,\mathrm{e}^{-\mathrm{i}\boldsymbol{j}\cdot\boldsymbol{\theta}}\mathrm{d}\boldsymbol{\theta}=\overline{\frac{1}{(2\pi)^d}\int_{[-\pi,\pi]^d}f(\boldsymbol{\theta})\,\mathrm{e}^{\mathrm{i}\boldsymbol{j}\cdot\boldsymbol{\theta}}\mathrm{d}\boldsymbol{\theta}}=\overline{f_{-\boldsymbol{j}}},$$

which is satisfied for all $\boldsymbol{j}\in\mathbb{Z}^d$. Hence, for all $\boldsymbol{i},\boldsymbol{j}=\mathbf{1},\ldots,\boldsymbol{n}$,

$$\left[T_{\boldsymbol{n}}(\overline{f})\right]_{\boldsymbol{ij}}=(\overline{f})_{\boldsymbol{i}-\boldsymbol{j}}=\overline{f_{\boldsymbol{j}-\boldsymbol{i}}}=\left[(T_{\boldsymbol{n}}(f))^*\right]_{\boldsymbol{ij}},$$

i.e.,

$$(T_{\boldsymbol{n}}(f))^*=T_{\boldsymbol{n}}(\overline{f}),\qquad f\in L^1([-\pi,\pi]^d).\qquad(3.13)$$

In particular, if f is real a.e. then all the Toeplitz matrices $T_{\boldsymbol{n}}(f)$ are Hermitian.

The next lemma provides an integral expression for the quantity $\mathbf{u}^*T_{\boldsymbol{n}}(f)\mathbf{v}$, where $\mathbf{u},\mathbf{v}\in\mathbb{C}^{N(\boldsymbol{n})}$. This expression will be used in Theorem 3.1 to obtain a localization of the spectrum of $T_{\boldsymbol{n}}(f)$ in the case where f is real a.e. (and hence $T_{\boldsymbol{n}}(f)$ is Hermitian).

Lemma 3.2 *For every $f\in L^1([-\pi,\pi]^d)$ and every $\boldsymbol{n}\in\mathbb{N}^d$,*

$$\mathbf{u}^*T_{\boldsymbol{n}}(f)\mathbf{v}=\frac{1}{(2\pi)^d}\int_{[-\pi,\pi]^d}f(\boldsymbol{\theta})\,\mathbf{u}^*U(\boldsymbol{\theta})\mathbf{v}\,\mathrm{d}\boldsymbol{\theta},\qquad \mathbf{u},\mathbf{v}\in\mathbb{C}^{N(\boldsymbol{n})},\qquad(3.14)$$

where $U(\boldsymbol{\theta})=\left[\mathrm{e}^{-\mathrm{i}(\boldsymbol{i}-\boldsymbol{j})\cdot\boldsymbol{\theta}}\right]_{\boldsymbol{i},\boldsymbol{j}=\mathbf{1}}^{\boldsymbol{n}}$. The matrix $U(\boldsymbol{\theta})$ satisfies

$$\mathbf{u}^*U(\boldsymbol{\theta})\mathbf{u}=\left|\sum_{\boldsymbol{j}=\mathbf{1}}^{\boldsymbol{n}}u_{\boldsymbol{j}}\,\mathrm{e}^{\mathrm{i}\boldsymbol{j}\cdot\boldsymbol{\theta}}\right|^2,\qquad \boldsymbol{\theta}\in[-\pi,\pi]^d,\qquad \mathbf{u}=[u_{\boldsymbol{j}}]_{\boldsymbol{j}=\mathbf{1}}^{\boldsymbol{n}}\in\mathbb{C}^{N(\boldsymbol{n})},\quad(3.15)$$

hence $U(\boldsymbol{\theta})$ is HPSD for every $\boldsymbol{\theta}\in[-\pi,\pi]^d$. Moreover

$$\frac{1}{(2\pi)^d}\int_{[-\pi,\pi]^d}\mathbf{u}^*U(\boldsymbol{\theta})\mathbf{u}\,\mathrm{d}\boldsymbol{\theta}=\|\mathbf{u}\|^2,\qquad \mathbf{u}\in\mathbb{C}^{N(\boldsymbol{n})}.\qquad(3.16)$$

Proof If we index the entries of any vector $\mathbf{y} \in \mathbb{C}^{N(\boldsymbol{n})}$ by a d-index $\boldsymbol{j} = \mathbf{1}, \ldots, \boldsymbol{n}$ and the entries of any matrix $A \in \mathbb{C}^{N(\boldsymbol{n})\times N(\boldsymbol{n})}$ by a pair of d-indices $\boldsymbol{i}, \boldsymbol{j} = \mathbf{1}, \ldots \boldsymbol{n}$ (so that $y = [y_{\boldsymbol{j}}]_{\boldsymbol{j}=\mathbf{1}}^{\boldsymbol{n}}$ and $A = [a_{\boldsymbol{ij}}]_{\boldsymbol{i},\boldsymbol{j}=\mathbf{1}}^{\boldsymbol{n}}$), then we have

$$\mathbf{u}^* A\mathbf{v} = \sum_{\boldsymbol{i},\boldsymbol{j}=\mathbf{1}}^{\boldsymbol{n}} a_{\boldsymbol{ij}}\overline{u_{\boldsymbol{i}}}v_{\boldsymbol{j}}, \qquad \mathbf{u}, \mathbf{v} \in \mathbb{C}^{N(\boldsymbol{n})}, \qquad A \in \mathbb{C}^{N(\boldsymbol{n})\times N(\boldsymbol{n})};$$

see also Example 2.6. For every $\mathbf{u}, \mathbf{v} \in \mathbb{C}^{N(\boldsymbol{n})}$,

$$\begin{aligned}
\mathbf{u}^* T_{\boldsymbol{n}}(f)\mathbf{v} &= \sum_{\boldsymbol{i},\boldsymbol{j}=\mathbf{1}}^{\boldsymbol{n}} f_{\boldsymbol{i}-\boldsymbol{j}}\, \overline{u_{\boldsymbol{i}}}\, v_{\boldsymbol{j}} = \sum_{\boldsymbol{i},\boldsymbol{j}=\mathbf{1}}^{\boldsymbol{n}} \left(\frac{1}{(2\pi)^d}\int_{[-\pi,\pi]^d} f(\boldsymbol{\theta})\,\mathrm{e}^{-\mathrm{i}(\boldsymbol{i}-\boldsymbol{j})\cdot\boldsymbol{\theta}}\mathrm{d}\boldsymbol{\theta}\right)\overline{u_{\boldsymbol{i}}}\, v_{\boldsymbol{j}} \\
&= \frac{1}{(2\pi)^d}\int_{[-\pi,\pi]^d} f(\boldsymbol{\theta}) \sum_{\boldsymbol{i},\boldsymbol{j}=\mathbf{1}}^{\boldsymbol{n}} \mathrm{e}^{-\mathrm{i}(\boldsymbol{i}-\boldsymbol{j})\cdot\boldsymbol{\theta}}\, \overline{u_{\boldsymbol{i}}}\, v_{\boldsymbol{j}}\, \mathrm{d}\boldsymbol{\theta} \\
&= \frac{1}{(2\pi)^d}\int_{[-\pi,\pi]^d} f(\boldsymbol{\theta})\, \mathbf{u}^* U(\boldsymbol{\theta})\mathbf{v}\, \mathrm{d}\boldsymbol{\theta},
\end{aligned}$$

where $U(\boldsymbol{\theta})$ is defined in the statement of the lemma. This proves (3.14). Equation (3.16) follows from (3.14) by taking $\mathbf{v} = \mathbf{u}$ and $f = 1$ (recall that $T_{\boldsymbol{n}}(1) = I_{N(\boldsymbol{n})}$). Finally,

$$\mathbf{u}^* U(\boldsymbol{\theta})\mathbf{u} = \sum_{\boldsymbol{i},\boldsymbol{j}=\mathbf{1}}^{\boldsymbol{n}} \mathrm{e}^{-\mathrm{i}(\boldsymbol{i}-\boldsymbol{j})\cdot\boldsymbol{\theta}}\, \overline{u_{\boldsymbol{i}}}\, u_{\boldsymbol{j}} = \sum_{\boldsymbol{i}=\mathbf{1}}^{\boldsymbol{n}} \overline{u_{\boldsymbol{i}}}\, \mathrm{e}^{-\mathrm{i}\boldsymbol{i}\cdot\boldsymbol{\theta}} \sum_{\boldsymbol{j}=\mathbf{1}}^{\boldsymbol{n}} u_{\boldsymbol{j}}\, \mathrm{e}^{\mathrm{i}\boldsymbol{j}\cdot\boldsymbol{\theta}} = \left|\sum_{\boldsymbol{j}=\mathbf{1}}^{\boldsymbol{n}} u_{\boldsymbol{j}}\, \mathrm{e}^{\mathrm{i}\boldsymbol{j}\cdot\boldsymbol{\theta}}\right|^2,$$

which proves (3.15). □

Theorem 3.1 *Assume that $f \in L^1([-\pi, \pi]^d)$ is real a.e. and let*

$$m_f = \operatorname*{ess\,inf}_{\boldsymbol{\theta}\in[-\pi,\pi]^d} f(\boldsymbol{\theta}), \qquad M_f = \operatorname*{ess\,sup}_{\boldsymbol{\theta}\in[-\pi,\pi]^d} f(\boldsymbol{\theta}).$$

Then

$$\Lambda(T_{\boldsymbol{n}}(f)) \subseteq [m_f, M_f], \qquad \boldsymbol{n} \in \mathbb{N}^d.$$

If moreover $m_f < M_f$, then

$$\Lambda(T_{\boldsymbol{n}}(f)) \subset (m_f, M_f), \qquad \boldsymbol{n} \in \mathbb{N}^d.$$

Note that the case $m_f = M_f$ is trivial, because $f = m_f$ a.e. by [22, Lemma 2.1] and $T_{\boldsymbol{n}}(f) = m_f\, I_{N(\boldsymbol{n})}$. Whenever f is not constant a.e., we have $m_f < M_f$ and the spectrum of the Toeplitz matrices $T_{\boldsymbol{n}}(f)$ is contained in the open interval (m_f, M_f).

Proof By Lemma 3.2, for every $\mathbf{u} \in \mathbb{C}^{N(\boldsymbol{n})}$ such that $\|\mathbf{u}\| = 1$ we have

$$\mathbf{u}^* T_{\boldsymbol{n}}(f)\mathbf{u} = \frac{1}{(2\pi)^d}\int_{[-\pi,\pi]^d} f(\boldsymbol{\theta})\,\mathbf{u}^* U(\boldsymbol{\theta})\mathbf{u}\,\mathrm{d}\boldsymbol{\theta}, \qquad \frac{1}{(2\pi)^d}\int_{[-\pi,\pi]^d} \mathbf{u}^* U(\boldsymbol{\theta})\mathbf{u}\,\mathrm{d}\boldsymbol{\theta} = 1.$$

Since $U(\boldsymbol{\theta})$ is HPSD for all $\boldsymbol{\theta} \in [-\pi,\pi]^d$ and $m_f \le f \le M_f$ a.e. by [22, Lemma 2.1], we obtain

$$m_f \le \mathbf{u}^* T_{\boldsymbol{n}}(f)\mathbf{u} \le M_f,$$

and the inclusion $\Lambda(T_{\boldsymbol{n}}(f)) \subseteq [m_f, M_f]$ follows from the minimax principle for eigenvalues [22, Theorem 2.8]. Suppose now that $m_f < M_f$. In this case, we show that, for all $\mathbf{u} \in \mathbb{C}^{N(\boldsymbol{n})}$ satisfying $\|\mathbf{u}\| = 1$,

$$m_f < \mathbf{u}^* T_{\boldsymbol{n}}(f)\mathbf{u} < M_f. \tag{3.17}$$

Once this is proved, the inclusion $\Lambda(T_{\boldsymbol{n}}(f)) \subset (m_f, M_f)$ follows again from the minimax principle for eigenvalues. We only prove the left inequality in (3.17), because the proof of the right inequality is completely analogous. By contradiction, suppose there exists $\hat{\mathbf{u}} \in \mathbb{C}^{N(\boldsymbol{n})}$ such that $\|\hat{\mathbf{u}}\| = 1$ and $\hat{\mathbf{u}}^* T_{\boldsymbol{n}}(f)\hat{\mathbf{u}} = m_f$. By Lemma 3.2,

$$\begin{aligned} 0 = \hat{\mathbf{u}}^* T_{\boldsymbol{n}}(f)\hat{\mathbf{u}} - m_f &= \frac{1}{(2\pi)^d}\int_{[-\pi,\pi]^d} (f(\boldsymbol{\theta}) - m_f)\,\hat{\mathbf{u}}^* U(\boldsymbol{\theta})\hat{\mathbf{u}}\,\mathrm{d}\boldsymbol{\theta} \\ &= \frac{1}{(2\pi)^d}\int_{[-\pi,\pi]^d} (f(\boldsymbol{\theta}) - m_f)\left|\sum_{\boldsymbol{j}=\boldsymbol{1}}^{\boldsymbol{n}} \hat{u}_{\boldsymbol{j}}\,\mathrm{e}^{\mathrm{i}\boldsymbol{j}\cdot\boldsymbol{\theta}}\right|^2 \mathrm{d}\boldsymbol{\theta}. \end{aligned} \tag{3.18}$$

Considering that the integrand in (3.18) is nonnegative a.e. (because $f \ge m_f$ a.e.) and that $\left|\sum_{\boldsymbol{j}=\boldsymbol{1}}^{\boldsymbol{n}} \hat{u}_{\boldsymbol{j}}\,\mathrm{e}^{\mathrm{i}\boldsymbol{j}\cdot\boldsymbol{\theta}}\right| > 0$ a.e. (Lemma 2.2), it follows from (3.18) and the vanishing property [22, Eq. (2.1)] that $f - m_f = 0$ a.e. (a contradiction to the hypothesis $m_f < M_f$). □

Two important corollaries of Theorem 3.1 are reported below. The first follows from Theorem 3.1 and the observation that every nonnegative function f which does not vanish a.e. satisfies $m_f \ge 0$ and $M_f > 0$. The second follows from Theorem 3.1 and the linearity of the map $T_{\boldsymbol{n}}(\cdot)$; see (3.11). Throughout the book, if X, Y are square matrices of the same size, the notation $X \ge Y$ (resp., $X > Y$) means that X, Y are Hermitian and $X - Y$ is HPSD (resp., HPD).

Corollary 3.1 *Assume that $f \in L^1([-\pi,\pi]^d)$ is nonnegative and not a.e. equal to 0. Then $T_{\boldsymbol{n}}(f) > O_{N(\boldsymbol{n})}$ for all $\boldsymbol{n} \in \mathbb{N}^d$.*

Corollary 3.2 *For any fixed $\boldsymbol{n} \in \mathbb{N}^d$, the map $T_{\boldsymbol{n}}(\cdot) : L^1([-\pi,\pi]^d) \to \mathbb{C}^{N(\boldsymbol{n})\times N(\boldsymbol{n})}$ is a Linear Positive Operator (LPO), i.e., it is linear and satisfies $T_{\boldsymbol{n}}(f) \ge O_{N(\boldsymbol{n})}$ whenever $f \ge 0$ a.e. In particular, $T_{\boldsymbol{n}}(\cdot)$ is monotone, i.e.,*

$$f \ge g \text{ a.e.} \quad \Longrightarrow \quad T_{\boldsymbol{n}}(f) \ge T_{\boldsymbol{n}}(g).$$

The last result of this section provides an important relation between tensor products and multilevel Toeplitz matrices. Recall that if $f_1, \ldots, f_d \in L^1([-\pi, \pi])$ then $f_1 \otimes \cdots \otimes f_d \in L^1([-\pi, \pi]^d)$ by Fubini's theorem.

Lemma 3.3 *Let* $f_1, \ldots, f_d \in L^1([-\pi, \pi])$ *and* $\boldsymbol{n} \in \mathbb{N}^d$. *Then*

$$T_{\boldsymbol{n}}(f_1 \otimes \cdots \otimes f_d) = T_{n_1}(f_1) \otimes \cdots \otimes T_{n_d}(f_d). \tag{3.19}$$

Proof The proof is very simple if we use the crucial property (2.29). By Fubini's theorem, the Fourier coefficients of $f_1 \otimes \cdots \otimes f_d$ are given by

$$(f_1 \otimes \cdots \otimes f_d)_{\boldsymbol{k}} = (f_1)_{k_1} \cdots (f_d)_{k_d}, \qquad \boldsymbol{k} \in \mathbb{Z}^d.$$

Hence, for all $\boldsymbol{i}, \boldsymbol{j} = \boldsymbol{1}, \ldots, \boldsymbol{n}$,

$$\begin{aligned}[T_{n_1}(f_1) \otimes \cdots \otimes T_{n_d}(f_d)]_{\boldsymbol{i}\boldsymbol{j}} &= [T_{n_1}(f_1)]_{i_1 j_1} \cdots [T_{n_d}(f_d)]_{i_d j_d} \\ &= (f_1)_{i_1 - j_1} \cdots (f_d)_{i_d - j_d} \\ &= (f_1 \otimes \cdots \otimes f_d)_{\boldsymbol{i}-\boldsymbol{j}} \\ &= [T_{\boldsymbol{n}}(f_1 \otimes \cdots \otimes f_d)]_{\boldsymbol{i}\boldsymbol{j}}. \end{aligned}$$ □

3.3 Schatten *p*-Norms of Multilevel Toeplitz Matrices

Important inequalities involving multilevel Toeplitz matrices and Schatten p-norms are provided in Theorem 3.2. They originally appeared in [37, Corollary 4.2] and were generalized in [34, Corollary 3.5]. To prove Theorem 3.2, we need a couple of intermediate lemmas, which are interesting also in themselves. They combine results from [37, Theorems 3.2 and 3.3], which hold in the more general context of "LPOs and unitarily invariant norms". We recall that the map $T_{\boldsymbol{n}}(\cdot)$ is an LPO (Corollary 3.2) and that the Schatten p-norms are unitarily invariant.

Lemma 3.4 *For every* $f \in L^1([-\pi, \pi]^d)$ *and every* $\boldsymbol{n} \in \mathbb{N}^d$,

$$\begin{aligned}|\mathbf{u}^* T_{\boldsymbol{n}}(f)\mathbf{v}| &\le \sqrt{\big(\mathbf{u}^* T_{\boldsymbol{n}}(|f|)\mathbf{u}\big)\big(\mathbf{v}^* T_{\boldsymbol{n}}(|f|)\mathbf{v}\big)} \\ &\le \frac{1}{2}\mathbf{u}^* T_{\boldsymbol{n}}(|f|)\mathbf{u} + \frac{1}{2}\mathbf{v}^* T_{\boldsymbol{n}}(|f|)\mathbf{v}, \qquad \mathbf{u}, \mathbf{v} \in \mathbb{C}^{N(\boldsymbol{n})}.\end{aligned}$$

Proof The proof is based on Lemma 3.2 and on the Cauchy–Schwarz inequality applied first in $\mathbb{C}^{N(\boldsymbol{n})}$ and then in $L^2([-\pi, \pi]^d)$. Using these ingredients, we obtain

$$|\mathbf{u}^* T_{\boldsymbol{n}}(f)\mathbf{v}|^2 = \left|\frac{1}{(2\pi)^d} \int_{[-\pi,\pi]^d} f(\boldsymbol{\theta})\, \mathbf{u}^* U(\boldsymbol{\theta})\mathbf{v}\, \mathrm{d}\boldsymbol{\theta}\right|^2$$

$$
\begin{aligned}
&\leq \left(\frac{1}{(2\pi)^d} \int_{[-\pi,\pi]^d} |f(\boldsymbol{\theta})|\, |\mathbf{u}^* U(\boldsymbol{\theta}) \mathbf{v}|\, \mathrm{d}\boldsymbol{\theta} \right)^2 \\
&= \left(\frac{1}{(2\pi)^d} \int_{[-\pi,\pi]^d} |f(\boldsymbol{\theta})| \left| \big(U(\boldsymbol{\theta})^{1/2}\mathbf{u}\big)^* \big(U(\boldsymbol{\theta})^{1/2}\mathbf{v}\big) \right| \mathrm{d}\boldsymbol{\theta} \right)^2 \\
&\leq \left(\frac{1}{(2\pi)^d} \int_{[-\pi,\pi]^d} |f(\boldsymbol{\theta})| \sqrt{\mathbf{u}^* U(\boldsymbol{\theta}) \mathbf{u}}\, \sqrt{\mathbf{v}^* U(\boldsymbol{\theta}) \mathbf{v}}\, \mathrm{d}\boldsymbol{\theta} \right)^2 \\
&= \left(\frac{1}{(2\pi)^d} \int_{[-\pi,\pi]^d} \sqrt{|f(\boldsymbol{\theta})|\, \mathbf{u}^* U(\boldsymbol{\theta}) \mathbf{u}}\, \sqrt{|f(\boldsymbol{\theta})|\, \mathbf{v}^* U(\boldsymbol{\theta}) \mathbf{v}}\, \mathrm{d}\boldsymbol{\theta} \right)^2 \\
&\leq \frac{1}{(2\pi)^d} \int_{[-\pi,\pi]^d} |f(\boldsymbol{\theta})|\, \mathbf{u}^* U(\boldsymbol{\theta}) \mathbf{u}\, \mathrm{d}\boldsymbol{\theta}\; \frac{1}{(2\pi)^d} \int_{[-\pi,\pi]^d} |f(\boldsymbol{\theta})|\, \mathbf{v}^* U(\boldsymbol{\theta}) \mathbf{v}\, \mathrm{d}\boldsymbol{\theta} \\
&= \big(\mathbf{u}^* T_{\boldsymbol{n}}(|f|)\mathbf{u}\big)\big(\mathbf{v}^* T_{\boldsymbol{n}}(|f|)\mathbf{v}\big).
\end{aligned}
$$

This proves the first inequality of the lemma. The second one is just the geometric–arithmetic mean inequality; see, e.g., [32, p. 63]. □

Lemma 3.5 *Let $f \in L^1([-\pi,\pi]^d)$, $\boldsymbol{n} \in \mathbb{N}^d$ and $1 \leq p \leq \infty$. Then*

$$
\|T_{\boldsymbol{n}}(f)\|_p \leq \|T_{\boldsymbol{n}}(|f|)\|_p. \tag{3.20}
$$

Proof By Lemma 3.4, for any pair of orthonormal bases $\{\mathbf{u}_i\}_{i=1}^{N(\boldsymbol{n})}$, $\{\mathbf{v}_i\}_{i=1}^{N(\boldsymbol{n})}$ of $\mathbb{C}^{N(\boldsymbol{n})}$ we have

$$
\begin{aligned}
\left|\left[\mathbf{u}_i^* T_{\boldsymbol{n}}(f)\mathbf{v}_i\right]_{i=1}^{N(\boldsymbol{n})}\right|_p &\leq \left|\left[\frac{1}{2}\mathbf{u}_i^* T_{\boldsymbol{n}}(|f|)\mathbf{u}_i + \frac{1}{2}\mathbf{v}_i^* T_{\boldsymbol{n}}(|f|)\mathbf{v}_i\right]_{i=1}^{N(\boldsymbol{n})}\right|_p \\
&\leq \frac{1}{2}\left|\left[\mathbf{u}_i^* T_{\boldsymbol{n}}(|f|)\mathbf{u}_i\right]_{i=1}^{N(\boldsymbol{n})}\right|_p + \frac{1}{2}\left|\left[\mathbf{v}_i^* T_{\boldsymbol{n}}(|f|)\mathbf{v}_i\right]_{i=1}^{N(\boldsymbol{n})}\right|_p.
\end{aligned}
$$

Passing to the supremum over all pairs of orthonormal bases $\{\mathbf{u}_i\}_{i=1}^{N(\boldsymbol{n})}$, $\{\mathbf{v}_i\}_{i=1}^{N(\boldsymbol{n})}$, and taking into account that $T_{\boldsymbol{n}}(|f|)$ is HPSD by Corollary 3.2, from [22, Lemma 2.11] we obtain (3.20). □

In the statement of Theorem 3.2 and throughout the book, we use the natural convention $C/\infty = 0$ for all numbers C.

Theorem 3.2 *Let $f \in L^p([-\pi,\pi]^d)$, $\boldsymbol{n} \in \mathbb{N}^d$ and $1 \leq p \leq \infty$. Then*

$$
\|T_{\boldsymbol{n}}(f)\|_p \leq \frac{N(\boldsymbol{n})^{1/p}}{(2\pi)^{d/p}} \|f\|_{L^p}. \tag{3.21}
$$

In particular, for $p = \infty$ we have

$$
\|T_{\boldsymbol{n}}(f)\| = \|T_{\boldsymbol{n}}(f)\|_\infty \leq \|f\|_{L^\infty}. \tag{3.22}
$$

Proof In view of Lemma 3.5, it suffices to prove (3.21) in the case where $f \ge 0$. In this case, we know from Corollary 3.2 that $T_{\boldsymbol{n}}(f)$ is HPSD. In particular, $\|T_{\boldsymbol{n}}(f)\| = \lambda_{\max}(T_{\boldsymbol{n}}(f))$ and (3.22) follows directly from Theorem 3.1.

Suppose now that $1 \le p < \infty$. By Lemma 3.2 and the inequality in [22, Eq. (2.3)], for every $\mathbf{u} \in \mathbb{C}^{N(\boldsymbol{n})}$ such that $\|\mathbf{u}\| = 1$ we have

$$\begin{aligned}\big(\mathbf{u}^* T_{\boldsymbol{n}}(f)\mathbf{u}\big)^p &= \left(\frac{1}{(2\pi)^d}\int_{[-\pi,\pi]^d} f(\boldsymbol{\theta})\,\mathbf{u}^* U(\boldsymbol{\theta})\mathbf{u}\,\mathrm{d}\boldsymbol{\theta}\right)^p \\ &\le \frac{1}{(2\pi)^d}\int_{[-\pi,\pi]^d} f(\boldsymbol{\theta})^p\,\mathbf{u}^* U(\boldsymbol{\theta})\mathbf{u}\,\mathrm{d}\boldsymbol{\theta} = \mathbf{u}^* T_{\boldsymbol{n}}(f^p)\mathbf{u}.\end{aligned}$$

Hence, by [22, Lemma 2.11],

$$\begin{aligned}\|T_{\boldsymbol{n}}(f)\|_p^p &= \sup_{\substack{\{\mathbf{u}_i\}_{i=1}^{N(\boldsymbol{n})} \text{orthonormal} \\ \text{basis of } \mathbb{C}^{N(\boldsymbol{n})}}} \left|\left[\mathbf{u}_i^* T_{\boldsymbol{n}}(f)\mathbf{u}_i\right]_{i=1}^{N(\boldsymbol{n})}\right|_p^p \\ &\le \sup_{\substack{\{\mathbf{u}_i\}_{i=1}^{N(\boldsymbol{n})} \text{orthonormal} \\ \text{basis of } \mathbb{C}^{N(\boldsymbol{n})}}} \left|\left[\mathbf{u}_i^* T_{\boldsymbol{n}}(f^p)\mathbf{u}_i\right]_{i=1}^{N(\boldsymbol{n})}\right|_1 = \|T_{\boldsymbol{n}}(f^p)\|_1.\end{aligned}$$

To conclude, we observe that, since $T_{\boldsymbol{n}}(f^p)$ is HPSD, its singular values coincide with its eigenvalues. Thus,

$$\begin{aligned}\|T_{\boldsymbol{n}}(f^p)\|_1 &= \mathrm{trace}(T_{\boldsymbol{n}}(f^p)) = N(\boldsymbol{n})(f^p)_{\mathbf{0}} \\ &= \frac{N(\boldsymbol{n})}{(2\pi)^d}\int_{[-\pi,\pi]^d} f(\boldsymbol{\theta})^p \mathrm{d}\boldsymbol{\theta} = \frac{N(\boldsymbol{n})}{(2\pi)^d}\|f\|_{L^p}^p.\end{aligned}$$ □

The last result of this section shows that $\|T_{\boldsymbol{n}}(f)T_{\boldsymbol{n}}(g) - T_{\boldsymbol{n}}(fg)\|_1 = o(N(\boldsymbol{n}))$ as $\boldsymbol{n} \to \infty$ in the case where $f \in L^p([-\pi,\pi]^d)$ and $g \in L^q([-\pi,\pi]^d)$, with p, q conjugate exponents. Note that in this case $f, g, fg \in L^1([-\pi,\pi]^d)$ by Hölder's inequality, so the matrices $T_{\boldsymbol{n}}(f)$, $T_{\boldsymbol{n}}(g)$, $T_{\boldsymbol{n}}(fg)$ are defined.

Lemma 3.6 *Let $f, g \in L^1([-\pi,\pi]^d)$, where $f(\boldsymbol{\theta}) = \sum_{\boldsymbol{j}=-\boldsymbol{r}}^{\boldsymbol{r}} f_{\boldsymbol{j}}\,\mathrm{e}^{\mathrm{i}\boldsymbol{j}\cdot\boldsymbol{\theta}}$ is a d-variate trigonometric polynomial, and let $\boldsymbol{n} \in \mathbb{N}^d$. Then, the rows of $T_{\boldsymbol{n}}(f)T_{\boldsymbol{n}}(g) - T_{\boldsymbol{n}}(fg)$ corresponding to indices $\boldsymbol{k} \in \{\boldsymbol{r}+\mathbf{1}, \ldots, \boldsymbol{n}-\boldsymbol{r}\}$ are zero. In particular,*

$$\mathrm{rank}(T_{\boldsymbol{n}}(f)T_{\boldsymbol{n}}(g) - T_{\boldsymbol{n}}(fg)) \le N(\boldsymbol{n}) - N(\boldsymbol{n} - 2\boldsymbol{r}).$$

Proof The Fourier coefficients of fg are given by

$$\begin{aligned}(fg)_{\boldsymbol{k}} &= \frac{1}{(2\pi)^d}\int_{[-\pi,\pi]^d} f(\boldsymbol{\theta})g(\boldsymbol{\theta})\,\mathrm{e}^{-\mathrm{i}\boldsymbol{k}\cdot\boldsymbol{\theta}}\mathrm{d}\boldsymbol{\theta} \\ &= \sum_{\boldsymbol{j}=-\boldsymbol{r}}^{\boldsymbol{r}} f_{\boldsymbol{j}}\,\frac{1}{(2\pi)^d}\int_{[-\pi,\pi]^d} g(\boldsymbol{\theta})\,\mathrm{e}^{-\mathrm{i}(\boldsymbol{k}-\boldsymbol{j})\cdot\boldsymbol{\theta}}\mathrm{d}\boldsymbol{\theta} = \sum_{\boldsymbol{j}=-\boldsymbol{r}}^{\boldsymbol{r}} f_{\boldsymbol{j}}\,g_{\boldsymbol{k}-\boldsymbol{j}}, \qquad \boldsymbol{k} \in \mathbb{Z}^d.\end{aligned}$$

For all $\boldsymbol{k}, \boldsymbol{\ell} = \boldsymbol{1}, \ldots, \boldsymbol{n}$,

$$(T_{\boldsymbol{n}}(fg))_{\boldsymbol{k}\boldsymbol{\ell}} = (fg)_{\boldsymbol{k}-\boldsymbol{\ell}} = \sum_{\boldsymbol{j}=-\boldsymbol{r}}^{\boldsymbol{r}} f_{\boldsymbol{j}}\, g_{\boldsymbol{k}-\boldsymbol{\ell}-\boldsymbol{j}} \tag{3.23}$$

and

$$\begin{aligned}(T_{\boldsymbol{n}}(f)T_{\boldsymbol{n}}(g))_{\boldsymbol{k}\boldsymbol{\ell}} &= \sum_{\boldsymbol{s}=\boldsymbol{1}}^{\boldsymbol{n}}(T_{\boldsymbol{n}}(f))_{\boldsymbol{k}\boldsymbol{s}}(T_{\boldsymbol{n}}(g))_{\boldsymbol{s}\boldsymbol{\ell}} = \sum_{\boldsymbol{s}=\boldsymbol{1}}^{\boldsymbol{n}} f_{\boldsymbol{k}-\boldsymbol{s}}\, g_{\boldsymbol{s}-\boldsymbol{\ell}} \\ &= \sum_{\boldsymbol{j}=\boldsymbol{k}-\boldsymbol{n}}^{\boldsymbol{k}-\boldsymbol{1}} f_{\boldsymbol{j}}\, g_{\boldsymbol{k}-\boldsymbol{j}-\boldsymbol{\ell}}. \end{aligned} \tag{3.24}$$

Considering that $f_{\boldsymbol{j}} = 0$ for $\boldsymbol{j} \notin \{-\boldsymbol{r}, \ldots, \boldsymbol{r}\}$, it is clear that (3.23) and (3.24) coincide for $\boldsymbol{r} + \boldsymbol{1} \le \boldsymbol{k} \le \boldsymbol{n} - \boldsymbol{r}$, because in this case we have $\boldsymbol{k} - \boldsymbol{n} \le -\boldsymbol{r}$ and $\boldsymbol{k} - \boldsymbol{1} \ge \boldsymbol{r}$. □

Theorem 3.3 *Let $f \in L^p([-\pi, \pi]^d)$ and $g \in L^q([-\pi, \pi]^d)$, where $1 \le p, q \le \infty$ are conjugate exponents. Then*

$$\lim_{\boldsymbol{n}\to\infty} \frac{\|T_{\boldsymbol{n}}(f)T_{\boldsymbol{n}}(g) - T_{\boldsymbol{n}}(fg)\|_1}{N(\boldsymbol{n})} = 0. \tag{3.25}$$

Proof We first prove (3.25) under the assumption that $f, g \in L^\infty([-\pi, \pi]^d)$. Let $f_m(\boldsymbol{\theta}) = \sum_{\boldsymbol{j}=-\boldsymbol{r}_m}^{\boldsymbol{r}_m} (f_m)_{\boldsymbol{j}}\, \mathrm{e}^{\mathrm{i}\boldsymbol{j}\cdot\boldsymbol{\theta}}$ be a sequence of d-variate trigonometric polynomials such that

$$\|f_m\|_\infty \le \|f\|_{L^\infty}, \qquad \|f - f_m\|_{L^1} \to 0;$$

note that such a sequence exists by Lemma 2.3. For every m and every $\boldsymbol{n} \in \mathbb{N}^d$,

$$\begin{aligned}\|T_{\boldsymbol{n}}(f)T_{\boldsymbol{n}}(g) - T_{\boldsymbol{n}}(fg)\|_1 \le\ & \|T_{\boldsymbol{n}}(f)T_{\boldsymbol{n}}(g) - T_{\boldsymbol{n}}(f_m)T_{\boldsymbol{n}}(g)\|_1 \\ &+ \|T_{\boldsymbol{n}}(f_m)T_{\boldsymbol{n}}(g) - T_{\boldsymbol{n}}(f_m g)\|_1 \\ &+ \|T_{\boldsymbol{n}}(f_m g) - T_{\boldsymbol{n}}(fg)\|_1. \end{aligned} \tag{3.26}$$

Using the linearity of $T_{\boldsymbol{n}}(\cdot)$, the Hölder-type inequality (2.18) and Theorem 3.2, we can bound the first and last term in the right-hand side of (3.26) as follows:

$$\begin{aligned}\|T_{\boldsymbol{n}}(f)T_{\boldsymbol{n}}(g) - T_{\boldsymbol{n}}(f_m)T_{\boldsymbol{n}}(g)\|_1 &\le \|T_{\boldsymbol{n}}(f - f_m)\|_1 \|T_{\boldsymbol{n}}(g)\| \\ &\le N(\boldsymbol{n})\|f - f_m\|_{L^1}\|g\|_{L^\infty}, \qquad (3.27)\\ \|T_{\boldsymbol{n}}(f_m g) - T_{\boldsymbol{n}}(fg)\|_1 &\le N(\boldsymbol{n})\|f_m g - fg\|_{L^1} \\ &\le N(\boldsymbol{n})\|f_m - f\|_{L^1}\|g\|_{L^\infty}. \qquad (3.28)\end{aligned}$$

To bound the second term in the right-hand side of (3.26), we use Lemma 3.6 and the trace-norm inequality (2.19) in combination with Theorem 3.2. We have

$$\begin{aligned}
&\|T_{\boldsymbol{n}}(f_m)T_{\boldsymbol{n}}(g) - T_{\boldsymbol{n}}(f_m g)\|_1 \\
&\le \operatorname{rank}(T_{\boldsymbol{n}}(f_m)T_{\boldsymbol{n}}(g) - T_{\boldsymbol{n}}(f_m g))\|T_{\boldsymbol{n}}(f_m)T_{\boldsymbol{n}}(g) - T_{\boldsymbol{n}}(f_m g)\| \\
&\le (N(\boldsymbol{n}) - N(\boldsymbol{n} - 2\boldsymbol{r}_m))\big(\|T_{\boldsymbol{n}}(f_m)\|\,\|T_{\boldsymbol{n}}(g)\| + \|T_{\boldsymbol{n}}(f_m g)\|\big) \\
&\le (N(\boldsymbol{n}) - N(\boldsymbol{n} - 2\boldsymbol{r}_m))\big(\|f_m\|_\infty\|g\|_{L^\infty} + \|f_m g\|_{L^\infty}\big) \\
&\le 2(N(\boldsymbol{n}) - N(\boldsymbol{n} - 2\boldsymbol{r}_m))\|f\|_{L^\infty}\|g\|_{L^\infty}. \qquad (3.29)
\end{aligned}$$

Putting together (3.26) and (3.27)–(3.29), we get

$$\frac{\|T_{\boldsymbol{n}}(f)T_{\boldsymbol{n}}(g) - T_{\boldsymbol{n}}(fg)\|_1}{N(\boldsymbol{n})} \le 2\|f - f_m\|_{L^1}\|g\|_{L^\infty} + \frac{2(N(\boldsymbol{n}) - N(\boldsymbol{n} - 2\boldsymbol{r}_m))\|f\|_{L^\infty}\|g\|_{L^\infty}}{N(\boldsymbol{n})}.$$

Passing to the limit as $\boldsymbol{n} \to \infty$, we obtain

$$\limsup_{\boldsymbol{n}\to\infty} \frac{\|T_{\boldsymbol{n}}(f)T_{\boldsymbol{n}}(g) - T_{\boldsymbol{n}}(fg)\|_1}{N(\boldsymbol{n})} \le 2\|f - f_m\|_{L^1}\|g\|_{L^\infty}.$$

Passing to the limit as $m \to \infty$, we obtain (3.25). This concludes the proof of (3.25) in the case where $f, g \in L^\infty([-\pi, \pi]^d)$.

Suppose now that $f \in L^p([-\pi, \pi]^d)$ and $g \in L^q([-\pi, \pi]^d)$, with $1 \le p, q \le \infty$ conjugate exponents. Take two sequences $\{f_m\}_m$ and $\{g_m\}_m$ such that $f_m, g_m \in L^\infty([-\pi, \pi]^d)$ for all m, $f_m \to f$ in $L^p([-\pi, \pi]^d)$ and $g_m \to g$ in $L^q([-\pi, \pi]^d)$; for example, one can choose $f_m = f\,\chi_{\{|f|\le m\}}$ and $g_m = g\,\chi_{\{|g|\le m\}}$. By the linearity of $T_{\boldsymbol{n}}(\cdot)$, the Hölder-type inequality (2.18) and Theorem 3.2, for every m and every $\boldsymbol{n} \in \mathbb{N}^d$ we have

$$\begin{aligned}
&\|T_{\boldsymbol{n}}(f)T_{\boldsymbol{n}}(g) - T_{\boldsymbol{n}}(fg)\|_1 \\
&\le \|T_{\boldsymbol{n}}(f - f_m)T_{\boldsymbol{n}}(g)\|_1 + \|T_{\boldsymbol{n}}(f_m)T_{\boldsymbol{n}}(g - g_m)\|_1 \\
&\quad + \|T_{\boldsymbol{n}}(f_m)T_{\boldsymbol{n}}(g_m) - T_{\boldsymbol{n}}(f_m g_m)\|_1 + \|T_{\boldsymbol{n}}(f_m g_m - fg)\|_1 \\
&\le N(\boldsymbol{n})^{1/p}\|f - f_m\|_{L^p}\,N(\boldsymbol{n})^{1/q}\|g\|_{L^q} + N(\boldsymbol{n})^{1/p}\|f_m\|_{L^p}\,N(\boldsymbol{n})^{1/q}\|g - g_m\|_{L^q} \\
&\quad + \|T_{\boldsymbol{n}}(f_m)T_{\boldsymbol{n}}(g_m) - T_{\boldsymbol{n}}(f_m g_m)\|_1 + N(\boldsymbol{n})\|f_m g_m - fg\|_{L^1} \\
&\le N(\boldsymbol{n})\bigg[\|f - f_m\|_{L^p}\|g\|_{L^q} + \sup_i \|f_i\|_{L^p}\|g - g_m\|_{L^q} \\
&\quad + \frac{\|T_{\boldsymbol{n}}(f_m)T_{\boldsymbol{n}}(g_m) - T_{\boldsymbol{n}}(f_m g_m)\|_1}{N(\boldsymbol{n})} + \|f_m g_m - fg\|_{L^1}\bigg]. \qquad (3.30)
\end{aligned}$$

Note that $\sup_i \|f_i\|_{L^p} < \infty$, because $f_i \to f$ in $L^p([-\pi, \pi]^d)$ and hence $\|f_i\|_{L^p} \to \|f\|_{L^p}$. Since $f_m, g_m \in L^\infty([-\pi, \pi]^d)$, by the first part of the proof we have

$$\lim_{n\to\infty} \frac{\|T_{\boldsymbol{n}}(f_m)T_{\boldsymbol{n}}(g_m) - T_{\boldsymbol{n}}(f_m g_m)\|_1}{N(\boldsymbol{n})} = 0.$$

Dividing both sides of (3.30) by $N(\boldsymbol{n})$ and passing to the limit as $\boldsymbol{n} \to \infty$, we obtain

$$\begin{aligned}
&\limsup_{n\to\infty} \frac{\|T_{\boldsymbol{n}}(f)T_{\boldsymbol{n}}(g) - T_{\boldsymbol{n}}(fg)\|_1}{N(\boldsymbol{n})} \\
&\le \|f - f_m\|_{L^p}\|g\|_{L^q} + \sup_i \|f_i\|_{L^p}\|g - g_m\|_{L^q} + \|f_m g_m - fg\|_{L^1}. \qquad (3.31)
\end{aligned}$$

This relation holds for every m. As $m \to \infty$, $f_m \to f$ in $L^p([-\pi, \pi]^d)$ and $g_m \to g$ in $L^q([-\pi, \pi]^d)$ by construction. Moreover, $f_m g_m \to fg$ in $L^1([-\pi, \pi]^d)$ by Hölder's inequality, since

$$\begin{aligned}
\|fg - f_m g_m\|_{L^1} &\le \|(f - f_m)g\|_{L^1} + \|f_m(g - g_m)\|_{L^1} \\
&\le \|f - f_m\|_{L^p}\|g\|_{L^q} + \|f_m\|_{L^p}\|g - g_m\|_{L^q} \\
&\le \|f - f_m\|_{L^p}\|g\|_{L^q} + \sup_i \|f_i\|_{L^p}\|g - g_m\|_{L^q}.
\end{aligned}$$

Passing to the limit as $m \to \infty$ in (3.31), we obtain (3.25). □

3.4 Multilevel Circulant Matrices

Given a d-index $\boldsymbol{n} \in \mathbb{N}^d$, a matrix of the form

$$\left[a_{(\boldsymbol{i}-\boldsymbol{j}) \bmod \boldsymbol{n}}\right]_{\boldsymbol{i},\boldsymbol{j}=\boldsymbol{1}}^{\boldsymbol{n}} \in \mathbb{C}^{N(\boldsymbol{n})\times N(\boldsymbol{n})} \qquad (3.32)$$

is called a multilevel circulant matrix (or, more precisely, a d-level circulant matrix). Since the $(\boldsymbol{i}, \boldsymbol{j})$ entry $a_{(\boldsymbol{i}-\boldsymbol{j}) \bmod \boldsymbol{n}}$ depends only on the difference $\boldsymbol{i} - \boldsymbol{j}$, it is clear that any d-level circulant matrix is in particular a d-level Toeplitz matrix. In the case $d = 1$, the matrix (3.32) becomes

$$\left[a_{(i-j) \bmod n}\right]_{i,j=1}^{n} = \begin{bmatrix}
a_0 & a_{n-1} & a_{n-2} & \cdots & \cdots & a_1 \\
a_1 & \ddots & \ddots & \ddots & & \vdots \\
a_2 & \ddots & \ddots & \ddots & \ddots & \vdots \\
\vdots & \ddots & \ddots & \ddots & \ddots & a_{n-2} \\
\vdots & & \ddots & \ddots & \ddots & a_{n-1} \\
a_{n-1} & \cdots & \cdots & a_2 & a_1 & a_0
\end{bmatrix}, \qquad (3.33)$$

which is the expression of the generic unilevel circulant matrix. Unilevel circulant matrices are nothing else than classical circulant matrices, which have been the subject of [22, Sect. 6.4]. In the case $d = 2$, the matrix (3.32) can be written as

$$\left[a_{(\boldsymbol{i}-\boldsymbol{j}) \bmod \boldsymbol{n}}\right]_{\boldsymbol{i},\boldsymbol{j}=\boldsymbol{1}}^{\boldsymbol{n}} = \left[\left[a_{(i_1-j_1) \bmod n_1,(i_2-j_2) \bmod n_2}\right]_{i_2,j_2=1}^{n_2}\right]_{i_1,j_1=1}^{n_1}$$

$$= \left[\mathbf{a}_{(i_1-j_1) \bmod n_1}\right]_{i_1,j_1=1}^{n_1} = \begin{bmatrix} \mathbf{a}_0 & \mathbf{a}_{n_1-1} & \mathbf{a}_{n_1-2} & \cdots & \cdots & \mathbf{a}_1 \\ \mathbf{a}_1 & \ddots & \ddots & \ddots & & \vdots \\ \mathbf{a}_2 & \ddots & \ddots & \ddots & \ddots & \vdots \\ \vdots & \ddots & \ddots & \ddots & \ddots & \mathbf{a}_{n_1-2} \\ \vdots & & \ddots & \ddots & \ddots & \mathbf{a}_{n_1-1} \\ \mathbf{a}_{n_1-1} & \cdots & \cdots & \mathbf{a}_2 & \mathbf{a}_1 & \mathbf{a}_0 \end{bmatrix}, \tag{3.34}$$

where, for all $k = 0, \ldots, n_1 - 1$,

$$\mathbf{a}_k = \left[a_{k,(i_2-j_2) \bmod n_2}\right]_{i_2,j_2=1}^{n_2}$$

$$= \begin{bmatrix} a_{k,0} & a_{k,n_2-1} & a_{k,n_2-2} & \cdots & \cdots & a_{k,1} \\ a_{k,1} & \ddots & \ddots & \ddots & & \vdots \\ a_{k,2} & \ddots & \ddots & \ddots & \ddots & \vdots \\ \vdots & \ddots & \ddots & \ddots & \ddots & a_{k,n_2-2} \\ \vdots & & \ddots & \ddots & \ddots & a_{k,n_2-1} \\ a_{k,n_2-1} & \cdots & \cdots & a_{k,2} & a_{k,1} & a_{k,0} \end{bmatrix}. \tag{3.35}$$

A matrix of this form, which we call a 2-level circulant matrix, is also known as a block circulant matrix with circulant blocks, or BCCB matrix. In general, we can say that a d-level circulant matrix is a block circulant matrix with $(d-1)$-level circulant blocks.

Let C_n be the $n \times n$ matrix whose (i, j) entry equals 1 if $(i - j) \bmod n = 1$ and 0 otherwise:

$$C_n = \begin{bmatrix} 0 & & & & 1 \\ 1 & \ddots & & & \\ & \ddots & \ddots & & \\ & & \ddots & \ddots & \\ & & & \ddots & \ddots \\ & & & 1 & 0 \end{bmatrix}. \tag{3.36}$$

The matrix C_n is called the generator of classical circulant matrices of order n. This name is due to the fact that the powers of C_n are

$$C_n^2 = \begin{bmatrix} 0 & & & & 1 & 0 \\ 0 & \ddots & & & & 1 \\ 1 & \ddots & \ddots & & & \\ & \ddots & \ddots & \ddots & & \\ & & \ddots & \ddots & \ddots & \\ & & & 1 & 0 & 0 \end{bmatrix}, \quad C_n^3 = \begin{bmatrix} 0 & & & & 1 & 0 & 0 \\ 0 & \ddots & & & & 1 & 0 \\ 0 & \ddots & \ddots & & & & 1 \\ 1 & \ddots & \ddots & \ddots & & & \\ & \ddots & \ddots & \ddots & \ddots & & \\ & & & 1 & 0 & 0 & 0 \end{bmatrix}, \ldots, \quad C_n^n = I_n, \tag{3.37}$$

and so the classical circulant matrix (3.33) can be written as a linear combination of nonnegative powers of C_n:

$$\left[a_{(i-j) \bmod n}\right]_{i,j=1}^{n} = \sum_{k=0}^{n-1} a_k C_n^k. \tag{3.38}$$

For $\boldsymbol{n} \in \mathbb{N}^d$ and $\boldsymbol{k} \in \mathbb{Z}^d$, let

$$C_{\boldsymbol{n}}^{\boldsymbol{k}} = C_{n_1}^{k_1} \otimes C_{n_2}^{k_2} \otimes \cdots \otimes C_{n_d}^{k_d}. \tag{3.39}$$

Lemma 3.7, which is a "refined" version of Lemma 3.1 for multilevel circulant matrices, generalizes (3.38) to the multilevel case.

Lemma 3.7 *The d-level circulant matrix* (3.32) *admits the following expression:*

$$\left[a_{(\boldsymbol{i}-\boldsymbol{j}) \bmod \boldsymbol{n}}\right]_{\boldsymbol{i},\boldsymbol{j}=\boldsymbol{1}}^{\boldsymbol{n}} = \sum_{\boldsymbol{k}=\boldsymbol{0}}^{\boldsymbol{n}-\boldsymbol{1}} a_{\boldsymbol{k}} C_{\boldsymbol{n}}^{\boldsymbol{k}}, \tag{3.40}$$

where $C_{\boldsymbol{n}}^{\boldsymbol{k}}$ is defined in (3.39).

Proof The proof follows the same pattern as the proof of Lemma 3.1. Equation (3.40) is proved componentwise, by showing that the $(\boldsymbol{i}, \boldsymbol{j})$ entry of the matrix in the right-hand side is equal to $a_{(\boldsymbol{i}-\boldsymbol{j}) \bmod \boldsymbol{n}}$. Let $\delta_r = 1$ if $r = 0$ and $\delta_r = 0$ otherwise. By (3.36) and (3.37), for every $n \in \mathbb{N}$ and $k \in \mathbb{Z}$ we can express the (i, j) entry of C_n^k as follows:

$$(C_n^k)_{ij} = \delta_{(i-j-k) \bmod n}, \qquad i, j = 1, \ldots, n. \tag{3.41}$$

By the crucial property (2.29), for all $\boldsymbol{i}, \boldsymbol{j} = \boldsymbol{1}, \ldots, \boldsymbol{n}$ we have

$$\begin{aligned} (C_{\boldsymbol{n}}^{\boldsymbol{k}})_{\boldsymbol{i}\boldsymbol{j}} &= (C_{n_1}^{k_1})_{i_1 j_1} (C_{n_2}^{k_2})_{i_2 j_2} \cdots (C_{n_d}^{k_d})_{i_d j_d} \\ &= \delta_{(i_1-j_1-k_1) \bmod n_1} \delta_{(i_2-j_2-k_2) \bmod n_2} \cdots \delta_{(i_d-j_d-k_d) \bmod n_d} = \delta_{(\boldsymbol{i}-\boldsymbol{j}-\boldsymbol{k}) \bmod \boldsymbol{n}}, \end{aligned}$$

where $\delta_{\boldsymbol{r}} = 1$ if $\boldsymbol{r} = \boldsymbol{0}$ and $\delta_{\boldsymbol{r}} = 0$ otherwise. Therefore,

$$\Big(\sum_{k=0}^{\boldsymbol{n}-\boldsymbol{1}} a_{\boldsymbol{k}} C_{\boldsymbol{n}}^{\boldsymbol{k}}\Big)_{\boldsymbol{ij}} = \sum_{\boldsymbol{k}=\boldsymbol{0}}^{\boldsymbol{n}-\boldsymbol{1}} a_{\boldsymbol{k}} (C_{\boldsymbol{n}}^{\boldsymbol{k}})_{\boldsymbol{ij}} = \sum_{\boldsymbol{k}=\boldsymbol{0}}^{\boldsymbol{n}-\boldsymbol{1}} a_{\boldsymbol{k}} \delta_{(\boldsymbol{i}-\boldsymbol{j}-\boldsymbol{k}) \bmod \boldsymbol{n}} = a_{(\boldsymbol{i}-\boldsymbol{j}) \bmod \boldsymbol{n}},$$

and (3.40) is proved. □

As a consequence of Lemma 3.7, any linear combination of the form $\sum_{\boldsymbol{k}=-\boldsymbol{r}}^{\boldsymbol{r}} c_{\boldsymbol{k}} C_{\boldsymbol{n}}^{\boldsymbol{k}}$ is a d-level circulant matrix, because it can be written in the form (3.40) by using the identity $C_n^{-1} = C_n^{n-1}$ (see (3.37)) and the properties of tensor products (see Sect. 2.5).

Theorem 3.4 provides the spectral decomposition of any linear combination of the form $\sum_{\boldsymbol{k}=-\boldsymbol{r}}^{\boldsymbol{r}} c_{\boldsymbol{k}} C_{\boldsymbol{n}}^{\boldsymbol{k}}$ and hence of any multilevel circulant matrix. Let

$$F_n = \frac{1}{\sqrt{n}}\big(\mathrm{e}^{-2\pi \mathrm{i} jk/n}\big)_{j,k=0}^{n-1} = \frac{1}{\sqrt{n}}\big(\mathrm{e}^{-2\pi \mathrm{i}(j-1)(k-1)/n}\big)_{j,k=1}^{n}. \tag{3.42}$$

The matrix F_n is the so-called unitary discrete Fourier transform of order n. Using the equation

$$\sum_{k=0}^{N} z^k = \begin{cases} \frac{z^{N+1}-1}{z-1}, & \text{if } z \neq 1, \\ N+1, & \text{if } z = 1, \end{cases}$$

it is easy to check that F_n is indeed unitary: $F_n^* F_n = I_n$. For $\boldsymbol{n} \in \mathbb{N}^d$, let

$$F_{\boldsymbol{n}} = F_{n_1} \otimes \cdots \otimes F_{n_d}. \tag{3.43}$$

Theorem 3.4 *Let $\boldsymbol{n}, \boldsymbol{r} \in \mathbb{N}^d$ and $c_{-\boldsymbol{r}}, \ldots, c_{\boldsymbol{r}} \in \mathbb{C}$. Then,*

$$\sum_{\boldsymbol{k}=-\boldsymbol{r}}^{\boldsymbol{r}} c_{\boldsymbol{k}} C_{\boldsymbol{n}}^{\boldsymbol{k}} = F_{\boldsymbol{n}}\Big(\operatorname*{diag}_{\boldsymbol{j}=\boldsymbol{0},\ldots,\boldsymbol{n}-\boldsymbol{1}} c\Big(\frac{2\pi \boldsymbol{j}}{\boldsymbol{n}}\Big)\Big) F_{\boldsymbol{n}}^*,$$

where $c(\boldsymbol{\theta}) = \sum_{\boldsymbol{k}=-\boldsymbol{r}}^{\boldsymbol{r}} c_{\boldsymbol{k}} \mathrm{e}^{\mathrm{i}\boldsymbol{k}\cdot\boldsymbol{\theta}}$ and $F_{\boldsymbol{n}}$ is defined in (3.43). In particular, $\sum_{\boldsymbol{k}=-\boldsymbol{r}}^{\boldsymbol{r}} c_{\boldsymbol{k}} C_{\boldsymbol{n}}^{\boldsymbol{k}}$ is a normal matrix whose spectrum is given by

$$\Lambda\Big(\sum_{\boldsymbol{k}=-\boldsymbol{r}}^{\boldsymbol{r}} c_{\boldsymbol{k}} C_{\boldsymbol{n}}^{\boldsymbol{k}}\Big) = \Big\{c\Big(\frac{2\pi \boldsymbol{j}}{\boldsymbol{n}}\Big) : \boldsymbol{j} = \boldsymbol{0}, \ldots, \boldsymbol{n}-\boldsymbol{1}\Big\}.$$

Proof The spectral decomposition of C_n is known and is given by

$$C_n = F_n D_n F_n^*, \qquad D_n = \operatorname*{diag}_{j=0,\ldots,n-1} \mathrm{e}^{2\pi \mathrm{i} j/n} = \operatorname*{diag}_{j=1,\ldots,n} \mathrm{e}^{2\pi \mathrm{i}(j-1)/n}.$$

This can be verified by direct computation: for all $i, j = 1, \ldots, n$,

$$(F_n^* C_n)_{ij} = \frac{1}{\sqrt{n}} \sum_{\ell=1}^{n} \mathrm{e}^{2\pi \mathrm{i}(i-1)(\ell-1)/n} (C_n)_{\ell j} = \frac{1}{\sqrt{n}} \mathrm{e}^{2\pi \mathrm{i}(i-1)j/n} = (D_n F_n^*)_{ij}.$$

Therefore, also the spectral decomposition of the matrix $C_{\boldsymbol{n}}^{\boldsymbol{k}}$ is known. Indeed, using the properties of tensor products (see Sect. 2.5), we obtain

$$\begin{aligned}
C_{\boldsymbol{n}}^{\boldsymbol{k}} &= C_{n_1}^{k_1} \otimes C_{n_2}^{k_2} \otimes \cdots \otimes C_{n_d}^{k_d} = (F_{n_1} D_{n_1}^{k_1} F_{n_1}^*) \otimes (F_{n_2} D_{n_2}^{k_2} F_{n_2}^*) \otimes \cdots \otimes (F_{n_d} D_{n_d}^{k_d} F_{n_d}^*) \\
&= (F_{n_1} \otimes F_{n_2} \otimes \cdots \otimes F_{n_d})(D_{n_1}^{k_1} \otimes D_{n_2}^{k_2} \otimes \cdots \otimes D_{n_d}^{k_d})(F_{n_1} \otimes F_{n_2} \otimes \cdots \otimes F_{n_d})^* \\
&= F_{\boldsymbol{n}} D_{\boldsymbol{n}}^{\boldsymbol{k}} F_{\boldsymbol{n}}^*,
\end{aligned}$$

where

$$D_{\boldsymbol{n}}^{\boldsymbol{k}} = D_{n_1}^{k_1} \otimes D_{n_2}^{k_2} \otimes \cdots \otimes D_{n_d}^{k_d} = \operatorname*{diag}_{\boldsymbol{j}=\boldsymbol{0},\ldots,\boldsymbol{n}-\boldsymbol{1}} (\mathrm{e}^{2\pi\mathrm{i}\sum_{r=1}^d j_r k_r/n_r}) = \operatorname*{diag}_{\boldsymbol{j}=\boldsymbol{0},\ldots,\boldsymbol{n}-\boldsymbol{1}} (\mathrm{e}^{2\pi\mathrm{i}(\boldsymbol{j}/\boldsymbol{n})\cdot\boldsymbol{k}});$$

note that the second equality follows from the crucial property (2.29). Thus,

$$\begin{aligned}
\sum_{\boldsymbol{k}=-\boldsymbol{r}}^{\boldsymbol{r}} c_{\boldsymbol{k}} C_{\boldsymbol{n}}^{\boldsymbol{k}} &= \sum_{\boldsymbol{k}=-\boldsymbol{r}}^{\boldsymbol{r}} c_{\boldsymbol{k}} F_{\boldsymbol{n}} D_{\boldsymbol{n}}^{\boldsymbol{k}} F_{\boldsymbol{n}}^* = F_{\boldsymbol{n}} \Bigg(\sum_{\boldsymbol{k}=-\boldsymbol{r}}^{\boldsymbol{r}} c_{\boldsymbol{k}} D_{\boldsymbol{n}}^{\boldsymbol{k}} \Bigg) F_{\boldsymbol{n}}^* \\
&= F_{\boldsymbol{n}} \Bigg(\sum_{\boldsymbol{k}=-\boldsymbol{r}}^{\boldsymbol{r}} c_{\boldsymbol{k}} \operatorname*{diag}_{\boldsymbol{j}=\boldsymbol{0},\ldots,\boldsymbol{n}-\boldsymbol{1}} (\mathrm{e}^{2\pi\mathrm{i}(\boldsymbol{j}/\boldsymbol{n})\cdot\boldsymbol{k}}) \Bigg) F_{\boldsymbol{n}}^* \\
&= F_{\boldsymbol{n}} \operatorname*{diag}_{\boldsymbol{j}=\boldsymbol{0},\ldots,\boldsymbol{n}-\boldsymbol{1}} \Bigg(\sum_{\boldsymbol{k}=-\boldsymbol{r}}^{\boldsymbol{r}} c_{\boldsymbol{k}} \mathrm{e}^{2\pi\mathrm{i}(\boldsymbol{j}/\boldsymbol{n})\cdot\boldsymbol{k}} \Bigg) F_{\boldsymbol{n}}^*,
\end{aligned}$$

and the thesis follows from the identity $\sum_{\boldsymbol{k}=-\boldsymbol{r}}^{\boldsymbol{r}} c_{\boldsymbol{k}} \mathrm{e}^{2\pi\mathrm{i}(\boldsymbol{j}/\boldsymbol{n})\cdot\boldsymbol{k}} = c(2\pi\boldsymbol{j}/\boldsymbol{n})$. □

3.5 Singular Value and Spectral Distribution of Multilevel Toeplitz Sequences: An a.c.s.-Based Proof

This section is devoted to the proof of Theorem 3.5, which comprises the multivariate L^1 versions of Szegő's first limit theorem and the Avram–Parter theorem. It provides the singular value distribution of multilevel Toeplitz sequences generated by a function $f \in L^1([-\pi,\pi]^d)$ and the spectral distribution of multilevel Toeplitz sequences generated by a real function $f \in L^1([-\pi,\pi]^d)$. For the eigenvalues it goes back to Szegő [27], and for the singular values it was established by Avram [2] and Parter [29]. They assumed $d = 1$ and $f \in L^\infty([-\pi,\pi]^d)$; see [8, Sect. 5] and [9, Sect. 10.14] for more on the subject in the case of L^∞ generating functions. The extension to $d \ge 1$ and $f \in L^1([-\pi,\pi]^d)$ was performed by Tyrtyshnikov and Zamarashkin [41–43] and Tilli [38]. For the eigenvalues in the case of $d \ge 1$ and $f \in L^\infty([-\pi,\pi]^d)$, Theorem 3.5 can also be derived from [7, Corollary 22]. In this section, we are going to see a proof of Theorem 3.5 based on the notion of a.c.s. It is essentially the same as the proof of [22, Theorem 6.5], with the only difference that in [22] we focused on the case $d = 1$, while here we will address the general case $d \geq 1$.

Theorem 3.5 *If $f \in L^1([-\pi,\pi]^d)$ and $\{\boldsymbol{n} = \boldsymbol{n}(n)\}_n \subseteq \mathbb{N}^d$ is any sequence such that $\boldsymbol{n} \to \infty$ as $n \to \infty$, then $\{T_{\boldsymbol{n}}(f)\}_n \sim_\sigma f$. If moreover f is real, then $\{T_{\boldsymbol{n}}(f)\}_n \sim_\lambda f$.*

Proof The proof consists of two steps. In the first step, we show that the theorem holds if f is a d-variate trigonometric polynomial. In the second step, using an approximation argument based on the concept of a.c.s., we show that the theorem holds for every $f \in L^1([-\pi,\pi]^d)$.

Step 1. Suppose f is a d-variate trigonometric polynomial, so that

$$f(\boldsymbol{\theta}) = \sum_{\boldsymbol{k}=-\boldsymbol{r}}^{\boldsymbol{r}} f_{\boldsymbol{k}}\, \mathrm{e}^{\mathrm{i}\boldsymbol{k}\cdot\boldsymbol{\theta}}$$

for some $\boldsymbol{r} \in \mathbb{N}^d$. Consider the d-level circulant matrix

$$C_{\boldsymbol{n}}(f) = \sum_{\boldsymbol{k}=-\boldsymbol{r}}^{\boldsymbol{r}} f_{\boldsymbol{k}}\, C_{\boldsymbol{n}}^{\boldsymbol{k}}, \tag{3.44}$$

where $C_{\boldsymbol{n}}^{\boldsymbol{k}} = C_{n_1}^{k_1} \otimes C_{n_2}^{k_2} \otimes \cdots \otimes C_{n_d}^{k_d}$ as in (3.39) and C_n is defined in (3.36). Note that $C_{\boldsymbol{n}}(f)$ is Hermitian whenever f is real, by Theorem 3.4. We are going to show that

$$\{C_{\boldsymbol{n}}(f)\}_n \sim_{\sigma,\lambda} f \tag{3.45}$$

and

$$\{C_{\boldsymbol{n}}(f)\}_n \xrightarrow{\text{a.c.s.}} \{T_{\boldsymbol{n}}(f)\}_n. \tag{3.46}$$

Once this is done, the singular value distribution $\{T_{\boldsymbol{n}}(f)\}_n \sim_\sigma f$ follows from Corollary 2.4, and the spectral distribution $\{T_{\boldsymbol{n}}(f)\}_n \sim_\lambda f$ follows from Corollary 2.5 under the assumption that f is real (in which case both $T_{\boldsymbol{n}}(f)$ and $C_{\boldsymbol{n}}(f)$ are Hermitian).

To prove (3.45) we use Theorem 3.4. Since $C_{\boldsymbol{n}}(f)$ is normal, it is enough to show that $\{C_{\boldsymbol{n}}(f)\}_n \sim_\lambda f$ (see Remark 2.2). By Theorem 3.4, the eigenvalues of $C_{\boldsymbol{n}}(f)$ are given by $f(2\pi\boldsymbol{j}/\boldsymbol{n})$, $\boldsymbol{j} = \boldsymbol{0},\ldots,\boldsymbol{n}-\boldsymbol{1}$. Hence, for every $F \in C_c(\mathbb{C})$,

$$\begin{aligned}\lim_{n\to\infty} \frac{1}{N(\boldsymbol{n})} \sum_{j=1}^{N(\boldsymbol{n})} F(\lambda_j(C_{\boldsymbol{n}}(f))) &= \lim_{n\to\infty} \frac{1}{N(\boldsymbol{n})} \sum_{\boldsymbol{j}=\boldsymbol{0}}^{\boldsymbol{n}-\boldsymbol{1}} F\Big(f\Big(\frac{2\pi\boldsymbol{j}}{\boldsymbol{n}}\Big)\Big) \\ = \frac{1}{(2\pi)^d} \int_{[0,2\pi]^d} F(f(\boldsymbol{\theta}))\mathrm{d}\boldsymbol{\theta} &= \frac{1}{(2\pi)^d} \int_{[-\pi,\pi]^d} F(f(\boldsymbol{\theta}))\mathrm{d}\boldsymbol{\theta}.\end{aligned}$$

Note that the last equality holds because f is a d-variate trigonometric polynomial (so it is periodic in each direction with period 2π), while the second equality is due to the fact that $\frac{(2\pi)^d}{N(\boldsymbol{n})}\sum_{\boldsymbol{j}=\boldsymbol{0}}^{\boldsymbol{n}-\boldsymbol{1}} F(f(\frac{2\pi \boldsymbol{j}}{\boldsymbol{n}}))$ is a Riemann sum for $\int_{[0,2\pi]^d} F(f(\boldsymbol{\theta}))\mathrm{d}\boldsymbol{\theta}$ and converges to this integral as $\boldsymbol{n}\to\infty$, because the function $F(f(\boldsymbol{\theta}))$ is continuous (and hence Riemann-integrable). Thus, $\{C_{\boldsymbol{n}}(f)\}_n \sim_\lambda f$ and the proof of (3.45) is complete.

To prove (3.46), it is enough to show that, for $\boldsymbol{n}\ge \boldsymbol{r}+\boldsymbol{1}$,

$$\mathrm{rank}(T_{\boldsymbol{n}}(f)-C_{\boldsymbol{n}}(f)) \le N(2\boldsymbol{r}+\boldsymbol{1})N(\boldsymbol{n})\sum_{i=1}^d \frac{r_i}{n_i}. \tag{3.47}$$

This implies that the matrix-sequence $\{T_{\boldsymbol{n}}(f)-C_{\boldsymbol{n}}(f)\}_n$ is zero-distributed by Theorem 2.2, hence $d_{\text{a.c.s.}}(\{T_{\boldsymbol{n}}(f)\}_n,\{C_{\boldsymbol{n}}(f)\}_n)=0$ and (3.46) is met. By (3.9) and the fact that $f_{\boldsymbol{k}}=0$ if $\boldsymbol{k}\notin\{-\boldsymbol{r},\dots,\boldsymbol{r}\}$, for $\boldsymbol{n}\ge\boldsymbol{r}+\boldsymbol{1}$ we have

$$T_{\boldsymbol{n}}(f)=\sum_{\boldsymbol{k}=-(\boldsymbol{n}-\boldsymbol{1})}^{\boldsymbol{n}-\boldsymbol{1}} f_{\boldsymbol{k}}\,J_{\boldsymbol{n}}^{(\boldsymbol{k})}=\sum_{\boldsymbol{k}=-\boldsymbol{r}}^{\boldsymbol{r}} f_{\boldsymbol{k}}\,J_{\boldsymbol{n}}^{(\boldsymbol{k})}, \tag{3.48}$$

where $J_{\boldsymbol{n}}^{(\boldsymbol{k})}=J_{n_1}^{(k_1)}\otimes J_{n_2}^{(k_2)}\otimes\cdots\otimes J_{n_d}^{(k_d)}$ as in (3.5) and $J_n^{(k)}$ is defined in (3.4). Since it is clear from (3.4) and (3.37) that the nonzero rows of $C_n^k - J_n^{(k)}$ are at most $|k|$, we have

$$\mathrm{rank}(C_n^k - J_n^{(k)})\le |k|,\qquad k\in\mathbb{Z},\qquad n\in\mathbb{N}.$$

Hence, by (2.31),

$$\mathrm{rank}(C_{\boldsymbol{n}}^{\boldsymbol{k}} - J_{\boldsymbol{n}}^{(\boldsymbol{k})})\le N(\boldsymbol{n})\sum_{i=1}^d \frac{|k_i|}{n_i},\qquad \boldsymbol{k}\in\mathbb{Z}^d,\qquad \boldsymbol{n}\in\mathbb{N}^d.$$

Comparing (3.44) and (3.48), we see that, for $\boldsymbol{n}\ge\boldsymbol{r}+\boldsymbol{1}$,

$$\begin{aligned}\mathrm{rank}(C_{\boldsymbol{n}}(f)-T_{\boldsymbol{n}}(f)) &\le \sum_{\boldsymbol{k}=-\boldsymbol{r}}^{\boldsymbol{r}} \mathrm{rank}(C_{\boldsymbol{n}}^{\boldsymbol{k}} - J_{\boldsymbol{n}}^{(\boldsymbol{k})}) \le \sum_{\boldsymbol{k}=-\boldsymbol{r}}^{\boldsymbol{r}} N(\boldsymbol{n})\sum_{i=1}^d \frac{|k_i|}{n_i}\\ &\le N(2\boldsymbol{r}+\boldsymbol{1})N(\boldsymbol{n})\sum_{i=1}^d\frac{r_i}{n_i},\end{aligned}$$

and (3.47) is proved.

Step 2. Let $f\in L^1([-\pi,\pi]^d)$. Since the set of d-variate trigonometric polynomials is dense in $L^1([-\pi,\pi]^d)$ by, e.g., [22, Lemma 2.2] or Lemma 2.3, there exists a sequence of d-variate trigonometric polynomials $\{f_m\}_m$ such that $f_m\to f$ in $L^1([-\pi,\pi]^d)$. By replacing f_m with $\Re(f_m)$ (if necessary), we may assume that f_m is real if f is real. In this way, all the matrices $T_{\boldsymbol{n}}(f)$ and $T_{\boldsymbol{n}}(f_m)$ are Hermitian if f is real. By Step 1,

$$\{T_{\boldsymbol{n}}(f_m)\}_n \sim_\sigma f_m$$

and, if f is real,

$$\{T_{\boldsymbol{n}}(f_m)\}_n \sim_\lambda f_m.$$

Moreover,

$$\{T_{\boldsymbol{n}}(f_m)\}_n \xrightarrow{\text{a.c.s.}} \{T_{\boldsymbol{n}}(f)\}_n$$

by Theorem 2.10, because, by Theorem 3.2,

$$\|T_{\boldsymbol{n}}(f) - T_{\boldsymbol{n}}(f_m)\|_1 = \|T_{\boldsymbol{n}}(f - f_m)\|_1 \leq N(\boldsymbol{n})\|f - f_m\|_{L^1}.$$

Thus, the relations $\{T_{\boldsymbol{n}}(f)\}_n \sim_{\sigma,\lambda} f$ follow from Corollaries 2.4 and 2.5. □

A noteworthy extension of the spectral distribution result $\{T_{\boldsymbol{n}}(f)\}_n \sim_\lambda f$ to the case where f is not real has been performed in [15], on the basis of Tilli's pioneering paper [40]. It was proved in [15] that the relation $\{T_{\boldsymbol{n}}(f)\}_n \sim_\lambda f$ holds whenever f satisfies the following conditions:

- $f \in L^\infty([-\pi, \pi]^d)$;
- the essential range $\mathcal{ER}(f)$ has empty interior and does not disconnect the complex plane $\mathbb{C}$.

The class of functions $f \in L^\infty([-\pi, \pi]^d)$ whose essential range has empty interior and does not disconnect $\mathbb{C}$ is sometimes referred to as the *Tilli class*. The hypothesis of "being in the Tilli class" has been used not only in [15] but also in the recent work [13], which provided an extension of both Theorem 3.5 and the above theorem from [15]. For more on this subject, see [13].

3.6 Extreme Eigenvalues of Hermitian Multilevel Toeplitz Matrices

Suppose that $f \in L^1([-\pi, \pi]^d)$ is real a.e. In this case, each Toeplitz matrix $T_{\boldsymbol{n}}(f)$ is Hermitian and $\Lambda(T_{\boldsymbol{n}}(f)) \subseteq [m_f, M_f]$, where $m_f = \operatorname{ess\,inf}_{\boldsymbol{\theta}\in[-\pi,\pi]^d} f(\boldsymbol{\theta})$ and $M_f = \operatorname{ess\,sup}_{\boldsymbol{\theta}\in[-\pi,\pi]^d} f(\boldsymbol{\theta})$; see Theorem 3.1. The next theorem shows that, for each fixed $j \geq 1$, the j smallest eigenvalues of $T_{\boldsymbol{n}}(f)$ converge to m_f and the j largest eigenvalues of $T_{\boldsymbol{n}}(f)$ converge to M_f as $\boldsymbol{n} \to \infty$.

Theorem 3.6 *Let $f \in L^1([-\pi, \pi]^d)$ be real a.e., let*

$$m_f = \operatorname*{ess\,inf}_{\boldsymbol{\theta}\in[-\pi,\pi]^d} f(\boldsymbol{\theta}), \qquad M_f = \operatorname*{ess\,sup}_{\boldsymbol{\theta}\in[-\pi,\pi]^d} f(\boldsymbol{\theta}),$$

and let $\lambda_1(T_{\boldsymbol{n}}(f)) \geq \cdots \geq \lambda_{N(\boldsymbol{n})}(T_{\boldsymbol{n}}(f))$ be the eigenvalues of $T_{\boldsymbol{n}}(f)$ sorted in non-increasing order. Then, for each fixed $j \geq 1$,

$$\lim_{\boldsymbol{n}\to\infty} \lambda_j(T_{\boldsymbol{n}}(f)) = M_f, \qquad \lim_{\boldsymbol{n}\to\infty} \lambda_{N(\boldsymbol{n})-j+1}(T_{\boldsymbol{n}}(f)) = m_f. \tag{3.49}$$

Proof We only prove the left limit in (3.49) as the proof of the right limit is conceptually identical. To prove the left limit, we show that

$$\lim_{n\to\infty} \lambda_j(T_{\boldsymbol{n}}(f)) = M_f$$

for all sequences $\{\boldsymbol{n} = \boldsymbol{n}(n)\}_n \subseteq \mathbb{N}^d$ such that $\boldsymbol{n} \to \infty$ as $n \to \infty$. The proof of the latter assertion is based on the observation that, by Theorems 2.1 and 3.5, each point of the essential range $\mathcal{ER}(f) \subseteq [m_f, M_f]$ strongly attracts the spectrum $\Lambda(T_{\boldsymbol{n}}(f))$ with infinite order.

If M_f is finite, then it belongs to the essential range $\mathcal{ER}(f)$ because $M_f = \sup \mathcal{ER}(f)$ by definition and $\mathcal{ER}(f)$ is closed by [22, Lemma 2.1]. As a consequence, M_f strongly attracts the spectrum $\Lambda(T_{\boldsymbol{n}}(f))$ with infinite order, so the left limit in (3.49) holds for each fixed $j \geq 1$ by Definition 2.3.

To prove that this limit continues to hold even if M_f is not finite, suppose by contradiction that there exists a fixed $j \geq 1$ such that

$$\liminf_{n\to\infty} \lambda_j(T_{\boldsymbol{n}}(f)) = \ell < M_f.$$

Passing, if necessary, to a subsequence of $\{T_{\boldsymbol{n}}(f)\}_n$, we may assume that

$$\lim_{n\to\infty} \lambda_j(T_{\boldsymbol{n}}(f)) = \ell < M_f.$$

This means that all the eigenvalues of $T_{\boldsymbol{n}}(f)$ (except possibly for the largest $j-1$) are eventually smaller than $\ell + \epsilon$ for some positive ϵ such that $\ell + \epsilon < M_f$. By definition of M_f, we can find a point $y \in \mathcal{ER}(f)$ which lies in $(\ell + \epsilon,\ M_f]$. Clearly, the point y cannot strongly attract $\Lambda(T_{\boldsymbol{n}}(f))$ with infinite order because only the largest $j-1$ eigenvalues of $T_{\boldsymbol{n}}(f)$ can converge to y as $n \to \infty$. This contradiction proves the left limit in (3.49). □

Chapter 4
Multilevel Locally Toeplitz Sequences

The theory of Locally Toeplitz (LT) sequences dates back to Tilli's pioneering paper [39]. It was then carried forward by the second author [35, 36], and it was finally reviewed and extended in [20]. Following [20], in this chapter we address the multivariate version of the theory of LT sequences, also known as the theory of multilevel LT sequences. The topic is presented here on an abstract level, whereas for motivations and insights we refer the reader to [22, Sect. 7.1]. Needless to say, the theory of multilevel LT sequences is the keystone of the theory of multilevel GLT sequences, which will be the subject of Chap. 5. We stress that *many results and proofs contained in this chapter have an exact analog in* [22, Chap. 7], where we dealt with classical (unilevel) LT sequences. However, we decided to reproduce here all the proofs without omitting details, in order to help the reader become familiar with the multilevel language (especially, the multi-index notation). In this regard, the reader is invited to compare the "multivariate proofs" presented in this chapter with the corresponding "univariate proofs" in [22, Chap. 7], in order to learn the way in which the multilevel language allows one to transfer many results from the univariate to the multivariate case by simply turning some letters (n, i, j, x, θ, etc.) in boldface ($\boldsymbol{n}, \boldsymbol{i}, \boldsymbol{j}, \mathbf{x}, \boldsymbol{\theta}$, etc.).

4.1 Multilevel LT Operator

Just as the theory of LT sequences [22, Chap. 7] begins with the notion of LT operator, the theory of multilevel LT sequences begins with the notion of multilevel LT operator. After introducing the multilevel LT operator in Sect. 4.1.1, we study its properties in Sect. 4.1.2.

C. Garoni and S. Serra-Capizzano, *Generalized Locally Toeplitz Sequences: Theory and Applications*, https://doi.org/10.1007/978-3-030-02233-4_4

4.1.1 Definition of Multilevel LT Operator

Throughout this book, if $\boldsymbol{n} \in \mathbb{N}^d$ and $a : [0,1]^d \to \mathbb{C}$, the $\boldsymbol{n}$th (d-level) diagonal sampling matrix generated by a is denoted by $D_{\boldsymbol{n}}(a)$ and is defined as the following diagonal matrix of size $N(\boldsymbol{n})$:

$$D_{\boldsymbol{n}}(a) = \underset{\boldsymbol{i}=\boldsymbol{1},\ldots,\boldsymbol{n}}{\mathrm{diag}}\, a\Big(\frac{\boldsymbol{i}}{\boldsymbol{n}}\Big), \tag{4.1}$$

where we recall that $\boldsymbol{i}$ varies from $\boldsymbol{1}$ to $\boldsymbol{n}$ following the lexicographic ordering; see Sect. 2.1.2. Note that $D_{\boldsymbol{n}}(a)$ can also be defined through a recursive formula: if $d = 1$, then

$$D_n(a) = \underset{i=1,\ldots,n}{\mathrm{diag}}\, a\Big(\frac{i}{n}\Big);$$

if $d > 1$, then

$$D_{\boldsymbol{n}}(a) = D_{n_1,\ldots,n_d}(a) = \underset{i_1=1,\ldots,n_1}{\mathrm{diag}}\, D_{n_2,\ldots,n_d}\Big(a\Big(\frac{i_1}{n_1}, x_2, \ldots, x_d\Big)\Big), \tag{4.2}$$

where $a(i_1/n_1, x_2, \ldots, x_d)$ is the $(d-1)$-variate function defined as follows:

$$a\Big(\frac{i_1}{n_1}, x_2, \ldots, x_d\Big) : [0,1]^{d-1} \to \mathbb{C}, \qquad (x_2, \ldots, x_d) \mapsto a\Big(\frac{i_1}{n_1}, x_2, \ldots, x_d\Big).$$

If $u : D \to \mathbb{C}$ and $v : E \to \mathbb{C}$ are arbitrary functions, the tensor-product function $u \otimes v : D \times E \to \mathbb{C}$ is defined as

$$(u \otimes v)(\xi, \vartheta) = u(\xi)v(\vartheta), \qquad (\xi, \vartheta) \in D \times E.$$

Definition 4.1 (*multilevel locally Toeplitz operator*)

- Let $m, n \in \mathbb{N}$, let $a : [0,1] \to \mathbb{C}$, and let $f \in L^1([-\pi, \pi])$. In accordance with [22, Definition 7.1], the associated (1-level) Locally Toeplitz (LT) operator is defined as the following $n \times n$ matrix:

$$\begin{aligned} LT_n^m(a, f) &= D_m(a) \otimes T_{\lfloor n/m \rfloor}(f) \oplus O_{n \bmod m} \\ &= \underset{i=1,\ldots,m}{\mathrm{diag}} \Big[a\Big(\frac{i}{m}\Big) T_{\lfloor n/m \rfloor}(f)\Big] \oplus O_{n \bmod m} \\ &= \underset{i=1,\ldots,m}{\mathrm{diag}}\, a\Big(\frac{i}{m}\Big) T_{\lfloor n/m \rfloor}(f) \oplus O_{n \bmod m}. \end{aligned}$$

  It is understood that $LT_n^m(a, f) = O_n$ when $n < m$ and that the term $O_{n \bmod m}$ is not present when n is a multiple of m. Moreover, here and in what follows, the tensor product operation $\otimes$ is always applied before the direct sum $\oplus$, exactly as in the case of numbers, where multiplication is always applied before addition. Note also that in the last equality we intentionally removed the square brackets in order to illustrate a notation that will be used hereinafter to simplify the presentation (roughly speaking, we are assuming that the "diag operator" is applied before the direct sum $\oplus$).

- Let $\boldsymbol{m}, \boldsymbol{n} \in \mathbb{N}^d$, let $a : [0,1]^d \to \mathbb{C}$, and let $f_1, \dots, f_d \in L^1([-\pi, \pi])$. The associated ($d$-level) Locally Toeplitz (LT) operator is defined as the following $N(\boldsymbol{n}) \times N(\boldsymbol{n})$ matrix:

$$
\begin{aligned}
LT_{\boldsymbol{n}}^{\boldsymbol{m}}(a, f_1 \otimes \cdots \otimes f_d) &= LT_{n_1,\dots,n_d}^{m_1,\dots,m_d}(a(x_1,\dots,x_d), f_1 \otimes \cdots \otimes f_d) \\
&= \operatorname*{diag}_{i_1=1,\dots,m_1} T_{\lfloor n_1/m_1 \rfloor}(f_1) \otimes LT_{n_2,\dots,n_d}^{m_2,\dots,m_d}\left(a\left(\frac{i_1}{m_1}, x_2, \dots, x_d\right), f_2 \otimes \cdots \otimes f_d\right) \\
&\quad \oplus O_{(n_1 \bmod m_1) n_2 \cdots n_d}.
\end{aligned}
$$

This is a recursive definition, whose base case has been given in the previous item. For example, in the case $d = 2$ we have

$$
\begin{aligned}
&LT_{n_1,n_2}^{m_1,m_2}(a, f_1 \otimes f_2) \\
&= \operatorname*{diag}_{i_1=1,\dots,m_1} T_{\lfloor n_1/m_1 \rfloor}(f_1) \otimes \left[\operatorname*{diag}_{i_2=1,\dots,m_2} a\left(\frac{i_1}{m_1}, \frac{i_2}{m_2}\right) T_{\lfloor n_2/m_2 \rfloor}(f_2) \oplus O_{n_2 \bmod m_2} \right] \\
&\quad \oplus O_{(n_1 \bmod m_1) n_2}.
\end{aligned}
$$

In Definition 4.1, we have defined the multilevel LT operator $LT_{\boldsymbol{n}}^{\boldsymbol{m}}(a, f)$ in the case where f is a separable function of the form $f = f_1 \otimes \cdots \otimes f_d$ with $f_1, \dots, f_d \in L^1([-\pi, \pi])$. We are going to see in Definition 4.2 that $LT_{\boldsymbol{n}}^{\boldsymbol{m}}(a, f)$ is actually well-defined (in a unique way) for any $f \in L^1([-\pi, \pi]^d)$. The crucial result in view of Definition 4.2 is Theorem 4.1. It shows that $LT_{\boldsymbol{n}}^{\boldsymbol{m}}(a, f_1 \otimes \cdots \otimes f_d)$ coincides with $D_{\boldsymbol{m}}(a) \otimes T_{\lfloor \boldsymbol{n}/\boldsymbol{m} \rfloor}(f_1 \otimes \cdots \otimes f_d) \oplus O$ up to a permutation transformation $\Pi_{\boldsymbol{n}}^{\boldsymbol{m}}$ depending only on $\boldsymbol{m}, \boldsymbol{n}$ and not on the functions $a, f_1, \dots, f_d$.

Theorem 4.1 *For any $\boldsymbol{m}, \boldsymbol{n} \in \mathbb{N}^d$ there exists a permutation matrix $\Pi_{\boldsymbol{n}}^{\boldsymbol{m}}$ of size $N(\boldsymbol{n})$ such that*

$$
\begin{aligned}
&LT_{\boldsymbol{n}}^{\boldsymbol{m}}(a, f_1 \otimes \cdots \otimes f_d) = \\
&\Pi_{\boldsymbol{n}}^{\boldsymbol{m}} \Big[D_{\boldsymbol{m}}(a) \otimes T_{\lfloor \boldsymbol{n}/\boldsymbol{m} \rfloor}(f_1 \otimes \cdots \otimes f_d) \oplus O_{N(\boldsymbol{n}) - N(\boldsymbol{m}) N(\lfloor \boldsymbol{n}/\boldsymbol{m} \rfloor)} \Big] (\Pi_{\boldsymbol{n}}^{\boldsymbol{m}})^T
\end{aligned}
$$

for every $a : [0,1]^d \to \mathbb{C}$ and every $f_1, \dots, f_d \in L^1([-\pi, \pi])$.

Proof The proof is done by induction on d. For $d = 1$, the result holds with $\Pi_n^m = I_n$. For $d \geq 2$, set $\boldsymbol{\nu} = (n_2, \dots, n_d)$ and $\boldsymbol{\mu} = (m_2, \dots, m_d)$. By definition,

$$LT_{\boldsymbol{n}}^{\boldsymbol{m}}(a, f_1 \otimes \cdots \otimes f_d) = \underset{i_1=1,\ldots,m_1}{\operatorname{diag}} \; T_{\lfloor n_1/m_1 \rfloor}(f_1) \otimes LT_{\boldsymbol{\nu}}^{\boldsymbol{\mu}}\Big(a\Big(\frac{i_1}{m_1}, \cdot\Big), f_2 \otimes \cdots \otimes f_d\Big) \;\oplus\; O_{(n_1 \bmod m_1)n_2\cdots n_d}, \tag{4.3}$$

where $a(i_1/m_1, \cdot) : [0, 1]^{d-1} \to \mathbb{C}$ is the function $(x_2, \ldots, x_d) \mapsto a(i_1/m_1, x_2, \ldots, x_d)$. By induction hypothesis, setting $N(\boldsymbol{\nu}, \boldsymbol{\mu}) = N(\boldsymbol{\nu}) - N(\boldsymbol{\mu})N(\lfloor \boldsymbol{\nu}/\boldsymbol{\mu} \rfloor)$, we have

$$LT_{\boldsymbol{\nu}}^{\boldsymbol{\mu}}\Big(a\Big(\frac{i_1}{m_1}, \cdot\Big), f_2 \otimes \cdots \otimes f_d\Big) = \Pi_{\boldsymbol{\nu}}^{\boldsymbol{\mu}}\Big[D_{\boldsymbol{\mu}}\Big(a\Big(\frac{i_1}{m_1}, \cdot\Big)\Big) \otimes T_{\lfloor \boldsymbol{\nu}/\boldsymbol{\mu} \rfloor}(f_2 \otimes \cdots \otimes f_d) \;\oplus\; O_{N(\boldsymbol{\nu},\boldsymbol{\mu})}\Big](\Pi_{\boldsymbol{\nu}}^{\boldsymbol{\mu}})^T. \tag{4.4}$$

Let us work on the argument of the "diag operator" in (4.3). From Lemma 2.6, Eq. (4.4) and the properties of tensor products (see Sect. 2.5), we get

$$\begin{aligned}
&T_{\lfloor n_1/m_1 \rfloor}(f_1) \otimes LT_{\boldsymbol{\nu}}^{\boldsymbol{\mu}}\Big(a\Big(\frac{i_1}{m_1}, \cdot\Big), f_2 \otimes \cdots \otimes f_d\Big) \\
&= \Pi_{(N(\boldsymbol{\nu}), \lfloor n_1/m_1 \rfloor);[2,1]} \Big\{LT_{\boldsymbol{\nu}}^{\boldsymbol{\mu}}\Big(a\Big(\frac{i_1}{m_1}, \cdot\Big), f_2 \otimes \cdots \otimes f_d\Big) \otimes T_{\lfloor n_1/m_1 \rfloor}(f_1)\Big\} \\
&\quad \cdot (\Pi_{(N(\boldsymbol{\nu}), \lfloor n_1/m_1 \rfloor);[2,1]})^T \\
&= \Pi_{(N(\boldsymbol{\nu}), \lfloor n_1/m_1 \rfloor);[2,1]} \\
&\quad \cdot \Big\{\Pi_{\boldsymbol{\nu}}^{\boldsymbol{\mu}}\Big[D_{\boldsymbol{\mu}}\Big(a\Big(\frac{i_1}{m_1}, \cdot\Big)\Big) \otimes T_{\lfloor \boldsymbol{\nu}/\boldsymbol{\mu} \rfloor}(f_2 \otimes \cdots \otimes f_d) \;\oplus\; O_{N(\boldsymbol{\nu},\boldsymbol{\mu})}\Big](\Pi_{\boldsymbol{\nu}}^{\boldsymbol{\mu}})^T \\
&\qquad \otimes\; T_{\lfloor n_1/m_1 \rfloor}(f_1)\Big\} \\
&\quad \cdot (\Pi_{(N(\boldsymbol{\nu}), \lfloor n_1/m_1 \rfloor);[2,1]})^T \\
&= \Pi_{(N(\boldsymbol{\nu}), \lfloor n_1/m_1 \rfloor);[2,1]}(\Pi_{\boldsymbol{\nu}}^{\boldsymbol{\mu}} \otimes I_{\lfloor n_1/m_1 \rfloor}) \\
&\quad \cdot \Big\{\Big[D_{\boldsymbol{\mu}}\Big(a\Big(\frac{i_1}{m_1}, \cdot\Big)\Big) \otimes T_{\lfloor \boldsymbol{\nu}/\boldsymbol{\mu} \rfloor}(f_2 \otimes \cdots \otimes f_d) \;\oplus\; O_{N(\boldsymbol{\nu},\boldsymbol{\mu})}\Big] \otimes T_{\lfloor n_1/m_1 \rfloor}(f_1)\Big\} \\
&\quad \cdot (\Pi_{\boldsymbol{\nu}}^{\boldsymbol{\mu}} \otimes I_{\lfloor n_1/m_1 \rfloor})^T(\Pi_{(N(\boldsymbol{\nu}), \lfloor n_1/m_1 \rfloor);[2,1]})^T.
\end{aligned} \tag{4.5}$$

Using Eq. (2.35), Lemmas 2.6, 3.3 and the properties of tensor products and direct sums (see Sect. 2.5), we obtain

$$\begin{aligned}
&\Big[D_{\boldsymbol{\mu}}\Big(a\Big(\frac{i_1}{m_1}, \cdot\Big)\Big) \otimes T_{\lfloor \boldsymbol{\nu}/\boldsymbol{\mu} \rfloor}(f_2 \otimes \cdots \otimes f_d) \;\oplus\; O_{N(\boldsymbol{\nu},\boldsymbol{\mu})}\Big] \otimes T_{\lfloor n_1/m_1 \rfloor}(f_1) \\
&= D_{\boldsymbol{\mu}}\Big(a\Big(\frac{i_1}{m_1}, \cdot\Big)\Big) \otimes T_{\lfloor \boldsymbol{\nu}/\boldsymbol{\mu} \rfloor}(f_2 \otimes \cdots \otimes f_d) \otimes T_{\lfloor n_1/m_1 \rfloor}(f_1) \;\oplus\; O_{N(\boldsymbol{\nu},\boldsymbol{\mu})\lfloor n_1/m_1 \rfloor}
\end{aligned}$$

$$
\begin{aligned}
&= \Pi_{(N(\boldsymbol{\mu}),\lfloor n_1/m_1\rfloor,\,N(\lfloor \boldsymbol{\nu}/\boldsymbol{\mu}\rfloor));[1,3,2]} \\
&\quad \cdot \left[D_{\boldsymbol{\mu}}\Big(a\Big(\frac{i_1}{m_1},\cdot\Big)\Big) \otimes T_{\lfloor n_1/m_1\rfloor}(f_1) \otimes T_{\lfloor \boldsymbol{\nu}/\boldsymbol{\mu}\rfloor}(f_2\otimes\cdots\otimes f_d)\right] \\
&\quad \cdot (\Pi_{(N(\boldsymbol{\mu}),\lfloor n_1/m_1\rfloor,\,N(\lfloor \boldsymbol{\nu}/\boldsymbol{\mu}\rfloor));[1,3,2]})^T \ \oplus\ O_{N(\boldsymbol{\nu},\boldsymbol{\mu})\lfloor n_1/m_1\rfloor} \\
&= \Pi_{(N(\boldsymbol{\mu}),\lfloor n_1/m_1\rfloor,\,N(\lfloor \boldsymbol{\nu}/\boldsymbol{\mu}\rfloor));[1,3,2]} \\
&\quad \cdot \left[D_{\boldsymbol{\mu}}\Big(a\Big(\frac{i_1}{m_1},\cdot\Big)\Big) \otimes T_{\lfloor \boldsymbol{n}/\boldsymbol{m}\rfloor}(f_1\otimes\cdots\otimes f_d)\right] \\
&\quad \cdot (\Pi_{(N(\boldsymbol{\mu}),\lfloor n_1/m_1\rfloor,\,N(\lfloor \boldsymbol{\nu}/\boldsymbol{\mu}\rfloor));[1,3,2]})^T \ \oplus\ O_{N(\boldsymbol{\nu},\boldsymbol{\mu})\lfloor n_1/m_1\rfloor} \\
&= (\Pi_{(N(\boldsymbol{\mu}),\lfloor n_1/m_1\rfloor,\,N(\lfloor \boldsymbol{\nu}/\boldsymbol{\mu}\rfloor));[1,3,2]} \ \oplus\ I_{N(\boldsymbol{\nu},\boldsymbol{\mu})\lfloor n_1/m_1\rfloor}) \\
&\quad \cdot \left[D_{\boldsymbol{\mu}}\Big(a\Big(\frac{i_1}{m_1},\cdot\Big)\Big) \otimes T_{\lfloor \boldsymbol{n}/\boldsymbol{m}\rfloor}(f_1\otimes\cdots\otimes f_d) \ \oplus\ O_{N(\boldsymbol{\nu},\boldsymbol{\mu})\lfloor n_1/m_1\rfloor}\right] \\
&\quad \cdot (\Pi_{(N(\boldsymbol{\mu}),\lfloor n_1/m_1\rfloor,\,N(\lfloor \boldsymbol{\nu}/\boldsymbol{\mu}\rfloor));[1,3,2]} \ \oplus\ I_{N(\boldsymbol{\nu},\boldsymbol{\mu})\lfloor n_1/m_1\rfloor})^T. \qquad (4.6)
\end{aligned}
$$

Substituting (4.6) into (4.5), we arrive at

$$
\begin{aligned}
&T_{\lfloor n_1/m_1\rfloor}(f_1) \otimes LT_{\boldsymbol{\nu}}^{\boldsymbol{\mu}}\Big(a\Big(\frac{i_1}{m_1},\cdot\Big),\, f_2\otimes\cdots\otimes f_d\Big) \\
&= P_{\boldsymbol{n}}^{\boldsymbol{m}} \left[D_{\boldsymbol{\mu}}\Big(a\Big(\frac{i_1}{m_1},\cdot\Big)\Big) \otimes T_{\lfloor \boldsymbol{n}/\boldsymbol{m}\rfloor}(f_1\otimes\cdots\otimes f_d) \ \oplus\ O_{N(\boldsymbol{\nu},\boldsymbol{\mu})\lfloor n_1/m_1\rfloor}\right] (P_{\boldsymbol{n}}^{\boldsymbol{m}})^T, \qquad (4.7)
\end{aligned}
$$

where

$$
\begin{aligned}
P_{\boldsymbol{n}}^{\boldsymbol{m}} &= \Pi_{(N(\boldsymbol{\nu}),\lfloor n_1/m_1\rfloor);[2,1]}(\Pi_{\boldsymbol{\nu}}^{\boldsymbol{\mu}} \otimes I_{\lfloor n_1/m_1\rfloor}) \\
&\quad \cdot (\Pi_{(N(\boldsymbol{\mu}),\lfloor n_1/m_1\rfloor,\,N(\lfloor \boldsymbol{\nu}/\boldsymbol{\mu}\rfloor));[1,3,2]} \ \oplus\ I_{N(\boldsymbol{\nu},\boldsymbol{\mu})\lfloor n_1/m_1\rfloor}).
\end{aligned}
$$

Combining (4.7) and (4.3), we obtain

$$
\begin{aligned}
&LT_{\boldsymbol{n}}^{\boldsymbol{m}}(a, f_1\otimes\cdots\otimes f_d) \\
&= \Big(\bigoplus_{i_1=1}^{m_1} P_{\boldsymbol{n}}^{\boldsymbol{m}}\Big) \underset{i_1=1,\ldots,m_1}{\operatorname{diag}} \Big[D_{\boldsymbol{\mu}}\Big(a\Big(\frac{i_1}{m_1},\cdot\Big)\Big) \otimes T_{\lfloor \boldsymbol{n}/\boldsymbol{m}\rfloor}(f_1\otimes\cdots\otimes f_d) \\
&\qquad\qquad \oplus\ O_{N(\boldsymbol{\nu},\boldsymbol{\mu})\lfloor n_1/m_1\rfloor}\Big] \Big(\bigoplus_{i_1=1}^{m_1} P_{\boldsymbol{n}}^{\boldsymbol{m}}\Big)^T \ \oplus\ O_{(n_1 \bmod m_1)n_2\cdots n_d}.
\end{aligned}
$$

From Lemma 2.7 and Eqs. (2.35) and (4.2),

$$
\underset{i_1=1,\ldots,m_1}{\operatorname{diag}} \left[D_{\boldsymbol{\mu}}\Big(a\Big(\frac{i_1}{m_1},\cdot\Big)\Big) \otimes T_{\lfloor \boldsymbol{n}/\boldsymbol{m}\rfloor}(f_1\otimes\cdots\otimes f_d) \ \oplus\ O_{N(\boldsymbol{\nu},\boldsymbol{\mu})\lfloor n_1/m_1\rfloor}\right]
$$

$$= \bigoplus_{i_1=1}^{m_1} \Big[D_{\mu}\Big(a\Big(\frac{i_1}{m_1},\cdot\Big)\Big) \otimes T_{\lfloor \boldsymbol{n}/\boldsymbol{m} \rfloor}(f_1 \otimes \cdots \otimes f_d) \,\oplus\, O_{N(\boldsymbol{\nu},\boldsymbol{\mu})\lfloor n_1/m_1 \rfloor} \Big]$$
$$= V_{\boldsymbol{n}}^{\boldsymbol{m}} \left[\bigoplus_{i_1=1}^{m_1} \Big[D_{\mu}\Big(a\Big(\frac{i_1}{m_1},\cdot\Big)\Big) \otimes T_{\lfloor \boldsymbol{n}/\boldsymbol{m} \rfloor}(f_1 \otimes \cdots \otimes f_d) \Big] \,\oplus\, O_{N(\boldsymbol{\nu},\boldsymbol{\mu})\lfloor n_1/m_1 \rfloor m_1} \right] (V_{\boldsymbol{n}}^{\boldsymbol{m}})^T$$
$$= V_{\boldsymbol{n}}^{\boldsymbol{m}} \big[D_{\boldsymbol{m}}(a) \otimes T_{\lfloor \boldsymbol{n}/\boldsymbol{m} \rfloor}(f_1 \otimes \cdots \otimes f_d) \,\oplus\, O_{N(\boldsymbol{\nu},\boldsymbol{\mu})\lfloor n_1/m_1 \rfloor m_1} \big] (V_{\boldsymbol{n}}^{\boldsymbol{m}})^T,$$

where

$$\begin{aligned}
V_{\boldsymbol{n}}^{\boldsymbol{m}} &= V_{\boldsymbol{h}(\boldsymbol{m},\boldsymbol{n});\sigma}, \\
\sigma &= [1, m_1+1, 2, m_1+2, \ldots, m_1, 2m_1], \\
\boldsymbol{h}(\boldsymbol{m},\boldsymbol{n}) &= \big(\underbrace{N(\boldsymbol{\mu})N(\lfloor \boldsymbol{n}/\boldsymbol{m} \rfloor), \ldots, N(\boldsymbol{\mu})N(\lfloor \boldsymbol{n}/\boldsymbol{m} \rfloor)}_{m_1}, \\
&\qquad \underbrace{N(\boldsymbol{\nu},\boldsymbol{\mu})\lfloor n_1/m_1 \rfloor, \ldots, N(\boldsymbol{\nu},\boldsymbol{\mu})\lfloor n_1/m_1 \rfloor}_{m_1}\big).
\end{aligned}$$

Thus,

$$\begin{aligned}
&LT_{\boldsymbol{n}}^{\boldsymbol{m}}(a, f_1 \otimes \cdots \otimes f_d) \\
&= \Big(\bigoplus_{i_1=1}^{m_1} P_{\boldsymbol{n}}^{\boldsymbol{m}} \Big) V_{\boldsymbol{n}}^{\boldsymbol{m}} \\
&\quad \cdot \big[D_{\boldsymbol{m}}(a) \otimes T_{\lfloor \boldsymbol{n}/\boldsymbol{m} \rfloor}(f_1 \otimes \cdots \otimes f_d) \,\oplus\, O_{N(\boldsymbol{\nu},\boldsymbol{\mu})\lfloor n_1/m_1 \rfloor m_1} \big] \\
&\quad \cdot (V_{\boldsymbol{n}}^{\boldsymbol{m}})^T \Big(\bigoplus_{i_1=1}^{m_1} P_{\boldsymbol{n}}^{\boldsymbol{m}} \Big)^T \,\oplus\, O_{(n_1 \bmod m_1)n_2\cdots n_d} \\
&= \Big[\Big(\bigoplus_{i_1=1}^{m_1} P_{\boldsymbol{n}}^{\boldsymbol{m}} \Big) V_{\boldsymbol{n}}^{\boldsymbol{m}} \,\oplus\, I_{(n_1 \bmod m_1)n_2\cdots n_d} \Big] \\
&\quad \cdot \big[D_{\boldsymbol{m}}(a) \otimes T_{\lfloor \boldsymbol{n}/\boldsymbol{m} \rfloor}(f_1 \otimes \cdots \otimes f_d) \,\oplus\, O_{N(\boldsymbol{\nu},\boldsymbol{\mu})\lfloor n_1/m_1 \rfloor m_1 + (n_1 \bmod m_1)n_2\cdots n_d} \big] \\
&\quad \cdot \Big[(V_{\boldsymbol{n}}^{\boldsymbol{m}})^T \Big(\bigoplus_{i_1=1}^{m_1} P_{\boldsymbol{n}}^{\boldsymbol{m}} \Big)^T \,\oplus\, I_{(n_1 \bmod m_1)n_2\cdots n_d} \Big].
\end{aligned}$$

This concludes the proof; note that the permutation matrix $\Pi_{\boldsymbol{n}}^{\boldsymbol{m}}$ is given by

$$\Pi_{\boldsymbol{n}}^{\boldsymbol{m}} = \Big(\bigoplus_{i_1=1}^{m_1} P_{\boldsymbol{n}}^{\boldsymbol{m}} \Big) V_{\boldsymbol{n}}^{\boldsymbol{m}} \,\oplus\, I_{(n_1 \bmod m_1)n_2\cdots n_d}$$

and, moreover, the number $N(\boldsymbol{\nu},\boldsymbol{\mu})\lfloor n_1/m_1 \rfloor m_1 + (n_1 \bmod m_1)n_2\cdots n_d$ is equal to $N(\boldsymbol{n}) - N(\boldsymbol{m})N(\lfloor \boldsymbol{n}/\boldsymbol{m} \rfloor)$. □

Definition 4.2 (*multilevel locally Toeplitz operator*) Let $\boldsymbol{m}, \boldsymbol{n} \in \mathbb{N}^d$, let $a : [0, 1]^d \to \mathbb{C}$, and let $f \in L^1([-\pi, \pi]^d)$. The associated (d-level) Locally Toeplitz (LT) operator is defined as the following $N(\boldsymbol{n}) \times N(\boldsymbol{n})$ matrix:

$$LT_{\boldsymbol{n}}^{\boldsymbol{m}}(a, f) = \Pi_{\boldsymbol{n}}^{\boldsymbol{m}} \left[D_{\boldsymbol{m}}(a) \otimes T_{\lfloor \boldsymbol{n}/\boldsymbol{m} \rfloor}(f) \oplus O_{N(\boldsymbol{n})-N(\boldsymbol{m})N(\lfloor \boldsymbol{n}/\boldsymbol{m} \rfloor)} \right] (\Pi_{\boldsymbol{n}}^{\boldsymbol{m}})^T,$$

where $\Pi_{\boldsymbol{n}}^{\boldsymbol{m}}$ is the permutation matrix appearing in Theorem 4.1.

Remark 4.1 We note that $LT_{\boldsymbol{n}}^{\boldsymbol{m}}(a, f) = LT_{\boldsymbol{n}}^{\boldsymbol{m}}(a, g)$ whenever $f = g$ a.e. Moreover, suppose that $f = f_1 \otimes \cdots \otimes f_d$ a.e., with $f_1, \ldots, f_d \in L^1([-\pi, \pi])$; then $LT_{\boldsymbol{n}}^{\boldsymbol{m}}(a, f)$ is equal to $LT_{\boldsymbol{n}}^{\boldsymbol{m}}(a, f_1 \otimes \cdots \otimes f_d)$, as defined by Definition 4.1. This shows that Definition 4.2 is an extension of Definition 4.1.

4.1.2 Properties of the Multilevel LT Operator

We now investigate the properties of the multilevel LT operator $LT_{\boldsymbol{n}}^{\boldsymbol{m}}(a, f)$. Let $\mathbb{C}^{[0,1]^d}$ be the complex vector space of all functions $a : [0, 1]^d \to \mathbb{C}$. For each fixed $\boldsymbol{n}, \boldsymbol{m} \in \mathbb{N}^d$, the map

$$LT_{\boldsymbol{n}}^{\boldsymbol{m}}(a, \cdot) : L^1([-\pi, \pi]^d) \to \mathbb{C}^{N(\boldsymbol{n}) \times N(\boldsymbol{n})}, \qquad f \mapsto LT_{\boldsymbol{n}}^{\boldsymbol{m}}(a, f), \tag{4.8}$$

is linear for any $a : [0, 1]^d \to \mathbb{C}$, and the map

$$LT_{\boldsymbol{n}}^{\boldsymbol{m}}(\cdot, f) : \mathbb{C}^{[0,1]^d} \to \mathbb{C}^{N(\boldsymbol{n}) \times N(\boldsymbol{n})}, \qquad a \mapsto LT_{\boldsymbol{n}}^{\boldsymbol{m}}(a, f), \tag{4.9}$$

is linear for any $f \in L^1([-\pi, \pi]^d)$. This follows from Definition 4.2, the linearity of the maps $a \mapsto D_{\boldsymbol{m}}(a)$ and $f \mapsto T_{\lfloor \boldsymbol{n}/\boldsymbol{m} \rfloor}(f)$, and the bilinearity of tensor products. For any $\boldsymbol{n}, \boldsymbol{m} \in \mathbb{N}^d$ and any pair of functions $a : [0, 1]^d \to \mathbb{C}$ and $f \in L^1([-\pi, \pi]^d)$, we have

$$(LT_{\boldsymbol{n}}^{\boldsymbol{m}}(a, f))^* = LT_{\boldsymbol{n}}^{\boldsymbol{m}}(\overline{a}, \overline{f}) \tag{4.10}$$

and

$$\begin{aligned} \|LT_{\boldsymbol{n}}^{\boldsymbol{m}}(a, f)\|_p &= \|D_{\boldsymbol{m}}(a)\|_p \, \|T_{\lfloor \boldsymbol{n}/\boldsymbol{m} \rfloor}(f)\|_p \\ &= \left\| \left[a\left(\frac{\boldsymbol{i}}{\boldsymbol{m}}\right) \right]_{\boldsymbol{i}=\boldsymbol{1}}^{\boldsymbol{m}} \right\|_p \|T_{\lfloor \boldsymbol{n}/\boldsymbol{m} \rfloor}(f)\|_p \\ &= \begin{cases} \max_{\boldsymbol{i}=\boldsymbol{1},\ldots,\boldsymbol{m}} \left| a(\frac{\boldsymbol{i}}{\boldsymbol{m}}) \right| \, \|T_{\lfloor \boldsymbol{n}/\boldsymbol{m} \rfloor}(f)\|, & \text{if } p = \infty, \\ \left(\sum_{\boldsymbol{i}=\boldsymbol{1}}^{\boldsymbol{m}} \left| a(\frac{\boldsymbol{i}}{\boldsymbol{m}}) \right|^p \right)^{1/p} \|T_{\lfloor \boldsymbol{n}/\boldsymbol{m} \rfloor}(f)\|_p, & \text{if } 1 \le p < \infty. \end{cases} \end{aligned} \tag{4.11}$$

Equation (4.10) follows from Definition 4.2, the relations $(X \otimes Y)^* = X^* \otimes Y^*$ and $(X \oplus Y)^* = X^* \oplus Y^*$, and the identity $(T_{\boldsymbol{k}}(f))^* = T_{\boldsymbol{k}}(\overline{f})$ in (3.13). The equations in (4.11) follow from Definition 4.2, the invariance of $\|\cdot\|_p$ with respect to unitary transformations (such as permutations), and Eqs. (2.25)–(2.26).

Now, let $1 \le p, q \le \infty$ be conjugate exponents. If $f \in L^p([-\pi,\pi]^d)$ and $\tilde{f} \in L^q([-\pi,\pi]^d)$, then $f\tilde{f} \in L^1([-\pi,\pi]^d)$ by Hölder's inequality, so for any $a, \tilde{a} : [0,1]^d \to \mathbb{C}$ we can consider the matrices $LT_{\boldsymbol{n}}^{\boldsymbol{m}}(a, f)$, $LT_{\boldsymbol{n}}^{\boldsymbol{m}}(\tilde{a}, \tilde{f})$, $LT_{\boldsymbol{n}}^{\boldsymbol{m}}(a\tilde{a}, f\tilde{f})$. In Proposition 4.1 we show that $LT_{\boldsymbol{n}}^{\boldsymbol{m}}(a, f)LT_{\boldsymbol{n}}^{\boldsymbol{m}}(\tilde{a}, \tilde{f})$ is "close" to $LT_{\boldsymbol{n}}^{\boldsymbol{m}}(a\tilde{a}, f\tilde{f})$, as long as $a, \tilde{a}$ are bounded.

Proposition 4.1 *Let $a, \tilde{a} : [0,1]^d \to \mathbb{C}$ be bounded, and let $f \in L^p([-\pi,\pi]^d)$ and $\tilde{f} \in L^q([-\pi,\pi]^d)$, where $1 \le p, q \le \infty$ are conjugate exponents. Then, for every $\boldsymbol{n}, \boldsymbol{m} \in \mathbb{N}^d$,*

$$\|LT_{\boldsymbol{n}}^{\boldsymbol{m}}(a, f)\, LT_{\boldsymbol{n}}^{\boldsymbol{m}}(\tilde{a}, \tilde{f}) - LT_{\boldsymbol{n}}^{\boldsymbol{m}}(a\tilde{a}, f\tilde{f})\|_1 \le \varepsilon(\lfloor \boldsymbol{n}/\boldsymbol{m} \rfloor)\, N(\boldsymbol{n}), \tag{4.12}$$

where

$$\varepsilon(\boldsymbol{k}) = \|a\tilde{a}\|_\infty \frac{\|T_{\boldsymbol{k}}(f)T_{\boldsymbol{k}}(\tilde{f}) - T_{\boldsymbol{k}}(f\tilde{f})\|_1}{N(\boldsymbol{k})}$$

and $\lim_{\boldsymbol{k}\to\infty} \varepsilon(\boldsymbol{k}) = 0$ by Theorem 3.3. In particular, for every $\boldsymbol{m} \in \mathbb{N}^d$ there exists $\boldsymbol{n}_{\boldsymbol{m}} \in \mathbb{N}^d$ such that, for $\boldsymbol{n} \ge \boldsymbol{n}_{\boldsymbol{m}}$,

$$\|LT_{\boldsymbol{n}}^{\boldsymbol{m}}(a, f)LT_{\boldsymbol{n}}^{\boldsymbol{m}}(\tilde{a}, \tilde{f}) - LT_{\boldsymbol{n}}^{\boldsymbol{m}}(a\tilde{a}, f\tilde{f})\|_1 \le \frac{N(\boldsymbol{n})}{N(\boldsymbol{m})} \tag{4.13}$$

and

$$\begin{gathered} LT_{\boldsymbol{n}}^{\boldsymbol{m}}(a, f)\, LT_{\boldsymbol{n}}^{\boldsymbol{m}}(\tilde{a}, \tilde{f}) = LT_{\boldsymbol{n}}^{\boldsymbol{m}}(a\tilde{a}, f\tilde{f}) + R_{\boldsymbol{n},\boldsymbol{m}} + N_{\boldsymbol{n},\boldsymbol{m}}, \\ \operatorname{rank}(R_{\boldsymbol{n},\boldsymbol{m}}) \le \frac{N(\boldsymbol{n})}{\sqrt{N(\boldsymbol{m})}}, \qquad \|N_{\boldsymbol{n},\boldsymbol{m}}\| \le \frac{1}{\sqrt{N(\boldsymbol{m})}}. \end{gathered} \tag{4.14}$$

Proof By Definition 4.2 and the properties of tensor products and direct sums,

$$\begin{aligned} & LT_{\boldsymbol{n}}^{\boldsymbol{m}}(a, f)LT_{\boldsymbol{n}}^{\boldsymbol{m}}(\tilde{a}, \tilde{f}) - LT_{\boldsymbol{n}}^{\boldsymbol{m}}(a\tilde{a}, f\tilde{f}) \\ & = \Pi_{\boldsymbol{n}}^{\boldsymbol{m}} \Big[D_{\boldsymbol{m}}(a\tilde{a}) \otimes \Big(T_{\lfloor \boldsymbol{n}/\boldsymbol{m} \rfloor}(f) T_{\lfloor \boldsymbol{n}/\boldsymbol{m} \rfloor}(\tilde{f}) - T_{\lfloor \boldsymbol{n}/\boldsymbol{m} \rfloor}(f\tilde{f}) \Big) \oplus O \Big] (\Pi_{\boldsymbol{n}}^{\boldsymbol{m}})^T. \end{aligned}$$

Hence,

$$\begin{aligned} & \|LT_{\boldsymbol{n}}^{\boldsymbol{m}}(a, f)LT_{\boldsymbol{n}}^{\boldsymbol{m}}(\tilde{a}, \tilde{f}) - LT_{\boldsymbol{n}}^{\boldsymbol{m}}(a\tilde{a}, f\tilde{f})\|_1 \\ & = \|D_{\boldsymbol{m}}(a\tilde{a})\|_1\, \|T_{\lfloor \boldsymbol{n}/\boldsymbol{m} \rfloor}(f) T_{\lfloor \boldsymbol{n}/\boldsymbol{m} \rfloor}(\tilde{f}) - T_{\lfloor \boldsymbol{n}/\boldsymbol{m} \rfloor}(f\tilde{f})\|_1 \\ & \le N(\boldsymbol{n})\|a\tilde{a}\|_\infty \frac{\|T_{\lfloor \boldsymbol{n}/\boldsymbol{m} \rfloor}(f) T_{\lfloor \boldsymbol{n}/\boldsymbol{m} \rfloor}(\tilde{f}) - T_{\lfloor \boldsymbol{n}/\boldsymbol{m} \rfloor}(f\tilde{f})\|_1}{N(\lfloor \boldsymbol{n}/\boldsymbol{m} \rfloor)}, \end{aligned}$$

and (4.12) is proved. Since $\varepsilon(\boldsymbol{k}) \to 0$ as $\boldsymbol{k} \to \infty$, for every $\boldsymbol{m} \in \mathbb{N}^d$ there exists $\boldsymbol{n}_{\boldsymbol{m}} \in \mathbb{N}^d$ such that, for $\boldsymbol{n} \geq \boldsymbol{n}_{\boldsymbol{m}}$, (4.13) holds; and (4.14) follows from (4.13) and [22, Lemma 5.6]. □

Theorems 4.2 and 4.3 provide information about the asymptotic singular value and eigenvalue distribution of a finite sum of the form $\sum_{i=1}^{p} LT_{\boldsymbol{n}}^{\boldsymbol{m}}(a_i, f_i)$. As we shall see, they play a central role in the computation of the singular value and eigenvalue distribution of multilevel GLT sequences.

Theorem 4.2 *Let* $a_1, \ldots, a_p : [0, 1]^d \to \mathbb{C}$ *and let* $f_1, \ldots, f_p \in L^1([-\pi, \pi]^d)$. *Then, for every* $\boldsymbol{m} \in \mathbb{N}^d$ *and every* $F \in C_c(\mathbb{R})$,

$$\lim_{\boldsymbol{n}\to\infty} \frac{1}{N(\boldsymbol{n})} \sum_{r=1}^{N(\boldsymbol{n})} F\biggl(\sigma_r\Bigl(\sum_{i=1}^{p} LT_{\boldsymbol{n}}^{\boldsymbol{m}}(a_i, f_i)\Bigr)\biggr)$$
$$= \phi_{\boldsymbol{m}}(F) = \frac{1}{N(\boldsymbol{m})} \sum_{\boldsymbol{j}=\boldsymbol{1}}^{\boldsymbol{m}} \frac{1}{(2\pi)^d} \int_{[-\pi,\pi]^d} F\biggl(\Bigl|\sum_{i=1}^{p} a_i\Bigl(\frac{\boldsymbol{j}}{\boldsymbol{m}}\Bigr) f_i(\boldsymbol{\theta})\Bigr|\biggr) \mathrm{d}\boldsymbol{\theta}. \tag{4.15}$$

Moreover, if $a_1, \ldots, a_p$ *are Riemann-integrable, then, for every* $F \in C_c(\mathbb{R})$,

$$\lim_{\boldsymbol{m}\to\infty} \phi_{\boldsymbol{m}}(F) = \phi(F) = \frac{1}{(2\pi)^d} \int_{[0,1]^d\times[-\pi,\pi]^d} F\biggl(\Bigl|\sum_{i=1}^{p} a_i(\mathbf{x}) f_i(\boldsymbol{\theta})\Bigr|\biggr) \mathrm{d}\mathbf{x}\mathrm{d}\boldsymbol{\theta}. \tag{4.16}$$

Proof By Definition 4.2,

$$\begin{aligned}&(\Pi_{\boldsymbol{n}}^{\boldsymbol{m}})^T\Bigl(\sum_{i=1}^{p} LT_{\boldsymbol{n}}^{\boldsymbol{m}}(a_i, f_i)\Bigr)\Pi_{\boldsymbol{n}}^{\boldsymbol{m}}\\ &= \Bigl(\sum_{i=1}^{p} D_{\boldsymbol{m}}(a_i) \otimes T_{\lfloor \boldsymbol{n}/\boldsymbol{m}\rfloor}(f_i)\Bigr) \oplus O_{N(\boldsymbol{n})-N(\boldsymbol{m})N(\lfloor \boldsymbol{n}/\boldsymbol{m}\rfloor)}.\end{aligned} \tag{4.17}$$

Recalling that $D_{\boldsymbol{m}}(a_i) = \mathrm{diag}_{\boldsymbol{j}=\boldsymbol{1},\ldots,\boldsymbol{m}}\, a_i(\boldsymbol{j}/\boldsymbol{m})$, for every $\boldsymbol{j} = \boldsymbol{1}, \ldots, \boldsymbol{m}$ the $\boldsymbol{j}$th diagonal block of size $N(\lfloor \boldsymbol{n}/\boldsymbol{m}\rfloor)$ of the matrix (4.17) is given by

$$\sum_{i=1}^{p} a_i\Bigl(\frac{\boldsymbol{j}}{\boldsymbol{m}}\Bigr) T_{\lfloor \boldsymbol{n}/\boldsymbol{m}\rfloor}(f_i) = T_{\lfloor \boldsymbol{n}/\boldsymbol{m}\rfloor}\Bigl(\sum_{i=1}^{p} a_i\Bigl(\frac{\boldsymbol{j}}{\boldsymbol{m}}\Bigr) f_i\Bigr).$$

It follows that the singular values of $\sum_{i=1}^{p} LT_{\boldsymbol{n}}^{\boldsymbol{m}}(a_i, f_i)$ are

$$\sigma_k\biggl(T_{\lfloor \boldsymbol{n}/\boldsymbol{m}\rfloor}\Bigl(\sum_{i=1}^{p} a_i\Bigl(\frac{\boldsymbol{j}}{\boldsymbol{m}}\Bigr) f_i\Bigr)\biggr), \qquad k = 1, \ldots, N(\lfloor \boldsymbol{n}/\boldsymbol{m}\rfloor), \qquad \boldsymbol{j} = \boldsymbol{1}, \ldots, \boldsymbol{m},$$

plus further $N(\boldsymbol{n}) - N(\boldsymbol{m})N(\lfloor \boldsymbol{n}/\boldsymbol{m} \rfloor)$ singular values equal to 0; see (2.24). Therefore, by Theorem 3.5, for any $F \in C_c(\mathbb{R})$ we have

$$\begin{aligned}
&\lim_{\boldsymbol{n}\to\infty} \frac{1}{N(\boldsymbol{n})} \sum_{r=1}^{N(\boldsymbol{n})} F\left(\sigma_r\left(\sum_{i=1}^{p} LT_{\boldsymbol{n}}^{\boldsymbol{m}}(a_i, f_i)\right)\right)\\
&= \lim_{\boldsymbol{n}\to\infty} \frac{N(\boldsymbol{m})N(\lfloor \boldsymbol{n}/\boldsymbol{m} \rfloor)}{N(\boldsymbol{n})} \times\\
&\qquad \times \frac{1}{N(\boldsymbol{m})} \sum_{\boldsymbol{j}=\boldsymbol{1}}^{\boldsymbol{m}} \frac{1}{N(\lfloor \boldsymbol{n}/\boldsymbol{m} \rfloor)} \sum_{k=1}^{N(\lfloor \boldsymbol{n}/\boldsymbol{m} \rfloor)} F\left(\sigma_k\left(T_{\lfloor \boldsymbol{n}/\boldsymbol{m} \rfloor}\left(\sum_{i=1}^{p} a_i\left(\frac{\boldsymbol{j}}{\boldsymbol{m}}\right) f_i\right)\right)\right)\\
&= \frac{1}{N(\boldsymbol{m})} \sum_{\boldsymbol{j}=\boldsymbol{1}}^{\boldsymbol{m}} \frac{1}{(2\pi)^d} \int_{[-\pi,\pi]^d} F\left(\left|\sum_{i=1}^{p} a_i\left(\frac{\boldsymbol{j}}{\boldsymbol{m}}\right) f_i(\boldsymbol{\theta})\right|\right) \mathrm{d}\boldsymbol{\theta}. \qquad (4.18)
\end{aligned}$$

This proves (4.15).

If we assume that $a_1, \ldots, a_p$ are Riemann-integrable, then the function $\mathbf{x} \mapsto F\left(\left|\sum_{i=1}^{p} a_i(\mathbf{x}) f_i(\boldsymbol{\theta})\right|\right)$ is Riemann-integrable for each fixed $\boldsymbol{\theta} \in [-\pi, \pi]^d$, because it is the composition of a continuous function with a Riemann-integrable function. Hence, by Lemma 2.5, for each fixed $\boldsymbol{\theta} \in [-\pi, \pi]^d$ we have

$$\lim_{\boldsymbol{m}\to\infty} \frac{1}{N(\boldsymbol{m})} \sum_{\boldsymbol{j}=\boldsymbol{1}}^{\boldsymbol{m}} F\left(\left|\sum_{i=1}^{p} a_i\left(\frac{\boldsymbol{j}}{\boldsymbol{m}}\right) f_i(\boldsymbol{\theta})\right|\right) = \int_{[0,1]^d} F\left(\left|\sum_{i=1}^{p} a_i(\mathbf{x}) f_i(\boldsymbol{\theta})\right|\right) \mathrm{d}\mathbf{x}.$$

Passing to the limit as $\boldsymbol{m} \to \infty$ in (4.18), and using the dominated convergence theorem, we get (4.16). □

Theorem 4.3 *Let $a_1, \ldots, a_p : [0, 1]^d \to \mathbb{C}$ and let $f_1, \ldots, f_p \in L^1([-\pi, \pi]^d)$. Then, for every $\boldsymbol{m} \in \mathbb{N}^d$ and every $F \in C_c(\mathbb{C})$,*

$$\begin{aligned}
&\lim_{\boldsymbol{n}\to\infty} \frac{1}{N(\boldsymbol{n})} \sum_{r=1}^{N(\boldsymbol{n})} F\left(\lambda_r\left(\Re\left(\sum_{i=1}^{p} LT_{\boldsymbol{n}}^{\boldsymbol{m}}(a_i, f_i)\right)\right)\right)\\
&= \phi_{\boldsymbol{m}}(F) = \frac{1}{N(\boldsymbol{m})} \sum_{\boldsymbol{j}=\boldsymbol{1}}^{\boldsymbol{m}} \frac{1}{(2\pi)^d} \int_{[-\pi,\pi]^d} F\left(\Re\left(\sum_{i=1}^{p} a_i\left(\frac{\boldsymbol{j}}{\boldsymbol{m}}\right) f_i(\boldsymbol{\theta})\right)\right) \mathrm{d}\boldsymbol{\theta}. \qquad (4.19)
\end{aligned}$$

Moreover, if $a_1, \ldots, a_p$ are Riemann-integrable, then, for every $F \in C_c(\mathbb{C})$,

$$\lim_{\boldsymbol{m}\to\infty} \phi_{\boldsymbol{m}}(F) = \phi(F) = \frac{1}{(2\pi)^d} \int_{[0,1]^d\times[-\pi,\pi]^d} F\left(\Re\left(\sum_{i=1}^{p} a_i(\mathbf{x}) f_i(\boldsymbol{\theta})\right)\right) \mathrm{d}\mathbf{x}\mathrm{d}\boldsymbol{\theta}. \qquad (4.20)$$

Proof The proof follows the same pattern as the proof of Theorem 4.2. By (4.10) and Definition 4.2,

$$
\begin{aligned}
&(\Pi_{\boldsymbol{n}}^{\boldsymbol{m}})^T \left(\Re\Big(\sum_{i=1}^{p} LT_{\boldsymbol{n}}^{\boldsymbol{m}}(a_i, f_i) \Big) \right) \Pi_{\boldsymbol{n}}^{\boldsymbol{m}} \\
&= (\Pi_{\boldsymbol{n}}^{\boldsymbol{m}})^T \left(\frac{1}{2} \Big(\sum_{i=1}^{p} LT_{\boldsymbol{n}}^{\boldsymbol{m}}(a_i, f_i) + \sum_{i=1}^{p} LT_{\boldsymbol{n}}^{\boldsymbol{m}}(\overline{a_i}, \overline{f_i}) \Big) \right) \Pi_{\boldsymbol{n}}^{\boldsymbol{m}} \\
&= \frac{1}{2} \Big(\sum_{i=1}^{p} D_{\boldsymbol{m}}(a_i) \otimes T_{\lfloor \boldsymbol{n}/\boldsymbol{m} \rfloor}(f_i) + \sum_{i=1}^{p} D_{\boldsymbol{m}}(\overline{a_i}) \otimes T_{\lfloor \boldsymbol{n}/\boldsymbol{m} \rfloor}(\overline{f_i}) \Big) \\
&\quad \oplus O_{N(\boldsymbol{n})-N(\boldsymbol{m})N(\lfloor \boldsymbol{n}/\boldsymbol{m} \rfloor)}.
\end{aligned}
$$

For $\mathbf{1} \le \boldsymbol{j} \le \boldsymbol{m}$, the $\boldsymbol{j}$th block of this matrix is given by

$$
\begin{aligned}
&\frac{1}{2} \Big(\sum_{i=1}^{p} a_i\Big(\frac{\boldsymbol{j}}{\boldsymbol{m}}\Big) T_{\lfloor \boldsymbol{n}/\boldsymbol{m} \rfloor}(f_i) + \sum_{i=1}^{p} \overline{a_i}\Big(\frac{\boldsymbol{j}}{\boldsymbol{m}}\Big) T_{\lfloor \boldsymbol{n}/\boldsymbol{m} \rfloor}(\overline{f_i}) \Big) \\
&= T_{\lfloor \boldsymbol{n}/\boldsymbol{m} \rfloor} \left(\Re\Big(\sum_{i=1}^{p} a_i\Big(\frac{\boldsymbol{j}}{\boldsymbol{m}}\Big) f_i \Big) \right).
\end{aligned}
$$

It follows that the eigenvalues of $\Re(\sum_{i=1}^{p} LT_{\boldsymbol{n}}^{\boldsymbol{m}}(a_i, f_i))$ are

$$
\lambda_k \left(T_{\lfloor \boldsymbol{n}/\boldsymbol{m} \rfloor} \left(\Re\Big(\sum_{i=1}^{p} a_i\Big(\frac{\boldsymbol{j}}{\boldsymbol{m}}\Big) f_i \Big) \right) \right), \qquad k = 1, \ldots, N(\lfloor \boldsymbol{n}/\boldsymbol{m} \rfloor), \qquad \boldsymbol{j} = \mathbf{1}, \ldots, \boldsymbol{m},
$$

plus further $N(\boldsymbol{n}) - N(\boldsymbol{m})N(\lfloor \boldsymbol{n}/\boldsymbol{m} \rfloor)$ eigenvalues equal to 0; see (2.23). Therefore, by Theorem 3.5, for any $F \in C_c(\mathbb{C})$ we have

$$
\begin{aligned}
&\lim_{\boldsymbol{n}\to\infty} \frac{1}{N(\boldsymbol{n})} \sum_{r=1}^{N(\boldsymbol{n})} F\left(\lambda_r \left(\Re\Big(\sum_{i=1}^{p} LT_{\boldsymbol{n}}^{\boldsymbol{m}}(a_i, f_i) \Big) \right) \right) \\
&= \lim_{\boldsymbol{n}\to\infty} \frac{N(\boldsymbol{m})N(\lfloor \boldsymbol{n}/\boldsymbol{m} \rfloor)}{N(\boldsymbol{n})} \times \\
&\quad \times \frac{1}{N(\boldsymbol{m})} \sum_{\boldsymbol{j}=\mathbf{1}}^{\boldsymbol{m}} \frac{1}{N(\lfloor \boldsymbol{n}/\boldsymbol{m} \rfloor)} \sum_{k=1}^{N(\lfloor \boldsymbol{n}/\boldsymbol{m} \rfloor)} F\left(\lambda_k \left(T_{\lfloor \boldsymbol{n}/\boldsymbol{m} \rfloor} \left(\Re\Big(\sum_{i=1}^{p} a_i\Big(\frac{\boldsymbol{j}}{\boldsymbol{m}}\Big) f_i \Big) \right) \right) \right) \\
&= \frac{1}{N(\boldsymbol{m})} \sum_{\boldsymbol{j}=\mathbf{1}}^{\boldsymbol{m}} \frac{1}{(2\pi)^d} \int_{[-\pi,\pi]^d} F\left(\Re\Big(\sum_{i=1}^{p} a_i\Big(\frac{\boldsymbol{j}}{\boldsymbol{m}}\Big) f_i(\boldsymbol{\theta}) \Big) \right) \mathrm{d}\boldsymbol{\theta}. \qquad (4.21)
\end{aligned}
$$

This proves (4.19).

If we assume that $a_1, \ldots, a_p$ are Riemann-integrable, then the function $\mathbf{x} \mapsto F(\Re(\sum_{i=1}^{p} a_i(\mathbf{x}) f_i(\boldsymbol{\theta})))$ is Riemann-integrable for each fixed $\boldsymbol{\theta} \in [-\pi, \pi]^d$, and so, by Lemma 2.5,

$$\lim_{m\to\infty}\frac{1}{N(\boldsymbol{m})}\sum_{\boldsymbol{j}=\boldsymbol{1}}^{\boldsymbol{m}}F\left(\Re\left(\sum_{i=1}^{p}a_i\left(\frac{\boldsymbol{j}}{\boldsymbol{m}}\right)f_i(\boldsymbol{\theta})\right)\right)=\int_{[0,1]^d}F\left(\Re\left(\sum_{i=1}^{p}a_i(\mathbf{x})f_i(\boldsymbol{\theta})\right)\right)\mathrm{d}\mathbf{x}.$$

Passing to the limit as $\boldsymbol{m}\to\infty$ in (4.21), and using the dominated convergence theorem, we get (4.20). □

4.2 Definition of Multilevel LT and sLT Sequences

From a historical point of view, the theory of multilevel LT sequences started with the notion of multilevel separable LT (sLT) sequences. For convenience, however, we first introduce the notion of multilevel LT sequences and then, as a special case, we will obtain the notion of multilevel sLT sequences.

Definition 4.3 (*multilevel locally Toeplitz sequence*) Let $\{A_{\boldsymbol{n}}\}_n$ be a d-level matrix-sequence, let $a:[0,1]^d\to\mathbb{C}$ be Riemann-integrable and let $f\in L^1([-\pi,\pi]^d)$. We say that $\{A_{\boldsymbol{n}}\}_n$ is a (d-level) Locally Toeplitz (LT) sequence with *symbol* $a\otimes f$, and we write $\{A_{\boldsymbol{n}}\}_n\sim_{\rm LT}a\otimes f$, if

$$\{LT_{\boldsymbol{n}}^{\boldsymbol{m}}(a,f)\}_n\xrightarrow{\text{a.c.s.}}\{A_{\boldsymbol{n}}\}_n\text{ as }\boldsymbol{m}\to\infty.$$

The symbol $a\otimes f$ is sometimes called the *kernel* of $\{A_{\boldsymbol{n}}\}_n$. The functions a and f are, respectively, the *weight function* and the *generating function* of $\{A_{\boldsymbol{n}}\}_n$; we refer the reader to [22, Sect. 7.1] for the origin and the meaning of this terminology.

Definition 4.4 (*multilevel separable locally Toeplitz sequence*) Let $\{A_{\boldsymbol{n}}\}_n$ be a d-level matrix-sequence. We say that $\{A_{\boldsymbol{n}}\}_n$ is a (d-level) separable Locally Toeplitz (sLT) sequence if $\{A_{\boldsymbol{n}}\}_n\sim_{\rm LT}a\otimes f$ for some Riemann-integrable function $a:[0,1]^d\to\mathbb{C}$ and some *separable* function $f\in L^1([-\pi,\pi]^d)$. In this case, we write $\{A_{\boldsymbol{n}}\}_n\sim_{\rm sLT}a\otimes f$.

It is clear from the definition that an sLT sequence is just an LT sequence with separable generating function. From now on, if we write $\{A_{\boldsymbol{n}}\}_n\sim_{\rm LT}a\otimes f$ (resp., $\{A_{\boldsymbol{n}}\}_n\sim_{\rm sLT}a\otimes f$), it is understood that $a:[0,1]^d\to\mathbb{C}$ is Riemann-integrable and $f\in L^1([-\pi,\pi]^d)$ (resp., $f\in L^1([-\pi,\pi]^d)$ is separable).

4.3 Fundamental Examples of Multilevel LT Sequences

In this section we provide three fundamental examples of multilevel LT sequences: zero-distributed sequences, sequences of multilevel diagonal sampling matrices and multilevel Toeplitz sequences. These may be regarded as the "building blocks" of the theory of multilevel GLT sequences, because from them we can construct through algebraic operations a lot of other matrix-sequences which will turn out to be multilevel GLT sequences.

4.3.1 Zero-Distributed Sequences

We have introduced zero-distributed sequences in Sect. 2.6.3. We now show that any d-level zero-distributed sequence is a d-level LT sequence with symbol 0.

Theorem 4.4 *Let $\{Z_{\boldsymbol{n}}\}_n$ be a d-level matrix-sequence. The following conditions are equivalent.*

1. $\{Z_{\boldsymbol{n}}\}_n \sim_\sigma 0$.
2. $\{O_{N(\boldsymbol{n})}\}_n \xrightarrow{\text{a.c.s.}} \{Z_{\boldsymbol{n}}\}_n$.
3. $\{Z_{\boldsymbol{n}}\}_n \sim_{\text{sLT}} 0$.

Proof ($1 \iff 2$) By Theorem 2.5, we have $\{O_{N(\boldsymbol{n})}\}_n \xrightarrow{\text{a.c.s.}} \{Z_{\boldsymbol{n}}\}_n$ if and only if $d_{\text{a.c.s.}}(\{Z_{\boldsymbol{n}}\}_n, \{O_{N(\boldsymbol{n})}\}_n) \to 0$ if and only if $d_{\text{a.c.s.}}(\{Z_{\boldsymbol{n}}\}_n, \{O_{N(\boldsymbol{n})}\}_n) = 0$ if and only if $\{Z_{\boldsymbol{n}}\}_n \sim_\sigma 0$.

($2 \iff 3$) This follows from the definition of d-level sLT sequences (Definition 4.4) and the observation that $LT_{\boldsymbol{n}}^{\boldsymbol{m}}(0, 0) = O_{N(\boldsymbol{n})}$ and $0 \otimes 0 = 0$. □

4.3.2 Sequences of Multilevel Diagonal Sampling Matrices

To any $a : [0, 1]^d \to \mathbb{C}$ we associate the family of d-level diagonal sampling matrices generated by a, that is, the family $\{D_{\boldsymbol{n}}(a)\}_{\boldsymbol{n}\in\mathbb{N}^d}$ with $D_{\boldsymbol{n}}(a)$ defined as in (4.1). We are going to show that any d-level matrix-sequence of the form $\{D_{\boldsymbol{n}}(a)\}_n$ with $\boldsymbol{n} = \boldsymbol{n}(n) \to \infty$ as $n \to \infty$ is a d-level sLT sequence with symbol

$$(a \otimes 1)(\mathbf{x}, \boldsymbol{\theta}) = a(\mathbf{x}),$$

as long as a is Riemann-integrable. For the proof we need the following technical lemma.

Lemma 4.1 *Let $\mathcal{M}$ be any infinite subset of $\mathbb{N}$. For every $m \in \mathcal{M}$ let $\{x(m, \boldsymbol{k})\}_{\boldsymbol{k}\in\mathbb{N}^d}$ be a family of numbers such that $x(m, \boldsymbol{k}) \to x(m)$ as $\boldsymbol{k} \to \infty$, where $x(m) \to 0$ as $m \to \infty$. Then, there exists a family $\{m(\boldsymbol{k})\}_{\boldsymbol{k}\in\mathbb{N}^d} \subseteq \mathcal{M}$ such that $m(\boldsymbol{k}) \to \infty$ and $x(m(\boldsymbol{k}), \boldsymbol{k}) \to 0$ as $\boldsymbol{k} \to \infty$.*

Proof Since $x(m, \boldsymbol{k}) \to x(m)$ for every $m \in \mathcal{M} = \{m_1, m_2, m_3, \ldots\}$,

- for $m = m_1$ there exists $\boldsymbol{k}_1$ such that $|x(m_1, \boldsymbol{k}) - x(m_1)| \le 1/m_1$ for $\boldsymbol{k} \ge \boldsymbol{k}_1$;
- for $m = m_2$ there exists $\boldsymbol{k}_2 > \boldsymbol{k}_1$ such that $|x(m_2, \boldsymbol{k}) - x(m_2)| \le 1/m_2$ for $\boldsymbol{k} \ge \boldsymbol{k}_2$;
- for $m = m_3$ there exists $\boldsymbol{k}_3 > \boldsymbol{k}_2$ such that $|x(m_3, \boldsymbol{k}) - x(m_3)| \le 1/m_3$ for $\boldsymbol{k} \ge \boldsymbol{k}_3$;
- ...

Define

- $m(\boldsymbol{k}) = m_1$ for $\boldsymbol{k} \not\ge \boldsymbol{k}_2$;
- $m(\boldsymbol{k}) = m_2$ for $\boldsymbol{k} \ge \boldsymbol{k}_2 \ \wedge \ \boldsymbol{k} \not\ge \boldsymbol{k}_3$;

- $m(\boldsymbol{k}) = m_3$ for $\boldsymbol{k} \geq \boldsymbol{k}_3 \ \wedge \ \boldsymbol{k} \not\geq \boldsymbol{k}_4$;
- ...

By construction, $m(\boldsymbol{k}) \to \infty$ and $|x(m(\boldsymbol{k}), \boldsymbol{k}) - x(m(\boldsymbol{k}))| \leq 1/m(\boldsymbol{k})$ for $\boldsymbol{k} \geq \boldsymbol{k}_2$. Since $x(m(\boldsymbol{k})) \to 0$ as $\boldsymbol{k} \to \infty$, we conclude that $x(m(\boldsymbol{k}), \boldsymbol{k}) \to 0$ as well. □

Theorem 4.5 *If $a : [0, 1]^d \to \mathbb{C}$ is Riemann-integrable then $\{D_{\boldsymbol{n}}(a)\}_n \sim_{\text{sLT}} a \otimes 1$ for any sequence $\{\boldsymbol{n} = \boldsymbol{n}(n)\}_n \subseteq \mathbb{N}^d$ such that $\boldsymbol{n} \to \infty$ as $n \to \infty$.*

Proof The proof is organized in two steps: we first show that the thesis holds if a is continuous; then, by using an approximation argument, we show that it holds for any Riemann-integrable function a. As we shall see, the approximation argument heavily relies on the Riemann-integrability of a.

Step 1. We prove by induction on d that if $a \in C([0, 1]^d)$ and $\omega_a(\cdot)$ is the modulus of continuity of a, then

$$D_{\boldsymbol{n}}(a) = LT_{\boldsymbol{n}}^{\boldsymbol{m}}(a, 1) + R_{\boldsymbol{n},\boldsymbol{m}} + N_{\boldsymbol{n},\boldsymbol{m}},$$
$$\text{rank}(R_{\boldsymbol{n},\boldsymbol{m}}) \leq N(\boldsymbol{n}) \sum_{i=1}^{d} \frac{m_i}{n_i}, \qquad \|N_{\boldsymbol{n},\boldsymbol{m}}\| \leq \sum_{i=1}^{d} \omega_a\Big(\frac{1}{m_i} + \frac{m_i}{n_i}\Big). \tag{4.22}$$

Since $\omega_a(\delta) \to 0$ as $\delta \to 0$, the convergence $\{LT_{\boldsymbol{n}}^{\boldsymbol{m}}(a, 1)\}_n \xrightarrow{\text{a.c.s.}} \{D_{\boldsymbol{n}}(a)\}_n$ as $\boldsymbol{m} \to \infty$ (and hence the relation $\{D_{\boldsymbol{n}}(a)\}_n \sim_{\text{sLT}} a \otimes 1$) follows immediately from Definition 2.8 (take $n_{\boldsymbol{m}}$ such that $\boldsymbol{n} \geq \boldsymbol{m}^2$ for $n \geq n_{\boldsymbol{m}}$, and take $c(\boldsymbol{m}) = \sum_{i=1}^d 1/m_i$ and $\omega(\boldsymbol{m}) = \sum_{i=1}^d \omega_a(2/m_i)$).

In the case $d = 1$, we have $\boldsymbol{n} = \boldsymbol{n}(n) = (d_n)$ for some sequence of numbers $\{d_n\}_n$ such that $d_n \to \infty$ as $n \to \infty$, and Eq. (4.22) reduces to

$$D_{d_n}(a) = LT_{d_n}^{m}(a, 1) + R_{d_n,m} + N_{d_n,m},$$
$$\text{rank}(R_{d_n,m}) \leq m, \qquad \|N_{d_n,m}\| \leq \omega_a\Big(\frac{1}{m} + \frac{m}{d_n}\Big).$$

This is nothing else than Eq. (7.24) from [22] with d_n in place of n, and it was already proved in [22]. In the case $d > 1$, $LT_{\boldsymbol{n}}^{\boldsymbol{m}}(a, 1)$ is an $N(\boldsymbol{n}) \times N(\boldsymbol{n})$ diagonal matrix that can be written as follows:

$$\begin{aligned} LT_{\boldsymbol{n}}^{\boldsymbol{m}}(a, 1) &= \operatorname*{diag}_{j_1=1,\ldots,m_1} I_{\lfloor n_1/m_1 \rfloor} \otimes LT_{n_2,\ldots,n_d}^{m_2,\ldots,m_d}\Big(a\Big(\frac{j_1}{m_1}, \cdot\Big), 1\Big) \oplus O_{(n_1 \bmod m_1)n_2\cdots n_d} \\ &= \operatorname*{diag}_{j_1=1,\ldots,m_1} \Bigg[\operatorname*{diag}_{i_1=(j_1-1)\lfloor n_1/m_1 \rfloor+1,\ldots,j_1\lfloor n_1/m_1 \rfloor} LT_{n_2,\ldots,n_d}^{m_2,\ldots,m_d}\Big(a\Big(\frac{j_1}{m_1}, \cdot\Big), 1\Big) \Bigg] \\ &\quad \oplus O_{(n_1 \bmod m_1)n_2\cdots n_d}, \end{aligned} \tag{4.23}$$

where, for any $\hat{x}_1 \in [0, 1]$, the function $a(\hat{x}_1, \cdot)$ is defined as follows:

$$a(\hat{x}_1, \cdot) : [0, 1]^{d-1} \to \mathbb{C}, \qquad (x_2, \ldots, x_d) \mapsto a(\hat{x}_1, x_2, \ldots, x_d).$$

Moreover, by (4.2), $D_{\boldsymbol{n}}(a)$ is an $N(\boldsymbol{n}) \times N(\boldsymbol{n})$ diagonal matrix that can be written as follows:

$$\begin{aligned} D_{\boldsymbol{n}}(a) = & \operatorname*{diag}_{j_1=1,\ldots,m_1} \left[\operatorname*{diag}_{i_1=(j_1-1)\lfloor n_1/m_1 \rfloor+1,\ldots,j_1\lfloor n_1/m_1 \rfloor} D_{n_2,\ldots,n_d}\Big(a\Big(\frac{i_1}{n_1}, \cdot\Big)\Big) \right] \\ & \oplus \operatorname*{diag}_{i_1=m_1\lfloor n_1/m_1 \rfloor+1,\ldots,n_1} D_{n_2,\ldots,n_d}\Big(a\Big(\frac{i_1}{n_1}, \cdot\Big)\Big). \end{aligned} \tag{4.24}$$

For $j_1 = 1, \ldots, m_1$ and $i_1 = (j_1 - 1)\lfloor n_1/m_1 \rfloor + 1, \ldots, j_1\lfloor n_1/m_1 \rfloor$, by induction hypothesis we have

$$\begin{aligned} & LT_{n_2,\ldots,n_d}^{m_2,\ldots,m_d}\Big(a\Big(\frac{j_1}{m_1}, \cdot\Big), 1\Big) - D_{n_2,\ldots,n_d}\Big(a\Big(\frac{i_1}{n_1}, \cdot\Big)\Big) \\ & = \Big[D_{n_2,\ldots,n_d}\Big(a\Big(\frac{j_1}{m_1}, \cdot\Big)\Big) - D_{n_2,\ldots,n_d}\Big(a\Big(\frac{i_1}{n_1}, \cdot\Big)\Big)\Big] \\ & \quad + R_{n_2,\ldots,n_d,m_2,\ldots,m_d}^{[j_1/m_1]} + N_{n_2,\ldots,n_d,m_2,\ldots,m_d}^{[j_1/m_1]}, \end{aligned}$$

where

$$\operatorname{rank}(R_{n_2,\ldots,n_d,m_2,\ldots,m_d}^{[j_1/m_1]}) \le n_2 \cdots n_d \sum_{k=2}^{d} \frac{m_k}{n_k},$$

$$\|N_{n_2,\ldots,n_d,m_2,\ldots,m_d}^{[j_1/m_1]}\| \le \sum_{k=2}^{d} \omega_{a(j_1/m_1,\cdot)}\Big(\frac{1}{m_k} + \frac{m_k}{n_k}\Big) \le \sum_{k=2}^{d} \omega_a\Big(\frac{1}{m_k} + \frac{m_k}{n_k}\Big).$$

Moreover,

$$\Big\|D_{n_2,\ldots,n_d}\Big(a\Big(\frac{j_1}{m_1}, \cdot\Big)\Big) - D_{n_2,\ldots,n_d}\Big(a\Big(\frac{i_1}{n_1}, \cdot\Big)\Big)\Big\| \le \omega_a\Big(\frac{1}{m_1} + \frac{m_1}{n_1}\Big),$$

because

$$\left|\frac{j_1}{m_1} - \frac{i_1}{n_1}\right| \le \frac{j_1}{m_1} - \frac{(j_1-1)\lfloor n_1/m_1 \rfloor}{n_1} \le \frac{j_1}{m_1} - \frac{(j_1-1)(n_1/m_1 - 1)}{n_1} \le \frac{1}{m_1} + \frac{m_1}{n_1}.$$

Thus,

$$LT^{m_2,\dots,m_d}_{n_2,\dots,n_d}\Big(a\Big(\frac{j_1}{m_1},\cdot\Big),1\Big) - D_{n_2,\dots,n_d}\Big(a\Big(\frac{i_1}{n_1},\cdot\Big)\Big) = R^{[j_1/m_1]}_{n_2,\dots,n_d,m_2,\dots,m_d} + N^{[j_1/m_1,\,i_1/n_1]}_{\boldsymbol{n},\boldsymbol{m}}$$

$$\operatorname{rank}(R^{[j_1/m_1]}_{n_2,\dots,n_d,m_2,\dots,m_d}) \le n_2\cdots n_d\sum_{k=2}^{d}\frac{m_k}{n_k}, \tag{4.25}$$

$$\|N^{[j_1/m_1,\,i_1/n_1]}_{\boldsymbol{n},\boldsymbol{m}}\| \le \sum_{k=1}^{d}\omega_a\Big(\frac{1}{m_k}+\frac{m_k}{n_k}\Big).$$

Hence, by (4.23) and (4.24),

$$\begin{aligned}
&D_{\boldsymbol{n}}(a) - LT^{\boldsymbol{m}}_{\boldsymbol{n}}(a,1)\\
&= \operatorname*{diag}_{j_1=1,\dots,m_1}\Bigg[\operatorname*{diag}_{i_1=(j_1-1)\lfloor n_1/m_1\rfloor+1,\dots,j_1\lfloor n_1/m_1\rfloor}\Bigg[D_{n_2,\dots,n_d}\Big(a\Big(\frac{i_1}{n_1},\cdot\Big)\Big)\\
&\qquad\qquad - LT^{m_2,\dots,m_d}_{n_2,\dots,n_d}\Big(a\Big(\frac{j_1}{m_1},\cdot\Big),1\Big)\Bigg]\Bigg]\\
&\quad\oplus \operatorname*{diag}_{i_1=m_1\lfloor n_1/m_1\rfloor+1,\dots,n_1} D_{n_2,\dots,n_d}\Big(a\Big(\frac{i_1}{n_1},\cdot\Big)\Big)\\
&= \operatorname*{diag}_{j_1=1,\dots,m_1}\Bigg[\operatorname*{diag}_{i_1=(j_1-1)\lfloor n_1/m_1\rfloor+1,\dots,j_1\lfloor n_1/m_1\rfloor}\Big[-R^{[j_1/m_1]}_{n_2,\dots,n_d,m_2,\dots,m_d} - N^{[j_1/m_1,\,i_1/n_1]}_{\boldsymbol{n},\boldsymbol{m}}\Big]\Bigg]\\
&\quad\oplus \operatorname*{diag}_{i_1=m_1\lfloor n_1/m_1\rfloor+1,\dots,n_1} D_{n_2,\dots,n_d}\Big(a\Big(\frac{i_1}{n_1},\cdot\Big)\Big)\\
&= R_{\boldsymbol{n},\boldsymbol{m}} + N_{\boldsymbol{n},\boldsymbol{m}},
\end{aligned}$$

where

$$\begin{aligned}
R_{\boldsymbol{n},\boldsymbol{m}} &= \operatorname*{diag}_{j_1=1,\dots,m_1}\Bigg[\operatorname*{diag}_{i_1=(j_1-1)\lfloor n_1/m_1\rfloor+1,\dots,j_1\lfloor n_1/m_1\rfloor} -R^{[j_1/m_1]}_{n_2,\dots,n_d,m_2,\dots,m_d}\Bigg]\\
&\quad\oplus \operatorname*{diag}_{i_1=m_1\lfloor n_1/m_1\rfloor+1,\dots,n_1} D_{n_2,\dots,n_d}\Big(a\Big(\frac{i_1}{n_1},\cdot\Big)\Big),\\
N_{\boldsymbol{n},\boldsymbol{m}} &= \operatorname*{diag}_{j_1=1,\dots,m_1}\Bigg[\operatorname*{diag}_{i_1=(j_1-1)\lfloor n_1/m_1\rfloor+1,\dots,j_1\lfloor n_1/m_1\rfloor} -N^{[j_1/m_1,\,i_1/n_1]}_{\boldsymbol{n},\boldsymbol{m}}\Bigg]\\
&\quad\oplus O_{(n_1 \bmod m_1)n_2\cdots n_d}.
\end{aligned}$$

By (4.25), (2.26) and (2.28), we have

$$\begin{aligned}\operatorname{rank}(R_{\boldsymbol{n},\boldsymbol{m}}) &\le m_1 \left\lfloor \frac{n_1}{m_1} \right\rfloor n_2 \cdots n_d \sum_{k=2}^{d} \frac{m_k}{n_k} + (n_1 \bmod m_1) n_2 \cdots n_d \\ &\le n_1 n_2 \cdots n_d \sum_{k=2}^{d} \frac{m_k}{n_k} + m_1 n_2 \cdots n_d = N(\boldsymbol{n}) \sum_{k=1}^{d} \frac{m_k}{n_k}, \\ \|N_{\boldsymbol{n},\boldsymbol{m}}\| &\le \sum_{k=1}^{d} \omega_a\Big(\frac{1}{m_k} + \frac{m_k}{n_k}\Big),\end{aligned}$$

and (4.22) is proved.

Step 2. Let $a : [0, 1]^d \to \mathbb{C}$ be any Riemann-integrable function. Take any sequence of continuous functions $a_m : [0, 1]^d \to \mathbb{C}$ such that $a_m \to a$ in $L^1([0, 1]^d)$. Note that such a sequence exists because $C([0, 1]^d)$ is dense in $L^1([0, 1]^d)$. By Step 1, we have $\{D_{\boldsymbol{n}}(a_m)\}_n \sim_{\text{sLT}} a_m \otimes 1$ for every m. Hence, $\{LT_{\boldsymbol{n}}^{\boldsymbol{k}}(a_m, 1)\}_n \xrightarrow{\text{a.c.s.}} \{D_{\boldsymbol{n}}(a_m)\}_n$ as $\boldsymbol{k} \to \infty$, i.e., for every m and every $\boldsymbol{k} \in \mathbb{N}^d$ there is $n_{m,\boldsymbol{k}}$ such that, for $n \ge n_{m,\boldsymbol{k}}$,

$$\begin{gathered} D_{\boldsymbol{n}}(a_m) = LT_{\boldsymbol{n}}^{\boldsymbol{k}}(a_m, 1) + R_{\boldsymbol{n},m,\boldsymbol{k}} + N_{\boldsymbol{n},m,\boldsymbol{k}}, \\ \operatorname{rank}(R_{\boldsymbol{n},m,\boldsymbol{k}}) \le c(m, \boldsymbol{k}) N(\boldsymbol{n}), \qquad \|N_{\boldsymbol{n},m,\boldsymbol{k}}\| \le \omega(m, \boldsymbol{k}), \end{gathered}$$

where

$$\lim_{\boldsymbol{k}\to\infty} c(m, \boldsymbol{k}) = \lim_{\boldsymbol{k}\to\infty} \omega(m, \boldsymbol{k}) = 0.$$

Moreover, $\{D_{\boldsymbol{n}}(a_m)\}_n \xrightarrow{\text{a.c.s.}} \{D_{\boldsymbol{n}}(a)\}_n$. Indeed,

$$\|D_{\boldsymbol{n}}(a) - D_{\boldsymbol{n}}(a_m)\|_1 = \sum_{\boldsymbol{j}=\boldsymbol{1}}^{\boldsymbol{n}} \left| a\Big(\frac{\boldsymbol{j}}{\boldsymbol{n}}\Big) - a_m\Big(\frac{\boldsymbol{j}}{\boldsymbol{n}}\Big) \right|.$$

By the Riemann-integrability of $|a - a_m|$ and by the fact that $a_m \to a$ in $L^1([0, 1]^d)$, the quantity

$$\varepsilon(m, \boldsymbol{n}) = \frac{1}{N(\boldsymbol{n})} \sum_{\boldsymbol{j}=\boldsymbol{1}}^{\boldsymbol{n}} \left| a\Big(\frac{\boldsymbol{j}}{\boldsymbol{n}}\Big) - a_m\Big(\frac{\boldsymbol{j}}{\boldsymbol{n}}\Big) \right| \tag{4.26}$$

satisfies

$$\lim_{m\to\infty} \lim_{\boldsymbol{n}\to\infty} \varepsilon(m, \boldsymbol{n}) = \lim_{m\to\infty} \int_{[0,1]^d} |a(\mathbf{x}) - a_m(\mathbf{x})| \mathrm{d}\mathbf{x} = \lim_{m\to\infty} \|a - a_m\|_{L^1} = 0.$$

By Theorem 2.10, this implies that $\{D_{\boldsymbol{n}}(a_m)\}_n \xrightarrow{\text{a.c.s.}} \{D_{\boldsymbol{n}}(a)\}_n$. Thus, for every m there exists n_m such that, for $n \geq n_m$,

$$\begin{gathered} D_{\boldsymbol{n}}(a) = D_{\boldsymbol{n}}(a_m) + R_{\boldsymbol{n},m} + N_{\boldsymbol{n},m}, \\ \operatorname{rank}(R_{\boldsymbol{n},m}) \leq c(m)N(\boldsymbol{n}), \qquad \|N_{\boldsymbol{n},m}\| \leq \omega(m), \end{gathered}$$

where

$$\lim_{m\to\infty} c(m) = \lim_{m\to\infty} \omega(m) = 0.$$

It follows that, for every m, every $\boldsymbol{k} \in \mathbb{N}^d$, and every $n \geq \max(n_m, n_{m,\boldsymbol{k}})$,

$$\begin{aligned} &D_{\boldsymbol{n}}(a) = LT_{\boldsymbol{n}}^{\boldsymbol{k}}(a,1) + \big[LT_{\boldsymbol{n}}^{\boldsymbol{k}}(a_m,1) - LT_{\boldsymbol{n}}^{\boldsymbol{k}}(a,1)\big] \\ &\qquad\qquad + (R_{\boldsymbol{n},m} + R_{\boldsymbol{n},m,\boldsymbol{k}}) + (N_{\boldsymbol{n},m} + N_{\boldsymbol{n},m,\boldsymbol{k}}), \\ &\operatorname{rank}(R_{\boldsymbol{n},m} + R_{\boldsymbol{n},m,\boldsymbol{k}}) \leq (c(m) + c(m,\boldsymbol{k}))N(\boldsymbol{n}), \\ &\|N_{\boldsymbol{n},m} + N_{\boldsymbol{n},m,\boldsymbol{k}}\| \leq \omega(m) + \omega(m,\boldsymbol{k}), \\ &\|LT_{\boldsymbol{n}}^{\boldsymbol{k}}(a_m,1) - LT_{\boldsymbol{n}}^{\boldsymbol{k}}(a,1)\|_1 \leq \frac{N(\boldsymbol{n})}{N(\boldsymbol{k})} \sum_{\boldsymbol{j}=\boldsymbol{1}}^{\boldsymbol{k}} \left| a\left(\frac{\boldsymbol{j}}{\boldsymbol{k}}\right) - a_m\left(\frac{\boldsymbol{j}}{\boldsymbol{k}}\right)\right| = \varepsilon(m,\boldsymbol{k})N(\boldsymbol{n}), \end{aligned}$$

where in the last inequality we used the linearity of $LT_{\boldsymbol{n}}^{\boldsymbol{k}}(\cdot, 1)$ and (4.11); recall that $\varepsilon(m, \boldsymbol{k})$ is defined in (4.26). Let $\{m(\boldsymbol{k})\}_{\boldsymbol{k}\in\mathbb{N}^d}$ be a family of indices such that $m(\boldsymbol{k}) \to \infty$ as $\boldsymbol{k} \to \infty$ and

$$\lim_{\boldsymbol{k}\to\infty} \varepsilon(m(\boldsymbol{k}),\boldsymbol{k}) = \lim_{\boldsymbol{k}\to\infty} c(m(\boldsymbol{k}),\boldsymbol{k}) = \lim_{\boldsymbol{k}\to\infty} \omega(m(\boldsymbol{k}),\boldsymbol{k}) = 0.$$

Such a family exists by Lemma 4.1 (apply the lemma with $x(m,\boldsymbol{k}) = \varepsilon(m,\boldsymbol{k}) + c(m,\boldsymbol{k}) + \omega(m,\boldsymbol{k})$). Then, for every $\boldsymbol{k} \in \mathbb{N}^d$ and every $n \geq \max(n_{m(\boldsymbol{k})}, n_{m(\boldsymbol{k}),\boldsymbol{k}})$,

$$\begin{aligned} &D_{\boldsymbol{n}}(a) = LT_{\boldsymbol{n}}^{\boldsymbol{k}}(a,1) + \big[LT_{\boldsymbol{n}}^{\boldsymbol{k}}(a_{m(\boldsymbol{k})},1) - LT_{\boldsymbol{n}}^{\boldsymbol{k}}(a,1)\big] \\ &\qquad\qquad + (R_{\boldsymbol{n},m(\boldsymbol{k})} + R_{\boldsymbol{n},m(\boldsymbol{k}),\boldsymbol{k}}) + (N_{\boldsymbol{n},m(\boldsymbol{k})} + N_{\boldsymbol{n},m(\boldsymbol{k}),\boldsymbol{k}}), \\ &\operatorname{rank}(R_{\boldsymbol{n},m(\boldsymbol{k})} + R_{\boldsymbol{n},m(\boldsymbol{k}),\boldsymbol{k}}) \leq (c(m(\boldsymbol{k})) + c(m(\boldsymbol{k}),\boldsymbol{k}))N(\boldsymbol{n}), \\ &\|N_{\boldsymbol{n},m(\boldsymbol{k})} + N_{\boldsymbol{n},m(\boldsymbol{k}),\boldsymbol{k}}\| \leq \omega(m(\boldsymbol{k})) + \omega(m(\boldsymbol{k}),\boldsymbol{k}), \\ &\|LT_{\boldsymbol{n}}^{\boldsymbol{k}}(a_{m(\boldsymbol{k})},1) - LT_{\boldsymbol{n}}^{\boldsymbol{k}}(a,1)\|_1 \leq \varepsilon(m(\boldsymbol{k}),\boldsymbol{k})N(\boldsymbol{n}). \end{aligned}$$

By [22, Lemma 5.6], we can decompose the matrix $LT_{\boldsymbol{n}}^{\boldsymbol{k}}(a_{m(\boldsymbol{k})},1) - LT_{\boldsymbol{n}}^{\boldsymbol{k}}(a,1)$ as the sum of a small-rank term $\hat{R}_{\boldsymbol{n},\boldsymbol{k}}$, with rank bounded by $\sqrt{\varepsilon(m(\boldsymbol{k}),\boldsymbol{k})}\,N(\boldsymbol{n})$, plus a small-norm term $\hat{N}_{\boldsymbol{n},\boldsymbol{k}}$, with norm bounded by $\sqrt{\varepsilon(m(\boldsymbol{k}),\boldsymbol{k})}$. This shows that $\{LT_{\boldsymbol{n}}^{\boldsymbol{k}}(a,1)\}_n \xrightarrow{\text{a.c.s.}} \{D_{\boldsymbol{n}}(a)\}_n$ as $\boldsymbol{k} \to \infty$, hence $\{D_{\boldsymbol{n}}(a)\}_n \sim_{\text{sLT}} a \otimes 1$. □

4.3.3 Multilevel Toeplitz Sequences

Given $f \in L^1([-\pi,\pi]^d)$, we show that any d-level Toeplitz sequence of the form $\{T_{\boldsymbol{n}}(f)\}_n$ with $\boldsymbol{n} = \boldsymbol{n}(n) \to \infty$ as $n \to \infty$ is a d-level LT sequence with symbol

$$(1 \otimes f)(\mathbf{x}, \boldsymbol{\theta}) = f(\boldsymbol{\theta}).$$

In view of what follows, we recall from Corollary 2.1 that if f is a separable d-variate trigonometric polynomial then $f = f_1 \otimes \cdots \otimes f_d$ for some univariate trigonometric polynomials $f_1, \ldots, f_d$. We also recall that the degree of a univariate trigonometric polynomial g is the smallest nonnegative integer r such that the Fourier coefficients g_k are zero for $|k| > r$.

Theorem 4.6 *If $f \in L^1([-\pi,\pi]^d)$ then $\{T_{\boldsymbol{n}}(f)\}_n \sim_{\rm LT} 1 \otimes f$ for any sequence $\{\boldsymbol{n} = \boldsymbol{n}(n)\}_n \subseteq \mathbb{N}^d$ such that $\boldsymbol{n} \to \infty$ as $n \to \infty$.*

Proof The proof is organized in three steps: we first show that the thesis holds if f is a separable d-variate trigonometric polynomial; then, by linearity, we show that it holds if f is an arbitrary d-variate trigonometric polynomial; finally, using an approximation argument, we prove the theorem under the sole assumption that $f \in L^1([-\pi,\pi]^d)$.

Step 1. We show by induction on d that, if f is a separable d-variate trigonometric polynomial, say $f = f_1 \otimes \cdots \otimes f_d$ with $f_1, \ldots, f_d$ univariate trigonometric polynomials of degrees $r_1, \ldots, r_d$, respectively, then

$$T_{\boldsymbol{n}}(f) = LT_{\boldsymbol{n}}^{\boldsymbol{m}}(1, f) + R_{\boldsymbol{n},\boldsymbol{m}}, \qquad \mathrm{rank}(R_{\boldsymbol{n},\boldsymbol{m}}) \le N(\boldsymbol{n}) \sum_{i=1}^{d} \frac{(2r_i+1)m_i}{n_i}. \tag{4.27}$$

Once this is done, the convergence $\{LT_{\boldsymbol{n}}^{\boldsymbol{m}}(1,f)\}_n \xrightarrow{\text{a.c.s.}} \{T_{\boldsymbol{n}}(f)\}_n$ (and hence the relation $\{T_{\boldsymbol{n}}(f)\}_n \sim_{\rm LT} 1 \otimes f$) follows immediately from Definition 2.8 (take $n_{\boldsymbol{m}}$ such that $\boldsymbol{n} \ge \boldsymbol{m}^2$ for $n \ge n_{\boldsymbol{m}}$, and take $c(\boldsymbol{m}) = \sum_{i=1}^d (2r_i+1)/m_i$ and $\omega(\boldsymbol{m}) = 0$).

In the case $d = 1$, we have $\boldsymbol{n} = \boldsymbol{n}(n) = (d_n)$ for some sequence of numbers $\{d_n\}_n$ such that $d_n \to \infty$ as $n \to \infty$, and Eq. (4.27) reduces to

$$T_{d_n}(f) = LT_{d_n}^m(1, f) + R_{d_n,m}, \qquad \mathrm{rank}(R_{d_n,m}) \le (2r+1)m, \tag{4.28}$$

where r is the degree of f. This is nothing else than Eq. (7.26) from [22] with d_n in place of n, and it was already proved in [22]. In the case $d > 1$, let $f = f_1 \otimes \cdots \otimes f_d$ with $f_1, \ldots, f_d$ being univariate trigonometric polynomials of degrees $r_1, \ldots, r_d$, respectively. By induction hypothesis,

$$LT^{m_2,\ldots,m_d}_{n_2,\ldots,n_d}(1, f_2 \otimes \cdots \otimes f_d) = T_{n_2,\ldots,n_d}(f_2 \otimes \cdots \otimes f_d) - R_{n_2,\ldots,n_d,m_2,\ldots,m_d},$$

$$\mathrm{rank}(R_{n_2,\ldots,n_d,m_2,\ldots,m_d}) \le n_2 \cdots n_d \sum_{i=2}^{d} \frac{(2r_i+1)m_i}{n_i}.$$

From the definition of $LT^{\boldsymbol{m}}_{\boldsymbol{n}}(1, f)$ and the properties of tensor products and direct sums (see Sect. 2.5), we obtain

$$\begin{aligned}
&LT^{\boldsymbol{m}}_{\boldsymbol{n}}(1, f)\\
&= \operatorname*{diag}_{j_1=1,\ldots,m_1} T_{\lfloor n_1/m_1 \rfloor}(f_1) \otimes LT^{m_2,\ldots,m_d}_{n_2,\ldots,n_d}(1, f_2 \otimes \cdots \otimes f_d) \oplus O_{(n_1 \bmod m_1)n_2\cdots n_d}\\
&= \Big[\operatorname*{diag}_{j_1=1,\ldots,m_1} T_{\lfloor n_1/m_1 \rfloor}(f_1)\Big] \otimes \big[T_{n_2,\ldots,n_d}(f_2 \otimes \cdots \otimes f_d) - R_{n_2,\ldots,n_d,m_2,\ldots,m_d}\big]\\
&\quad \oplus O_{(n_1 \bmod m_1)n_2\cdots n_d}\\
&= \Big[\operatorname*{diag}_{j_1=1,\ldots,m_1} T_{\lfloor n_1/m_1 \rfloor}(f_1) \oplus O_{n_1 \bmod m_1}\Big]\\
&\quad \otimes \big[T_{n_2,\ldots,n_d}(f_2 \otimes \cdots \otimes f_d) - R_{n_2,\ldots,n_d,m_2,\ldots,m_d}\big]\\
&= LT^{m_1}_{n_1}(1, f_1) \otimes \big[T_{n_2,\ldots,n_d}(f_2 \otimes \cdots \otimes f_d) - R_{n_2,\ldots,n_d,m_2,\ldots,m_d}\big]\\
&= LT^{m_1}_{n_1}(1, f_1) \otimes T_{n_2,\ldots,n_d}(f_2 \otimes \cdots \otimes f_d) - \tilde{R}_{n_1,\ldots,n_d,m_1,\ldots,m_d},
\end{aligned}$$

where $\tilde{R}_{n_1,\ldots,n_d,m_1,\ldots,m_d} = LT^{m_1}_{n_1}(1, f_1) \otimes R_{n_2,\ldots,n_d,m_2,\ldots,m_d}$ satisfies

$$\mathrm{rank}(\tilde{R}_{n_1,\ldots,n_d,m_1,\ldots,m_d}) \le N(\boldsymbol{n}) \sum_{i=2}^{d} \frac{(2r_i+1)m_i}{n_i}.$$

Using (4.28) with $d_n = n_1(n) = n_1$, we can decompose $LT^{m_1}_{n_1}(1, f_1)$ into the sum of $T_{n_1}(f_1)$ plus a small-rank matrix $-R_{n_1,m_1}$, whose rank is bounded by $(2r_1+1)m_1$. Invoking Lemma 3.3, we obtain

$$\begin{aligned}
LT^{\boldsymbol{m}}_{\boldsymbol{n}}(1, f) &= \big(T_{n_1}(f_1) - R_{n_1,m_1}\big) \otimes T_{n_2,\ldots,n_d}(f_2 \otimes \cdots \otimes f_d) - \tilde{R}_{n_1,\ldots,n_d,m_1,\ldots,m_d}\\
&= T_{\boldsymbol{n}}(f) - R_{\boldsymbol{n},\boldsymbol{m}},
\end{aligned}$$

where $R_{\boldsymbol{n},\boldsymbol{m}} = R_{n_1,m_1} \otimes T_{n_2,\ldots,n_d}(f_2 \otimes \cdots \otimes f_d) + \tilde{R}_{n_1,\ldots,n_d,m_1,\ldots,m_d}$ satisfies

$$\begin{aligned}
\mathrm{rank}(R_{\boldsymbol{n},\boldsymbol{m}}) &\le (2r_1+1)m_1 n_2 \cdots n_d + N(\boldsymbol{n}) \sum_{i=2}^{d} \frac{(2r_i+1)m_i}{n_i}\\
&= N(\boldsymbol{n}) \sum_{i=1}^{d} \frac{(2r_i+1)m_i}{n_i}.
\end{aligned}$$

This completes the proof of (4.27).

Step 2. Let f be any d-variate trigonometric polynomial. By definition, f is a finite linear combination of the d-variate Fourier frequencies $\mathrm{e}^{\mathrm{i}\boldsymbol{j}\cdot\boldsymbol{\theta}}$, $\boldsymbol{j} \in \mathbb{Z}^d$, and so we can write $f(\boldsymbol{\theta}) = \sum_{\boldsymbol{j}=-\boldsymbol{r}}^{\boldsymbol{r}} f_{\boldsymbol{j}}\, \mathrm{e}^{\mathrm{i}\boldsymbol{j}\cdot\boldsymbol{\theta}}$ for some separable trigonometric polynomials $f_{\boldsymbol{j}}\, \mathrm{e}^{\mathrm{i}\boldsymbol{j}\cdot\boldsymbol{\theta}}$. By linearity,

$$T_{\boldsymbol{n}}(f) = \sum_{\boldsymbol{j}=-\boldsymbol{r}}^{\boldsymbol{r}} f_{\boldsymbol{j}}\, T_{\boldsymbol{n}}(\mathrm{e}^{\mathrm{i}\boldsymbol{j}\cdot\boldsymbol{\theta}}), \qquad LT_{\boldsymbol{n}}^{\boldsymbol{m}}(1, f) = \sum_{\boldsymbol{j}=-\boldsymbol{r}}^{\boldsymbol{r}} f_{\boldsymbol{j}}\, LT_{\boldsymbol{n}}^{\boldsymbol{m}}(1, \mathrm{e}^{\mathrm{i}\boldsymbol{j}\cdot\boldsymbol{\theta}}).$$

By Step 1, $\{T_{\boldsymbol{n}}(\mathrm{e}^{\mathrm{i}\boldsymbol{j}\cdot\boldsymbol{\theta}})\}_n \sim_{\mathrm{LT}} 1 \otimes \mathrm{e}^{\mathrm{i}\boldsymbol{j}\cdot\boldsymbol{\theta}}$, hence $\{LT_{\boldsymbol{n}}^{\boldsymbol{m}}(1, \mathrm{e}^{\mathrm{i}\boldsymbol{j}\cdot\boldsymbol{\theta}})\}_n \xrightarrow{\text{a.c.s.}} \{T_{\boldsymbol{n}}(\mathrm{e}^{\mathrm{i}\boldsymbol{j}\cdot\boldsymbol{\theta}})\}_n$ as $\boldsymbol{m} \to \infty$, and so $\{LT_{\boldsymbol{n}}^{\boldsymbol{m}}(1, f)\}_n \xrightarrow{\text{a.c.s.}} \{T_{\boldsymbol{n}}(f)\}_n$ as $\boldsymbol{m} \to \infty$ by Remark 2.6. Thus, $\{T_{\boldsymbol{n}}(f)\}_n \sim_{\mathrm{LT}} 1 \otimes f$.

Step 3. Let $f \in L^1([-\pi, \pi]^d)$. Since the set of d-variate trigonometric polynomials is dense in $L^1([-\pi, \pi]^d)$ by [22, Lemma 2.2] or Lemma 2.3, there is a sequence $\{f_m\}_m$ of d-variate trigonometric polynomials such that $f_m \to f$ in $L^1([-\pi, \pi]^d)$. By Step 2, $\{T_{\boldsymbol{n}}(f_m)\}_n \sim_{\mathrm{LT}} 1 \otimes f_m$. Hence, for every m and every $\boldsymbol{k} \in \mathbb{N}^d$ there is $n_{m,\boldsymbol{k}}$ such that, for $n \ge n_{m,\boldsymbol{k}}$,

$$T_{\boldsymbol{n}}(f_m) = LT_{\boldsymbol{n}}^{\boldsymbol{k}}(1, f_m) + R_{\boldsymbol{n},m,\boldsymbol{k}} + N_{\boldsymbol{n},m,\boldsymbol{k}},$$
$$\operatorname{rank}(R_{\boldsymbol{n},m,\boldsymbol{k}}) \le c(m, \boldsymbol{k}) N(\boldsymbol{n}), \qquad \|N_{\boldsymbol{n},m,\boldsymbol{k}}\| \le \omega(m, \boldsymbol{k}),$$

where

$$\lim_{\boldsymbol{k}\to\infty} c(m, \boldsymbol{k}) = \lim_{\boldsymbol{k}\to\infty} \omega(m, \boldsymbol{k}) = 0.$$

Moreover, by Theorem 3.2,

$$\|T_{\boldsymbol{n}}(f) - T_{\boldsymbol{n}}(f_m)\|_1 = \|T_{\boldsymbol{n}}(f - f_m)\|_1 \le N(\boldsymbol{n}) \|f - f_m\|_{L^1}$$

and so $\{T_{\boldsymbol{n}}(f_m)\}_n \xrightarrow{\text{a.c.s.}} \{T_{\boldsymbol{n}}(f)\}_n$ by Theorem 2.10: for every m there exists n_m such that, for $n \ge n_m$,

$$T_{\boldsymbol{n}}(f) = T_{\boldsymbol{n}}(f_m) + R_{\boldsymbol{n},m} + N_{\boldsymbol{n},m},$$
$$\operatorname{rank}(R_{\boldsymbol{n},m}) \le c(m) N(\boldsymbol{n}), \qquad \|N_{\boldsymbol{n},m}\| \le \omega(m),$$

where

$$\lim_{m\to\infty} c(m) = \lim_{m\to\infty} \omega(m) = 0.$$

It follows that, for every m, every $\boldsymbol{k} \in \mathbb{N}^d$ and every $n \ge \max(n_m, n_{m,\boldsymbol{k}})$,

$$\begin{aligned}
T_{\boldsymbol{n}}(f) &= LT_{\boldsymbol{n}}^{\boldsymbol{k}}(1, f) + \left[LT_{\boldsymbol{n}}^{\boldsymbol{k}}(1, f_m) - LT_{\boldsymbol{n}}^{\boldsymbol{k}}(1, f)\right] \\
&\quad + (R_{\boldsymbol{n},m} + R_{\boldsymbol{n},m,\boldsymbol{k}}) + (N_{\boldsymbol{n},m} + N_{\boldsymbol{n},m,\boldsymbol{k}}), \\
\operatorname{rank}(R_{\boldsymbol{n},m} &+ R_{\boldsymbol{n},m,\boldsymbol{k}}) \le (c(m) + c(m, \boldsymbol{k}))N(\boldsymbol{n}), \\
\|N_{\boldsymbol{n},m} &+ N_{\boldsymbol{n},m,\boldsymbol{k}}\| \le \omega(m) + \omega(m, \boldsymbol{k}), \\
\|LT_{\boldsymbol{n}}^{\boldsymbol{k}}(1, f_m) &- LT_{\boldsymbol{n}}^{\boldsymbol{k}}(1, f)\|_1 = \|LT_{\boldsymbol{n}}^{\boldsymbol{k}}(1, f_m - f)\|_1 \le N(\boldsymbol{n})\|f - f_m\|_{L^1},
\end{aligned}$$

where in the last inequality we used (4.11) and Theorem 3.2. Let $\{m(\boldsymbol{k})\}_{\boldsymbol{k}\in\mathbb{N}^d}$ be a family of indices such that $m(\boldsymbol{k}) \to \infty$ as $\boldsymbol{k} \to \infty$ and

$$\lim_{\boldsymbol{k}\to\infty} c(m(\boldsymbol{k}), \boldsymbol{k}) = \lim_{\boldsymbol{k}\to\infty} \omega(m(\boldsymbol{k}), \boldsymbol{k}) = 0.$$

Such a family exists by Lemma 4.1 (apply the lemma with $x(m, \boldsymbol{k}) = c(m, \boldsymbol{k}) + \omega(m, \boldsymbol{k})$). Then, for every $\boldsymbol{k} \in \mathbb{N}^d$ and every $\boldsymbol{n} \ge \max(\boldsymbol{n}_{m(\boldsymbol{k})}, \boldsymbol{n}_{m(\boldsymbol{k}),\boldsymbol{k}})$,

$$\begin{aligned}
T_{\boldsymbol{n}}(f) &= LT_{\boldsymbol{n}}^{\boldsymbol{k}}(1, f) + \left[LT_{\boldsymbol{n}}^{\boldsymbol{k}}(1, f_{m(\boldsymbol{k})}) - LT_{\boldsymbol{n}}^{\boldsymbol{k}}(1, f)\right] \\
&\quad + (R_{\boldsymbol{n},m(\boldsymbol{k})} + R_{\boldsymbol{n},m(\boldsymbol{k}),\boldsymbol{k}}) + (N_{\boldsymbol{n},m(\boldsymbol{k})} + N_{\boldsymbol{n},m(\boldsymbol{k}),\boldsymbol{k}}), \\
\operatorname{rank}(R_{\boldsymbol{n},m(\boldsymbol{k})} &+ R_{\boldsymbol{n},m(\boldsymbol{k}),\boldsymbol{k}}) \le (c(m(\boldsymbol{k})) + c(m(\boldsymbol{k}), \boldsymbol{k}))N(\boldsymbol{n}), \\
\|N_{\boldsymbol{n},m(\boldsymbol{k})} &+ N_{\boldsymbol{n},m(\boldsymbol{k}),\boldsymbol{k}}\| \le \omega(m(\boldsymbol{k})) + \omega(m(\boldsymbol{k}), \boldsymbol{k}), \\
\|LT_{\boldsymbol{n}}^{\boldsymbol{k}}(1, f_{m(\boldsymbol{k})}) &- LT_{\boldsymbol{n}}^{\boldsymbol{k}}(1, f)\|_1 \le N(\boldsymbol{n})\|f_{m(\boldsymbol{k})} - f\|_{L^1}.
\end{aligned}$$

By [22, Lemma 5.6], we can decompose the matrix $LT_{\boldsymbol{n}}^{\boldsymbol{k}}(1, f_{m(\boldsymbol{k})}) - LT_{\boldsymbol{n}}^{\boldsymbol{k}}(1, f)$ as the sum of a small-rank term $\hat{R}_{\boldsymbol{n},\boldsymbol{k}}$, with rank bounded by $\sqrt{\|f_{m(\boldsymbol{k})} - f\|_{L^1}}\, N(\boldsymbol{n})$, plus a small-norm term $\hat{N}_{\boldsymbol{n},\boldsymbol{k}}$, with norm bounded by $\sqrt{\|f_{m(\boldsymbol{k})} - f\|_{L^1}}$. This shows that $\{LT_{\boldsymbol{n}}^{\boldsymbol{k}}(a, 1)\}_n \xrightarrow{\text{a.c.s.}} \{T_{\boldsymbol{n}}(f)\}_n$ as $\boldsymbol{k} \to \infty$, hence $\{T_{\boldsymbol{n}}(f)\}_n \sim_{\text{LT}} 1 \otimes f$. □

It follows from Theorem 4.6 that $\{T_{\boldsymbol{n}}(f)\}_n \sim_{\text{sLT}} 1 \otimes f$ whenever $f \in L^1([-\pi, \pi]^d)$ is separable.

4.4 Singular Value and Spectral Distribution of a Finite Sum of Multilevel LT Sequences

Theorem 4.7 provides the singular value distribution of a finite sum of multilevel LT sequences.

Theorem 4.7 *If* $\{A_{\boldsymbol{n}}^{(i)}\}_n \sim_{\text{LT}} a_i \otimes f_i$, $i = 1, \ldots, p$, *then*

$$\left\{\sum_{i=1}^{p} A_{\boldsymbol{n}}^{(i)}\right\}_n \sim_\sigma \sum_{i=1}^{p} a_i \otimes f_i.$$

Proof Take any sequence $\{\boldsymbol{m} = \boldsymbol{m}(m)\}_m \subseteq \mathbb{N}^d$ such that $\boldsymbol{m} \to \infty$ as $m \to \infty$. By definition of multilevel LT sequences,

$$\{LT_{\boldsymbol{n}}^{\boldsymbol{m}}(a_i, f_i)\}_n \xrightarrow{\text{a.c.s.}} \{A_{\boldsymbol{n}}^{(i)}\}_n, \qquad i = 1, \dots, p.$$

By Theorem 2.9,

$$\left\{\sum_{i=1}^{p} LT_{\boldsymbol{n}}^{\boldsymbol{m}}(a_i, f_i)\right\}_n \xrightarrow{\text{a.c.s.}} \left\{\sum_{i=1}^{p} A_{\boldsymbol{n}}^{(i)}\right\}_n.$$

By Theorem 4.2, for every $F \in C_c(\mathbb{R})$ we have

$$\lim_{n\to\infty} \frac{1}{N(\boldsymbol{n})} \sum_{r=1}^{N(\boldsymbol{n})} F\left(\sigma_r\left(\sum_{i=1}^{p} LT_{\boldsymbol{n}}^{\boldsymbol{m}}(a_i, f_i)\right)\right) = \phi_{\boldsymbol{m}}(F),$$

$$\lim_{m\to\infty} \phi_{\boldsymbol{m}}(F) = \phi(F) = \frac{1}{(2\pi)^d} \int_{[0,1]^d\times[-\pi,\pi]^d} F\left(\left|\sum_{i=1}^{p} a_i(\mathbf{x}) f_i(\boldsymbol{\theta})\right|\right) \mathrm{d}\mathbf{x}\mathrm{d}\boldsymbol{\theta}.$$

By Theorem 2.6, we obtain

$$\left\{\sum_{i=1}^{p} A_{\boldsymbol{n}}^{(i)}\right\}_n \sim_\sigma \phi.$$

Since ϕ is just the functional $\phi_{|\sum_{i=1}^{p} a_i \otimes f_i|}$ associated with $|\sum_{i=1}^{p} a_i \otimes f_i|$ according to Eq. (2.2), the previous singular value distribution is equivalent to

$$\left\{\sum_{i=1}^{p} A_{\boldsymbol{n}}^{(i)}\right\}_n \sim_\sigma \sum_{i=1}^{p} a_i \otimes f_i,$$

and the thesis is proved. □

Using Theorem 4.7, we show in Proposition 4.2 that the symbol of a multilevel LT sequence is essentially unique. Afterwards, in Proposition 4.3, we show that the symbol of a multilevel LT sequence formed by Hermitian matrices is real a.e.

Proposition 4.2 *If* $\{A_{\boldsymbol{n}}\}_n \sim_{\text{LT}} a \otimes f$ *and* $\{A_{\boldsymbol{n}}\}_n \sim_{\text{LT}} \tilde{a} \otimes \tilde{f}$, *then* $a \otimes f = \tilde{a} \otimes \tilde{f}$ *a.e.*

Proof By Theorem 4.7,

$$\{O_{N(\boldsymbol{n})}\}_n = \{A_{\boldsymbol{n}} - A_{\boldsymbol{n}}\}_n \sim_\sigma a \otimes f - \tilde{a} \otimes \tilde{f}.$$

Hence, for every $F \in C_c(\mathbb{R})$,

$$F(0) = \frac{1}{(2\pi)^d} \int_{[0,1]^d\times[-\pi,\pi]^d} F(|a(\mathbf{x}) f(\boldsymbol{\theta}) - \tilde{a}(\mathbf{x}) \tilde{f}(\boldsymbol{\theta})|) \mathrm{d}\mathbf{x}\mathrm{d}\boldsymbol{\theta}.$$

This means that $\phi_{|a\otimes f-\tilde{a}\otimes\tilde{f}|}=\phi_0$ and so $|a\otimes f-\tilde{a}\otimes\tilde{f}|=0$ a.e. by [22, Remark 2.1]. □

Proposition 4.3 *If $\{A_{\boldsymbol{n}}\}_n\sim_{\rm LT} a\otimes f$ and the $A_{\boldsymbol{n}}$ are Hermitian, then $a\otimes f\in\mathbb{R}$ a.e.*

Proof It holds in general that $\{A_{\boldsymbol{n}}\}_n\sim_{\rm LT} a\otimes f$ implies $\{A_{\boldsymbol{n}}^*\}_n\sim_{\rm LT} \overline{a}\otimes\overline{f}$. This follows immediately from the definition of multilevel LT sequences in combination with (4.10) and Remark 2.6. If the matrices $A_{\boldsymbol{n}}$ are Hermitian, then, by Proposition 4.2, we have $a\otimes f=\overline{a}\otimes\overline{f}$ a.e., i.e., $a\otimes f\in\mathbb{R}$ a.e. □

Theorem 4.8 provides the spectral distribution of a finite sum of multilevel LT sequences formed by Hermitian matrices.

Theorem 4.8 *If $\{A_{\boldsymbol{n}}^{(i)}\}_n\sim_{\rm LT} a_i\otimes f_i$, $i=1,\ldots,p$, then*

$$\left\{\Re\left(\sum_{i=1}^{p}A_{\boldsymbol{n}}^{(i)}\right)\right\}_n\sim_\lambda \Re\left(\sum_{i=1}^{p}a_i\otimes f_i\right).$$

In particular, if the $A_{\boldsymbol{n}}^{(i)}$ are Hermitian,

$$\left\{\sum_{i=1}^{p}A_{\boldsymbol{n}}^{(i)}\right\}_n\sim_\lambda \sum_{i=1}^{p}a_i\otimes f_i.$$

Proof Take any sequence $\{\boldsymbol{m}=\boldsymbol{m}(m)\}_m\subseteq\mathbb{N}^d$ such that $\boldsymbol{m}\to\infty$ as $m\to\infty$. By definition of multilevel LT sequences,

$$\{LT_{\boldsymbol{n}}^{\boldsymbol{m}}(a_i,f_i)\}_n\xrightarrow{\text{a.c.s.}}\{A_{\boldsymbol{n}}^{(i)}\}_n,\qquad i=1,\ldots,p.$$

By Theorem 2.9,

$$\left\{\Re\left(\sum_{i=1}^{p}LT_{\boldsymbol{n}}^{\boldsymbol{m}}(a_i,f_i)\right)\right\}_n\xrightarrow{\text{a.c.s.}}\left\{\Re\left(\sum_{i=1}^{p}A_{\boldsymbol{n}}^{(i)}\right)\right\}_n.$$

By Theorem 4.3, for every $F\in C_c(\mathbb{C})$ we have

$$\lim_{n\to\infty}\frac{1}{N(\boldsymbol{n})}\sum_{r=1}^{N(\boldsymbol{n})}F\left(\lambda_r\left(\Re\left(\sum_{i=1}^{p}LT_{\boldsymbol{n}}^{\boldsymbol{m}}(a_i,f_i)\right)\right)\right)=\phi_{\boldsymbol{m}}(F),$$

$$\lim_{m\to\infty}\phi_{\boldsymbol{m}}(F)=\phi(F)=\frac{1}{(2\pi)^d}\int_{[0,1]^d\times[-\pi,\pi]^d}F\left(\Re\left(\sum_{i=1}^{p}a_i(\mathbf{x})f_i(\boldsymbol{\theta})\right)\right)\mathrm{d}\mathbf{x}\mathrm{d}\boldsymbol{\theta}.$$

By Theorem 2.7,

$$\left\{\Re\left(\sum_{i=1}^{p} A_{\boldsymbol{n}}^{(i)}\right)\right\}_n \sim_\lambda \phi.$$

Since $\phi = \phi_{\Re(\sum_{i=1}^p a_i \otimes f_i)}$, we obtain

$$\left\{\Re\left(\sum_{i=1}^{p} A_{\boldsymbol{n}}^{(i)}\right)\right\}_n \sim_\lambda \Re\left(\sum_{i=1}^{p} a_i \otimes f_i\right).$$

To conclude the proof, we note that if all the matrices $A_{\boldsymbol{n}}^{(i)}$ are Hermitian then we have $\Re(\sum_{i=1}^p A_{\boldsymbol{n}}^{(i)}) = \sum_{i=1}^p A_{\boldsymbol{n}}^{(i)}$ and $\Re(\sum_{i=1}^p a_i \otimes f_i) = \sum_{i=1}^p a_i \otimes f_i$ a.e. by Proposition 4.3. □

4.5 Algebraic Properties of Multilevel LT Sequences

Proposition 4.4 collects the most elementary algebraic properties of multilevel LT sequences, which follow from Remark 2.6, Eq. (4.10), and the bilinearity of the multilevel LT operator $LT_{\boldsymbol{n}}^{\boldsymbol{m}}(a, f)$ with respect to its arguments a and f (see Sect. 4.1.2, especially Eqs. (4.8) and (4.9)).

Proposition 4.4 *The following properties hold.*

- *If* $\{A_{\boldsymbol{n}}\}_n \sim_{\rm LT} a \otimes f$ *then* $\{A_{\boldsymbol{n}}^*\}_n \sim_{\rm LT} \overline{a} \otimes \overline{f}$.
- *If* $\{A_{\boldsymbol{n}}\}_n \sim_{\rm LT} a \otimes f$ *then* $\{\alpha A_{\boldsymbol{n}}\}_n \sim_{\rm LT} \alpha a \otimes f$ *for all* $\alpha \in \mathbb{C}$.
- *If* $\{A_{\boldsymbol{n}}^{(i)}\}_n \sim_{\rm LT} a \otimes f_i$, $i = 1, \dots, r$, *then* $\{\sum_{i=1}^r A_{\boldsymbol{n}}^{(i)}\}_n \sim_{\rm LT} a \otimes (\sum_{i=1}^r f_i)$.
- *If* $\{A_{\boldsymbol{n}}^{(i)}\}_n \sim_{\rm LT} a_i \otimes f$, $i = 1, \dots, r$, *then* $\{\sum_{i=1}^r A_{\boldsymbol{n}}^{(i)}\}_n \sim_{\rm LT} (\sum_{i=1}^r a_i) \otimes f$.

In Theorem 4.9, we show, under mild assumptions, that the product of multilevel LT sequences is again a multilevel LT sequence with symbol given by the product of the symbols.

Theorem 4.9 *Let* $\{A_{\boldsymbol{n}}\}_n \sim_{\rm LT} a \otimes f$ *and* $\{\tilde{A}_{\boldsymbol{n}}\}_n \sim_{\rm LT} \tilde{a} \otimes \tilde{f}$, *where* $f \in L^p([-\pi, \pi]^d)$, $\tilde{f} \in L^q([-\pi, \pi]^d)$, *and* $1 \le p, q \le \infty$ *are conjugate exponents. Then*

$$\{A_{\boldsymbol{n}}\tilde{A}_{\boldsymbol{n}}\}_n \sim_{\rm LT} a\tilde{a} \otimes f\tilde{f}.$$

Proof By Theorem 4.7 and Proposition 2.2, every multilevel LT sequence is s.u., so in particular $\{A_{\boldsymbol{n}}\}_n$ and $\{\tilde{A}_{\boldsymbol{n}}\}_n$ are s.u. Since, by definition of multilevel LT sequences,

$$\{LT_{\boldsymbol{n}}^{\boldsymbol{m}}(a, f)\}_n \xrightarrow{\text{a.c.s.}} \{A_{\boldsymbol{n}}\}_n \text{ as } \boldsymbol{m} \to \infty,$$
$$\{LT_{\boldsymbol{n}}^{\boldsymbol{m}}(\tilde{a}, \tilde{f})\}_n \xrightarrow{\text{a.c.s.}} \{\tilde{A}_{\boldsymbol{n}}\}_n \text{ as } \boldsymbol{m} \to \infty,$$

Remark 2.6 yields

$$\{LT_{\boldsymbol{n}}^{\boldsymbol{m}}(a, f)LT_{\boldsymbol{n}}^{\boldsymbol{m}}(\tilde{a}, \tilde{f})\}_n \xrightarrow{\text{a.c.s.}} \{A_{\boldsymbol{n}}\tilde{A}_{\boldsymbol{n}}\}_n \text{ as } \boldsymbol{m} \to \infty.$$

Using Proposition 4.1, especially (4.14), we obtain

$$\{LT_{\boldsymbol{n}}^{\boldsymbol{m}}(a\tilde{a}, f\tilde{f})\}_n \xrightarrow{\text{a.c.s.}} \{A_{\boldsymbol{n}}\tilde{A}_{\boldsymbol{n}}\}_n \text{ as } \boldsymbol{m} \to \infty,$$

hence $\{A_{\boldsymbol{n}}\tilde{A}_{\boldsymbol{n}}\}_n \sim_{\text{LT}} a\tilde{a} \otimes f\tilde{f}$. □

As a consequence of Theorems 4.5, 4.6 and 4.9, we immediately obtain the following result.

Theorem 4.10 *If $a : [0, 1]^d \to \mathbb{C}$ is Riemann-integrable and $f \in L^1([-\pi, \pi]^d)$ then $\{D_{\boldsymbol{n}}(a)T_{\boldsymbol{n}}(f)\}_n \sim_{\text{LT}} a \otimes f$ for any sequence $\{\boldsymbol{n} = \boldsymbol{n}(n)\}_n \subseteq \mathbb{N}^d$ such that $\boldsymbol{n} \to \infty$ as $n \to \infty$.*

4.6 Characterizations of Multilevel LT Sequences

Theorem 4.10 shows that, for any a, f as in Definition 4.3, there always exists a matrix-sequence $\{A_{\boldsymbol{n}}\}_n$ such that $\{A_{\boldsymbol{n}}\}_n \sim_{\text{LT}} a \otimes f$. Indeed, it suffices to take $A_{\boldsymbol{n}} = D_{\boldsymbol{n}}(a)T_{\boldsymbol{n}}(f)$. Theorem 4.11 shows that the sequences of the form $\{D_{\boldsymbol{n}}(a)T_{\boldsymbol{n}}(f)\}_n$ play a central role in the world of multilevel LT sequences. Indeed, $\{A_{\boldsymbol{n}}\}_n \sim_{\text{LT}} a \otimes f$ if and only if $A_{\boldsymbol{n}}$ equals $D_{\boldsymbol{n}}(a)T_{\boldsymbol{n}}(f)$ up to a small-rank plus small-norm correction. More precisely, any multilevel LT sequence $\{A_{\boldsymbol{n}}\}_n \sim_{\text{LT}} a \otimes f$ admits the fixed matrix-sequence $\{D_{\boldsymbol{n}}(a)T_{\boldsymbol{n}}(f)\}_n$ as an a.c.s., and, vice versa, any matrix-sequence $\{A_{\boldsymbol{n}}\}_n$ admitting $\{D_{\boldsymbol{n}}(a)T_{\boldsymbol{n}}(f)\}_n$ as an a.c.s. is a multilevel LT sequence with symbol $a \otimes f$. From a topological viewpoint, this means that

$$\{A_{\boldsymbol{n}}\}_n \sim_{\text{LT}} a \otimes f \iff d_{\text{a.c.s.}}(\{A_{\boldsymbol{n}}\}_n, \{D_{\boldsymbol{n}}(a)T_{\boldsymbol{n}}(f)\}_n) = 0.$$

Theorem 4.11 *Let $\{A_{\boldsymbol{n}}\}_n$ be a d-level matrix-sequence, let $a : [0, 1]^d \to \mathbb{C}$ be a Riemann-integrable function and let $f \in L^1([-\pi, \pi]^d)$. The following conditions are equivalent.*

1. $\{A_{\boldsymbol{n}}\}_n \sim_{\text{LT}} a \otimes f$.
2. *For all sequences $\{a_m\}_m$, $\{f_m\}_m$, $\{\{A_{\boldsymbol{n}}^{(m)}\}_n\}_m$ such that*
   - *$a_m : [0, 1]^d \to \mathbb{C}$ is Riemann-integrable and $a_m \to a$ in $L^1([0, 1]^d)$,*
   - *$f_m \in L^1([-\pi, \pi]^d)$ and $f_m \to f$ in $L^1([-\pi, \pi]^d)$,*
   - *$\{A_{\boldsymbol{n}}^{(m)}\}_n \sim_{\text{LT}} a_m \otimes f_m$,*

   we have $\{A_{\boldsymbol{n}}^{(m)}\}_n \xrightarrow{\text{a.c.s.}} \{A_{\boldsymbol{n}}\}_n$.

3. *There exist sequences* $\{a_m\}_m$, $\{f_m\}_m$ *such that*
   - $a_m : [0, 1]^d \to \mathbb{C}$ *is continuous with* $\|a_m\|_\infty \le \|a\|_{L^\infty}$ *for all* m *and* $a_m \to a$ *a.e.,*
   - $f_m : [-\pi, \pi]^d \to \mathbb{C}$ *is a* d*-variate trigonometric polynomial with* $\|f_m\|_\infty \le \operatorname{ess\,sup}_{[-\pi,\pi]^d}|f|$ *for all* m *and* $f_m \to f$ *a.e. and in* $L^1([-\pi, \pi]^d)$,
   - $\{D_{\boldsymbol{n}}(a_m)T_{\boldsymbol{n}}(f_m)\}_{\boldsymbol{n}} \xrightarrow{\text{a.c.s.}} \{A_{\boldsymbol{n}}\}_{\boldsymbol{n}}$.
4. *There exist sequences* $\{a_m\}_m$, $\{f_m\}_m$, $\{\{A_{\boldsymbol{n}}^{(m)}\}_{\boldsymbol{n}}\}_m$ *such that*
   - $a_m : [0, 1]^d \to \mathbb{C}$ *is Riemann-integrable and* $a_m \to a$ *in* $L^1([0, 1]^d)$,
   - $f_m \in L^1([-\pi, \pi]^d)$ *and* $f_m \to f$ *in* $L^1([-\pi, \pi]^d)$,
   - $\{A_{\boldsymbol{n}}^{(m)}\}_{\boldsymbol{n}} \sim_{\rm LT} a_m \otimes f_m$ *and* $\{A_{\boldsymbol{n}}^{(m)}\}_{\boldsymbol{n}} \xrightarrow{\text{a.c.s.}} \{A_{\boldsymbol{n}}\}_{\boldsymbol{n}}$.
5. $\{D_{\boldsymbol{n}}(a)T_{\boldsymbol{n}}(f)\}_{\boldsymbol{n}} \xrightarrow{\text{a.c.s.}} \{A_{\boldsymbol{n}}\}_{\boldsymbol{n}}$.
6. $A_{\boldsymbol{n}} = D_{\boldsymbol{n}}(a)T_{\boldsymbol{n}}(f) + Z_{\boldsymbol{n}}$ *for every* $\boldsymbol{n}$*, where* $\{Z_{\boldsymbol{n}}\}_{\boldsymbol{n}}$ *is zero-distributed.*

Proof (1 $\Longrightarrow$ 2) Let $\{a_m\}_m$, $\{f_m\}_m$, $\{\{A_{\boldsymbol{n}}^{(m)}\}_{\boldsymbol{n}}\}_m$ be sequences with the properties specified in item 2. Since $\{A_{\boldsymbol{n}}^{(m)}\}_{\boldsymbol{n}} \sim_{\rm LT} a_m \otimes f_m$, for every m and every $\boldsymbol{k} \in \mathbb{N}^d$ there is $n_{m,\boldsymbol{k}}$ such that, for $n \ge n_{m,\boldsymbol{k}}$,

$$\begin{gathered} A_{\boldsymbol{n}}^{(m)} = LT_{\boldsymbol{n}}^{\boldsymbol{k}}(a_m, f_m) + R_{\boldsymbol{n},m,\boldsymbol{k}} + N_{\boldsymbol{n},m,\boldsymbol{k}}, \\ \operatorname{rank}(R_{\boldsymbol{n},m,\boldsymbol{k}}) \le c(m, \boldsymbol{k})N(\boldsymbol{n}), \qquad \|N_{\boldsymbol{n},m,\boldsymbol{k}}\| \le \omega(m, \boldsymbol{k}), \end{gathered}$$

where

$$\lim_{\boldsymbol{k}\to\infty} c(m, \boldsymbol{k}) = \lim_{\boldsymbol{k}\to\infty} \omega(m, \boldsymbol{k}) = 0.$$

Moreover, since $\{A_{\boldsymbol{n}}\}_{\boldsymbol{n}} \sim_{\rm LT} a \otimes f$, for every $\boldsymbol{k} \in \mathbb{N}^d$ there is $n_{\boldsymbol{k}}$ such that, for $n \ge n_{\boldsymbol{k}}$,

$$\begin{gathered} A_{\boldsymbol{n}} = LT_{\boldsymbol{n}}^{\boldsymbol{k}}(a, f) + R_{\boldsymbol{n},\boldsymbol{k}} + N_{\boldsymbol{n},\boldsymbol{k}}, \\ \operatorname{rank}(R_{\boldsymbol{n},\boldsymbol{k}}) \le c(\boldsymbol{k})N(\boldsymbol{n}), \qquad \|N_{\boldsymbol{n},\boldsymbol{k}}\| \le \omega(\boldsymbol{k}), \end{gathered}$$

where

$$\lim_{\boldsymbol{k}\to\infty} c(\boldsymbol{k}) = \lim_{\boldsymbol{k}\to\infty} \omega(\boldsymbol{k}) = 0.$$

Hence, for every m, every $\boldsymbol{k} \in \mathbb{N}^d$ and every $n \ge \max(n_{m,\boldsymbol{k}}, n_{\boldsymbol{k}})$,

$$\begin{aligned} A_{\boldsymbol{n}} &= A_{\boldsymbol{n}}^{(m)} + \big[LT_{\boldsymbol{n}}^{\boldsymbol{k}}(a, f) - LT_{\boldsymbol{n}}^{\boldsymbol{k}}(a_m, f_m)\big] \\ &\quad + (R_{\boldsymbol{n},\boldsymbol{k}} - R_{\boldsymbol{n},m,\boldsymbol{k}}) + (N_{\boldsymbol{n},\boldsymbol{k}} - N_{\boldsymbol{n},m,\boldsymbol{k}}), \qquad (4.29) \\ &\operatorname{rank}(R_{\boldsymbol{n},\boldsymbol{k}} - R_{\boldsymbol{n},m,\boldsymbol{k}}) \le (c(\boldsymbol{k}) + c(m, \boldsymbol{k}))N(\boldsymbol{n}), \\ &\|N_{\boldsymbol{n},\boldsymbol{k}} - N_{\boldsymbol{n},m,\boldsymbol{k}}\| \le \omega(\boldsymbol{k}) + \omega(m, \boldsymbol{k}). \end{aligned}$$

Thanks to (4.11), the linearity of the maps (4.8)–(4.9) and Theorem 3.2, we have

$$\begin{aligned}
&\|LT_{\boldsymbol n}^{\boldsymbol k}(a,f)-LT_{\boldsymbol n}^{\boldsymbol k}(a_m,f_m)\|_1\\
&\le \|LT_{\boldsymbol n}^{\boldsymbol k}(a,f-f_m)\|_1+\|LT_{\boldsymbol n}^{\boldsymbol k}(a-a_m,f_m)\|_1\\
&=\sum_{\boldsymbol j=\boldsymbol 1}^{\boldsymbol k}\Bigl|a\Bigl(\frac{\boldsymbol j}{\boldsymbol k}\Bigr)\Bigr|\,\|T_{\lfloor \boldsymbol n/\boldsymbol k\rfloor}(f-f_m)\|_1+\sum_{\boldsymbol j=\boldsymbol 1}^{\boldsymbol k}\Bigl|a\Bigl(\frac{\boldsymbol j}{\boldsymbol k}\Bigr)-a_m\Bigl(\frac{\boldsymbol j}{\boldsymbol k}\Bigr)\Bigr|\,\|T_{\lfloor \boldsymbol n/\boldsymbol k\rfloor}(f_m)\|_1\\
&\le N(\boldsymbol n)\|a\|_\infty\|f-f_m\|_{L^1}+\|f_m\|_{L^1}\frac{N(\boldsymbol n)}{N(\boldsymbol k)}\sum_{\boldsymbol j=\boldsymbol 1}^{\boldsymbol k}\Bigl|a\Bigl(\frac{\boldsymbol j}{\boldsymbol k}\Bigr)-a_m\Bigl(\frac{\boldsymbol j}{\boldsymbol k}\Bigr)\Bigr|\\
&\le\left[\|a\|_\infty\|f-f_m\|_{L^1}+\sup_\ell\|f_\ell\|_{L^1}\frac{1}{N(\boldsymbol k)}\sum_{\boldsymbol j=\boldsymbol 1}^{\boldsymbol k}\Bigl|a\Bigl(\frac{\boldsymbol j}{\boldsymbol k}\Bigr)-a_m\Bigl(\frac{\boldsymbol j}{\boldsymbol k}\Bigr)\Bigr|\right]N(\boldsymbol n);
\end{aligned}\tag{4.30}$$

note that $\|f_\ell\|_{L^1}$ is uniformly bounded with respect to ℓ, because f_ℓ converges to f in $L^1([-\pi,\pi]^d)$. By the Riemann-integrability of $|a-a_m|$, and by the fact that $a_m\to a$ in $L^1([0,1]^d)$ and $f_m\to f$ in $L^1([-\pi,\pi]^d)$, the quantity

$$\varepsilon(m,\boldsymbol k)=\|a\|_\infty\|f-f_m\|_{L^1}+\sup_\ell\|f_\ell\|_{L^1}\frac{1}{N(\boldsymbol k)}\sum_{\boldsymbol j=\boldsymbol 1}^{\boldsymbol k}\Bigl|a\Bigl(\frac{\boldsymbol j}{\boldsymbol k}\Bigr)-a_m\Bigl(\frac{\boldsymbol j}{\boldsymbol k}\Bigr)\Bigr|\tag{4.31}$$

satisfies

$$\begin{aligned}
&\lim_{m\to\infty}\lim_{\boldsymbol k\to\infty}\varepsilon(m,\boldsymbol k)\\
&=\lim_{m\to\infty}\left(\|a\|_\infty\|f-f_m\|_{L^1}+\sup_\ell\|f_\ell\|_{L^1}\int_{[0,1]^d}|a(\mathbf x)-a_m(\mathbf x)|\mathrm d\mathbf x\right)=0.
\end{aligned}\tag{4.32}$$

Choose any sequence $\{\boldsymbol k(m)\}_m\subseteq\mathbb N^d$ such that $\boldsymbol k(m)\to\infty$ as $m\to\infty$ and

$$\lim_{m\to\infty}c(m,\boldsymbol k(m))=\lim_{m\to\infty}\omega(m,\boldsymbol k(m))=\lim_{m\to\infty}\varepsilon(m,\boldsymbol k(m))=0.$$

By (4.29)–(4.30), for every m and every $\boldsymbol n\ge\max(\boldsymbol n_{m,\boldsymbol k(m)},\boldsymbol n_{\boldsymbol k(m)})$ we have

$$\begin{aligned}
A_{\boldsymbol n}&=A_{\boldsymbol n}^{(m)}+\bigl[LT_{\boldsymbol n}^{\boldsymbol k(m)}(a,f)-LT_{\boldsymbol n}^{\boldsymbol k(m)}(a_m,f_m)\bigr]\\
&\qquad+(R_{\boldsymbol n,\boldsymbol k(m)}-R_{\boldsymbol n,m,\boldsymbol k(m)})+(N_{\boldsymbol n,\boldsymbol k(m)}-N_{\boldsymbol n,m,\boldsymbol k(m)}),\\
&\mathrm{rank}(R_{\boldsymbol n,\boldsymbol k(m)}-R_{\boldsymbol n,m,\boldsymbol k(m)})\le(c(\boldsymbol k(m))+c(m,\boldsymbol k(m)))N(\boldsymbol n),\\
&\|N_{\boldsymbol n,\boldsymbol k(m)}-N_{\boldsymbol n,m,\boldsymbol k(m)}\|\le\omega(\boldsymbol k(m))+\omega(m,\boldsymbol k(m)),\\
&\|LT_{\boldsymbol n}^{\boldsymbol k(m)}(a,f)-LT_{\boldsymbol n}^{\boldsymbol k(m)}(a_m,f_m)\|_1\le\varepsilon(m,\boldsymbol k(m))N(\boldsymbol n).
\end{aligned}\tag{4.33}$$

By [22, Lemma 5.6], we can decompose $LT_{\boldsymbol{n}}^{\boldsymbol{k}(m)}(a, f) - LT_{\boldsymbol{n}}^{\boldsymbol{k}(m)}(a_m, f_m)$ as the sum of a small-rank term $\hat{R}_{\boldsymbol{n},m}$, with rank bounded by $\sqrt{\varepsilon(m, \boldsymbol{k}(m))}\, N(\boldsymbol{n})$, plus a small-norm term $\hat{N}_{\boldsymbol{n},m}$, with norm bounded by $\sqrt{\varepsilon(m, \boldsymbol{k}(m))}$. We then infer from (4.33) that $\{A_{\boldsymbol{n}}^{(m)}\}_n \xrightarrow{\text{a.c.s.}} \{A_{\boldsymbol{n}}\}_n$. This concludes the proof of the implication $1 \Longrightarrow 2$.

($2 \Longrightarrow 3$) Since any Riemann-integrable function is bounded by definition, we have $a \in L^\infty([0, 1]^d)$. Hence, by [22, Theorem 2.2], there exists a sequence of continuous functions $a_m : [0, 1]^d \to \mathbb{C}$ such that $\|a_m\|_\infty \le \|a\|_{L^\infty}$ for all m and $a_m \to a$ a.e. The sequence $\{a_m\}_m$ satisfies the properties in item 3.

Since $f \in L^1([-\pi, \pi]^d)$, by Lemma 2.3 there exists a sequence of d-variate trigonometric polynomials $\{f_m\}_m$ such that $\|f_m\|_\infty \le \operatorname{ess\,sup}_{[-\pi,\pi]^d}|f|$ for all m and $f_m \to f$ a.e. and in $L^1([-\pi, \pi]^d)$. The sequence $\{f_m\}_m$ satisfies the properties in item 3.

By item 2 and Theorem 4.10, we have $\{D_{\boldsymbol{n}}(a_m)T_{\boldsymbol{n}}(f_m)\}_n \xrightarrow{\text{a.c.s.}} \{A_{\boldsymbol{n}}\}_n$, and the proof is complete.

($3 \Longrightarrow 4$) Simply note that, under the assumptions in item 3, $a_m \to a$ in $L^1([0, 1]^d)$ by the dominated convergence theorem and $\{D_{\boldsymbol{n}}(a_m)T_{\boldsymbol{n}}(f_m)\}_n \sim_{\rm LT} a_m \otimes f_m$ by Theorem 4.10.

($4 \Longrightarrow 1$) Since $\{A_{\boldsymbol{n}}^{(m)}\}_n \sim_{\rm LT} a_m \otimes f_m$, for every m and every $\boldsymbol{k} \in \mathbb{N}^d$ there is $n_{m,\boldsymbol{k}}$ such that, for $n \ge n_{m,\boldsymbol{k}}$,

$$\begin{gathered} A_{\boldsymbol{n}}^{(m)} = LT_{\boldsymbol{n}}^{\boldsymbol{k}}(a_m, f_m) + R_{\boldsymbol{n},m,\boldsymbol{k}} + N_{\boldsymbol{n},m,\boldsymbol{k}}, \\ \operatorname{rank}(R_{\boldsymbol{n},m,\boldsymbol{k}}) \le c(m, \boldsymbol{k})N(\boldsymbol{n}), \qquad \|N_{\boldsymbol{n},m,\boldsymbol{k}}\| \le \omega(m, \boldsymbol{k}), \end{gathered}$$

where

$$\lim_{\boldsymbol{k}\to\infty} c(m, \boldsymbol{k}) = \lim_{\boldsymbol{k}\to\infty} \omega(m, \boldsymbol{k}) = 0.$$

Since $\{A_{\boldsymbol{n}}^{(m)}\}_n \xrightarrow{\text{a.c.s.}} \{A_{\boldsymbol{n}}\}_n$, for every m there exists n_m such that, for $n \ge n_m$,

$$\begin{gathered} A_{\boldsymbol{n}} = A_{\boldsymbol{n}}^{(m)} + R_{\boldsymbol{n},m} + N_{\boldsymbol{n},m}, \\ \operatorname{rank}(R_{\boldsymbol{n},m}) \le c(m)N(\boldsymbol{n}), \qquad \|N_{\boldsymbol{n},m}\| \le \omega(m), \end{gathered}$$

where

$$\lim_{m\to\infty} c(m) = \lim_{m\to\infty} \omega(m) = 0.$$

Thus, for every m, every $\boldsymbol{k} \in \mathbb{N}^d$ and every $n \ge \max(n_m, n_{m,\boldsymbol{k}})$,

$$\begin{aligned} A_{\boldsymbol{n}} = LT_{\boldsymbol{n}}^{\boldsymbol{k}}(a, f) &+ \big[LT_{\boldsymbol{n}}^{\boldsymbol{k}}(a_m, f_m) - LT_{\boldsymbol{n}}^{\boldsymbol{k}}(a, f)\big] \\ &+ (R_{\boldsymbol{n},m} + R_{\boldsymbol{n},m,\boldsymbol{k}}) + (N_{\boldsymbol{n},m} + N_{\boldsymbol{n},m,\boldsymbol{k}}), \\ \operatorname{rank}(R_{\boldsymbol{n},m} &+ R_{\boldsymbol{n},m,\boldsymbol{k}}) \le (c(m) + c(m, \boldsymbol{k}))N(\boldsymbol{n}), \\ \|N_{\boldsymbol{n},m} &+ N_{\boldsymbol{n},m,\boldsymbol{k}}\| \le \omega(m) + \omega(m, \boldsymbol{k}), \\ \|LT_{\boldsymbol{n}}^{\boldsymbol{k}}(a_m, f_m) &- LT_{\boldsymbol{n}}^{\boldsymbol{k}}(a, f)\|_1 \le \varepsilon(m, \boldsymbol{k})N(\boldsymbol{n}), \end{aligned}$$

where in the last inequalities we used (4.30); the quantity $\varepsilon(m, \boldsymbol{k})$ is defined in (4.31) and satisfies (4.32). Let $\{m(\boldsymbol{k})\}_{\boldsymbol{k}\in\mathbb{N}^d}$ be a family of indices such that $m(\boldsymbol{k}) \to \infty$ as $\boldsymbol{k} \to \infty$ and

$$\lim_{\boldsymbol{k}\to\infty} \varepsilon(m(\boldsymbol{k}), \boldsymbol{k}) = \lim_{\boldsymbol{k}\to\infty} c(m(\boldsymbol{k}), \boldsymbol{k}) = \lim_{\boldsymbol{k}\to\infty} \omega(m(\boldsymbol{k}), \boldsymbol{k}) = 0.$$

Note that such a sequence exists by Lemma 4.1 (apply the lemma with $x(m, \boldsymbol{k}) = \varepsilon(m, \boldsymbol{k}) + c(m, \boldsymbol{k}) + \omega(m, \boldsymbol{k})$). Then, for every $\boldsymbol{k} \in \mathbb{N}^d$ and $n \geq \max(n_{m(\boldsymbol{k})}, n_{m(\boldsymbol{k}),\boldsymbol{k}})$,

$$\begin{aligned}
A_{\boldsymbol{n}} &= LT_{\boldsymbol{n}}^{\boldsymbol{k}}(a, f) + \left[LT_{\boldsymbol{n}}^{\boldsymbol{k}}(a_{m(\boldsymbol{k})}, f_{m(\boldsymbol{k})}) - LT_{\boldsymbol{n}}^{\boldsymbol{k}}(a, f)\right] \\
&\quad + (R_{\boldsymbol{n},m(\boldsymbol{k})} + R_{\boldsymbol{n},m(\boldsymbol{k}),\boldsymbol{k}}) + (N_{\boldsymbol{n},m(\boldsymbol{k})} + N_{\boldsymbol{n},m(\boldsymbol{k}),\boldsymbol{k}}), \\
&\mathrm{rank}(R_{\boldsymbol{n},m(\boldsymbol{k})} + R_{\boldsymbol{n},m(\boldsymbol{k}),\boldsymbol{k}}) \leq (c(m(\boldsymbol{k})) + c(m(\boldsymbol{k}), \boldsymbol{k}))N(\boldsymbol{n}), \\
&\|N_{\boldsymbol{n},m(\boldsymbol{k})} + N_{\boldsymbol{n},m(\boldsymbol{k}),\boldsymbol{k}}\| \leq \omega(m(\boldsymbol{k})) + \omega(m(\boldsymbol{k}), \boldsymbol{k}), \\
&\|LT_{\boldsymbol{n}}^{\boldsymbol{k}}(a_{m(\boldsymbol{k})}, f_{m(\boldsymbol{k})}) - LT_{\boldsymbol{n}}^{\boldsymbol{k}}(a, f)\|_1 \leq \varepsilon(m(\boldsymbol{k}), \boldsymbol{k})N(\boldsymbol{n}).
\end{aligned}$$

The use of [22, Lemma 5.6] allows one to decompose $LT_{\boldsymbol{n}}^{\boldsymbol{k}}(a_{m(\boldsymbol{k})}, f_{m(\boldsymbol{k})}) - LT_{\boldsymbol{n}}^{\boldsymbol{k}}(a, f)$ as the sum of a small-rank term $\hat{R}_{\boldsymbol{n},\boldsymbol{k}}$, with rank bounded by $\sqrt{\varepsilon(m(\boldsymbol{k}), \boldsymbol{k})}\, N(\boldsymbol{n})$, plus a small-norm term $\hat{N}_{\boldsymbol{n},\boldsymbol{k}}$, with norm bounded by $\sqrt{\varepsilon(m(\boldsymbol{k}), \boldsymbol{k})}$. We therefore infer that $\{LT_{\boldsymbol{n}}^{\boldsymbol{k}}(a, f)\}_n \xrightarrow{\text{a.c.s.}} \{A_{\boldsymbol{n}}\}_n$ as $\boldsymbol{k} \to \infty$, i.e., $\{A_{\boldsymbol{n}}\}_n \sim_{\mathrm{LT}} a \otimes f$. This concludes the proof of the implication $4 \Longrightarrow 1$.

($5 \Longleftrightarrow 6$) Item 5 is equivalent to saying that $\{O_{N(\boldsymbol{n})}\}_n \xrightarrow{\text{a.c.s.}} \{A_{\boldsymbol{n}} - D_{\boldsymbol{n}}(a)T_{\boldsymbol{n}}(f)\}_n$, which, by Theorem 4.4, is equivalent to saying that $\{A_{\boldsymbol{n}} - D_{\boldsymbol{n}}(a)T_{\boldsymbol{n}}(f)\}_n$ is zero-distributed.

($2 \Longrightarrow 5$) Obvious (take $a_m = a$, $f_m = f$ and $A_{\boldsymbol{n}}^{(m)} = D_{\boldsymbol{n}}(a)T_{\boldsymbol{n}}(f)$).

($5 \Longrightarrow 4$) Obvious (take $a_m = a$, $f_m = f$ and $A_{\boldsymbol{n}}^{(m)} = D_{\boldsymbol{n}}(a)T_{\boldsymbol{n}}(f)$). □

Chapter 5
Multilevel Generalized Locally Toeplitz Sequences

In this chapter we develop the multivariate version of the theory of GLT sequences, also known as the theory of multilevel GLT sequences, which dates back to the pioneering papers [35, 36] and was recently reviewed and extended in [21]. The topic is presented here on an abstract level, whereas for motivations and insights we refer the reader to [22, pp. 1–3]; see also Chap. 1. We stress that *essentially all the results and proofs contained in this chapter have an exact analog in* [22, Chap. 8], where we dealt with classical (unilevel) GLT sequences. However, we decided to reproduce here all the proofs without omitting details, in order to help the reader become familiar with the multilevel language (especially, the multi-index notation). In this regard, the reader is invited to compare the "multivariate proofs" presented in this chapter with the corresponding "univariate proofs" in [22, Chap. 8], in order to learn the way in which the multilevel language allows one to transfer many results from the univariate to the multivariate case by simply turning some letters (n, i, j, x, θ, etc.) in boldface ($\boldsymbol{n}, \boldsymbol{i}, \boldsymbol{j}, \mathbf{x}, \boldsymbol{\theta}$, etc.).

5.1 Equivalent Definitions of Multilevel GLT Sequences

We first report (a corrected version of) the original definition of multilevel GLT sequences. This definition is formulated in terms of a.c.s. parameterized by a positive $\varepsilon \to 0$ (see Sect. 2.7.5).

Definition 5.1 (*multilevel generalized locally Toeplitz sequence*) Let $\{A_{\boldsymbol{n}}\}_n$ be a d-level matrix-sequence and let $\kappa : [0, 1]^d \times [-\pi, \pi]^d \to \mathbb{C}$ be a measurable function. We say that $\{A_{\boldsymbol{n}}\}_n$ is a (d-level) Generalized Locally Toeplitz (GLT) sequence with *symbol* κ, and we write $\{A_{\boldsymbol{n}}\}_n \sim_{\text{GLT}} \kappa$, if the following condition is met.

C. Garoni and S. Serra-Capizzano, *Generalized Locally Toeplitz Sequences: Theory and Applications*, https://doi.org/10.1007/978-3-030-02233-4_5

For every $\varepsilon > 0$ there exists a finite number of d-level LT sequences $\{A_{\boldsymbol{n}}^{(i,\varepsilon)}\}_n \sim_{\rm LT} a_{i,\varepsilon} \otimes f_{i,\varepsilon}$, $i = 1, \ldots, N_\varepsilon$, such that

- $\sum_{i=1}^{N_\varepsilon} a_{i,\varepsilon} \otimes f_{i,\varepsilon} \to \kappa$ in measure as $\varepsilon \to 0$,
- $\{\sum_{i=1}^{N_\varepsilon} A_{\boldsymbol{n}}^{(i,\varepsilon)}\} \xrightarrow{\text{a.c.s.}} \{A_{\boldsymbol{n}}\}_n$ as $\varepsilon \to 0$.

The symbol κ is sometimes called the *kernel* of $\{A_{\boldsymbol{n}}\}_n$.

In what follows, whenever we write a relation such as $\{A_{\boldsymbol{n}}\}_n \sim_{\rm GLT} \kappa$, it is understood that $\{A_{\boldsymbol{n}}\}_n$ is a d-level matrix-sequence and $\kappa : [0, 1]^d \times [-\pi, \pi]^d \to \mathbb{C}$ is a measurable function, as in Definition 5.1.

Proposition 5.1 provides a straightforward characterization of multilevel GLT sequences, which may be taken as their definition instead of Definition 5.1. Actually, Proposition 5.1 is essentially the same as Definition 5.1, but it is easier to handle, because it is based on the standard a.c.s. notion (see Definition 2.6).

Proposition 5.1 *Let $\{A_{\boldsymbol{n}}\}_n$ be a d-level matrix-sequence and $\kappa : [0, 1]^d \times [-\pi, \pi]^d \to \mathbb{C}$ a measurable function. We have $\{A_{\boldsymbol{n}}\}_n \sim_{\rm GLT} \kappa$ if and only if the following condition is met.*

For every $m \in \mathbb{N}$ there exists a finite number of d-level LT sequences $\{A_{\boldsymbol{n}}^{(i,m)}\}_n \sim_{\rm LT} a_{i,m} \otimes f_{i,m}$, $i = 1, \ldots, N_m$, such that

- $\sum_{i=1}^{N_m} a_{i,m} \otimes f_{i,m} \to \kappa$ *in measure,*
- $\{\sum_{i=1}^{N_m} A_{\boldsymbol{n}}^{(i,m)}\}_n \xrightarrow{\text{a.c.s.}} \{A_{\boldsymbol{n}}\}_n$.

Proof If $\{A_{\boldsymbol{n}}\}_n \sim_{\rm GLT} \kappa$, then the condition of the proposition holds with

$$a_{i,m} = a_{i,\varepsilon(m)}, \qquad f_{i,m} = f_{i,\varepsilon(m)}, \qquad \{A_{\boldsymbol{n}}^{(i,m)}\}_n = \{A_{\boldsymbol{n}}^{(i,\varepsilon(m))}\}_n, \qquad N_m = N_{\varepsilon(m)},$$

where $a_{i,\varepsilon}$, $f_{i,\varepsilon}$, $\{A_{\boldsymbol{n}}^{(i,\varepsilon)}\}_n$, N_ε are as in Definition 5.1 and $\{\varepsilon(m)\}_m$ is any sequence of positive numbers such that $\varepsilon(m) \to 0$.

Conversely, suppose the condition of the proposition holds. Then $\{A_{\boldsymbol{n}}\}_n \sim_{\rm GLT} \kappa$, because the condition of Definition 5.1 holds with

$$a_{i,\varepsilon} = a_{i,m(\varepsilon)}, \qquad f_{i,\varepsilon} = f_{i,m(\varepsilon)}, \qquad \{A_{\boldsymbol{n}}^{(i,\varepsilon)}\}_n = \{A_{\boldsymbol{n}}^{(i,m(\varepsilon))}\}_n, \qquad N_\varepsilon = N_{m(\varepsilon)},$$

where $a_{i,m}$, $f_{i,m}$, $\{A_{\boldsymbol{n}}^{(i,m)}\}_n$, N_m are as in the proposition and $\{m(\varepsilon)\}_{\varepsilon>0} \subseteq \mathbb{N}$ is any family of indices such that $m(\varepsilon) \to \infty$ as $\varepsilon \to 0$. □

Remark 5.1 It is clear that any d-level LT sequence is a d-level GLT sequence. More precisely,

$$\{A_{\boldsymbol{n}}\}_n \sim_{\rm LT} a \otimes f \implies \{A_{\boldsymbol{n}}\}_n \sim_{\rm GLT} a \otimes f.$$

To see this, it suffices to take, in Proposition 5.1,

$$\{A_{\boldsymbol{n}}^{(1,m)}\}_n = \{A_{\boldsymbol{n}}\}_n, \qquad a_{1,m} = a, \qquad f_{1,m} = f, \qquad N_m = 1.$$

5.2 Singular Value and Spectral Distribution of Multilevel GLT Sequences

Theorem 5.1 provides the singular value distribution of a multilevel GLT sequence.

Theorem 5.1 *If* $\{A_{\boldsymbol{n}}\}_n \sim_{\mathrm{GLT}} \kappa$ *then* $\{A_{\boldsymbol{n}}\}_n \sim_\sigma \kappa$.

Proof By Proposition 5.1, there exist LT sequences

$$\{A_{\boldsymbol{n}}^{(i,m)}\}_n \sim_{\mathrm{LT}} a_{i,m} \otimes f_{i,m}, \qquad i = 1, \ldots, N_m,$$

such that

- $\sum_{i=1}^{N_m} a_{i,m} \otimes f_{i,m} \to \kappa$ in measure,
- $\left\{\sum_{i=1}^{N_m} A_{\boldsymbol{n}}^{(i,m)}\right\}_n \xrightarrow{\text{a.c.s.}} \{A_{\boldsymbol{n}}\}_n$.

By Theorem 4.7,

$$\left\{\sum_{i=1}^{N_m} A_{\boldsymbol{n}}^{(i,m)}\right\}_n \sim_\sigma \sum_{i=1}^{N_m} a_{i,m} \otimes f_{i,m}.$$

All the assumptions of Corollary 2.4 are then satisfied and $\{A_{\boldsymbol{n}}\}_n \sim_\sigma \kappa$. □

Using Theorem 5.1, we show in Proposition 5.2 that the symbol of a multilevel GLT sequence is essentially unique. Afterwards, in Proposition 5.3, we show that the symbol of a multilevel GLT sequence formed by Hermitian matrices is real a.e. For the proofs we need the following lemma, which is the most elementary result in the world of the algebraic properties possessed by multilevel GLT sequences. These properties will be investigated in Sect. 5.4 and give rise to the so-called multilevel GLT algebra.

Lemma 5.1 *Let* $\{A_{\boldsymbol{n}}\}_n \sim_{\mathrm{GLT}} \kappa$ *and* $\{B_{\boldsymbol{n}}\}_n \sim_{\mathrm{GLT}} \xi$. *Then*

- $\{A_{\boldsymbol{n}}^*\}_n \sim_{\mathrm{GLT}} \overline{\kappa}$,
- $\{\alpha A_{\boldsymbol{n}} + \beta B_{\boldsymbol{n}}\}_n \sim_{\mathrm{GLT}} \alpha\kappa + \beta\xi$ *for all* $\alpha, \beta \in \mathbb{C}$.

Proof It suffices to write the meaning of $\{A_{\boldsymbol{n}}\}_n \sim_{\mathrm{GLT}} \kappa$ and $\{B_{\boldsymbol{n}}\}_n \sim_{\mathrm{GLT}} \xi$ (using the characterization of Proposition 5.1), and to apply the first property of Proposition 4.4 in combination with Theorem 2.9. □

Proposition 5.2 *If* $\{A_{\boldsymbol{n}}\}_n \sim_{\mathrm{GLT}} \kappa$ *and* $\{A_{\boldsymbol{n}}\}_n \sim_{\mathrm{GLT}} \xi$ *then* $\kappa = \xi$ *a.e.*

Proof By Lemma 5.1 we have $\{O_{N(\boldsymbol{n})}\}_n \sim_{\mathrm{GLT}} \kappa - \xi$. Therefore, by Theorem 5.1, for all test functions $F \in C_c(\mathbb{R})$ we have

$$F(0) = \frac{1}{(2\pi)^d} \int_{[0,1]^d \times [-\pi,\pi]^d} F(|\kappa(\mathbf{x}, \boldsymbol{\theta}) - \xi(\mathbf{x}, \boldsymbol{\theta})|)\mathrm{d}\mathbf{x}\mathrm{d}\boldsymbol{\theta}.$$

This means that $\phi_{|\kappa-\xi|} = \phi_0$ and so $|\kappa - \xi| = 0$ a.e. by [22, Remark 2.1]. □

Proposition 5.3 *If* $\{A_{\boldsymbol{n}}\}_n \sim_{\rm GLT} \kappa$ *and the* $A_{\boldsymbol{n}}$ *are Hermitian then* $\kappa \in \mathbb{R}$ *a.e.*

Proof Since the matrices $A_{\boldsymbol{n}}$ are Hermitian, by Lemma 5.1 we have $\{A_{\boldsymbol{n}}\}_n \sim_{\rm GLT} \kappa$ and $\{A_{\boldsymbol{n}}\}_n \sim_{\rm GLT} \overline{\kappa}$. Thus, by Proposition 5.2, $\kappa = \overline{\kappa}$ a.e., i.e., $\kappa \in \mathbb{R}$ a.e. □

Theorem 5.2 provides the spectral distribution of a multilevel GLT sequence formed by Hermitian matrices.

Theorem 5.2 *If* $\{A_{\boldsymbol{n}}\}_n \sim_{\rm GLT} \kappa$ *and the* $A_{\boldsymbol{n}}$ *are Hermitian then* $\{A_{\boldsymbol{n}}\}_n \sim_\lambda \kappa$.

Proof By Proposition 5.1, there exist LT sequences

$$\{A_{\boldsymbol{n}}^{(i,m)}\}_n \sim_{\rm LT} a_{i,m} \otimes f_{i,m}, \qquad i = 1, \ldots, N_m,$$

such that

- $\sum_{i=1}^{N_m} a_{i,m} \otimes f_{i,m} \to \kappa$ in measure,
- $\big\{\sum_{i=1}^{N_m} A_{\boldsymbol{n}}^{(i,m)}\big\}_n \xrightarrow{\text{a.c.s.}} \{A_{\boldsymbol{n}}\}_n$.

Since the matrices $A_{\boldsymbol{n}}$ are Hermitian, we have $A_{\boldsymbol{n}} = \Re(A_{\boldsymbol{n}})$ and, consequently, by Theorem 2.9,

$$\left\{\Re\left(\sum_{i=1}^{N_m} A_{\boldsymbol{n}}^{(i,m)}\right)\right\}_n \xrightarrow{\text{a.c.s.}} \{A_{\boldsymbol{n}}\}_n.$$

By Theorem 4.8,

$$\left\{\Re\left(\sum_{i=1}^{N_m} A_{\boldsymbol{n}}^{(i,m)}\right)\right\}_n \sim_\lambda \Re\left(\sum_{i=1}^{N_m} a_{i,m} \otimes f_{i,m}\right).$$

The function κ is real a.e. by Proposition 5.3, so from the convergence in measure $\sum_{i=1}^{N_m} a_{i,m} \otimes f_{i,m} \to \kappa$ we get

$$\Re\left(\sum_{i=1}^{N_m} a_{i,m} \otimes f_{i,m}\right) \to \kappa \text{ in measure.}$$

All the assumptions of Corollary 2.5 are then satisfied and $\{A_{\boldsymbol{n}}\}_n \sim_\lambda \kappa$. □

Remark 5.2 In view of Lemma 5.1, it is clear that Theorem 4.7 is a particular case of Theorem 5.1, and Theorem 4.8 is a particular case of Theorem 5.2.

We end this section with a spectral distribution result for multilevel GLT sequences formed by perturbed Hermitian matrices.

Theorem 5.3 *Suppose* $\{A_{\boldsymbol{n}}\}_n \sim_{\rm GLT} \kappa$ *and* $A_{\boldsymbol{n}} = X_{\boldsymbol{n}} + Y_{\boldsymbol{n}}$, *where*

1. *every* $X_{\boldsymbol{n}}$ *is Hermitian,*
2. $\|X_{\boldsymbol{n}}\|, \|Y_{\boldsymbol{n}}\| \le C$ *for all* n, *where* C *is a constant independent of* n,
3. $\|Y_{\boldsymbol{n}}\|_1 = o(N(\boldsymbol{n}))$ *as* $n \to \infty$.

Then $\{A_{\boldsymbol{n}}\}_n \sim_\lambda \kappa$.

Proof $\{Y_{\boldsymbol{n}}\}_n$ is zero-distributed by Theorem 2.3, so $\{Y_{\boldsymbol{n}}\}_n \sim_{\rm GLT} 0$ by Theorem 4.4. Since $X_{\boldsymbol{n}} = A_{\boldsymbol{n}} - Y_{\boldsymbol{n}}$ and the matrices $X_{\boldsymbol{n}}$ are Hermitian, we have $\{X_{\boldsymbol{n}}\}_n \sim_{\rm GLT} \kappa$ by Lemma 5.1 and $\{X_{\boldsymbol{n}}\}_n \sim_\lambda \kappa$ by Theorem 5.2. All the assumptions of Corollary 2.3 are then satisfied and the thesis follows. □

5.3 Approximation Results for Multilevel GLT Sequences

Theorem 5.4 is the main approximation result for multilevel GLT sequences. It is the same as Corollaries 2.4 and 2.5 with "$\sim_\sigma$" and "$\sim_\lambda$" replaced by "$\sim_{\rm GLT}$". As we shall see, it is particularly useful to show that a given matrix-sequence $\{A_{\boldsymbol{n}}\}_n$ is a multilevel GLT sequence.

Theorem 5.4 *Let* $\{A_{\boldsymbol{n}}\}_n$ *be a* d*-level matrix-sequence and let* $\kappa : [0,1]^d \times [-\pi,\pi]^d \to \mathbb{C}$ *be a measurable function. Suppose that*

1. $\{B_{\boldsymbol{n},m}\}_n \sim_{\rm GLT} \kappa_m$ *for every* m,
2. $\{B_{\boldsymbol{n},m}\}_n \xrightarrow{\text{a.c.s.}} \{A_{\boldsymbol{n}}\}_n$,
3. $\kappa_m \to \kappa$ *in measure.*

Then $\{A_{\boldsymbol{n}}\}_n \sim_{\rm GLT} \kappa$.

Proof Since $\{B_{\boldsymbol{n},m}\}_n \sim_{\rm GLT} \kappa_m$, Proposition 5.1 ensures that, for every m, k, there exists a finite number of d-level LT sequences $\{A_{\boldsymbol{n},m}^{(i,k)}\}_n \sim_{\rm LT} a_{i,k,m} \otimes f_{i,k,m}$, $i = 1, \ldots, N_{k,m}$, such that

- $\sum_{i=1}^{N_{k,m}} a_{i,k,m} \otimes f_{i,k,m} \to \kappa_m$ in measure as $k \to \infty$,
- $\big\{\sum_{i=1}^{N_{k,m}} A_{\boldsymbol{n},m}^{(i,k)}\big\}_n \xrightarrow{\text{a.c.s.}} \{B_{\boldsymbol{n},m}\}_n$ as $k \to \infty$.

Hence, for every m, k there exists $n_{k,m}$ such that, for $n \ge n_{k,m}$,

$$B_{\boldsymbol{n},m} = \sum_{i=1}^{N_{k,m}} A_{\boldsymbol{n},m}^{(i,k)} + R_{\boldsymbol{n},k,m} + N_{\boldsymbol{n},k,m},$$

$$\text{rank}(R_{\boldsymbol{n},k,m}) \le c(k,m)N(\boldsymbol{n}), \qquad \|N_{\boldsymbol{n},k,m}\| \le \omega(k,m),$$

where

$$\lim_{k\to\infty} c(k,m) = \lim_{k\to\infty} \omega(k,m) = 0.$$

Let $\{\delta_m\}_m$ be a sequence of positive numbers such that $\delta_m \to 0$. Since $\sum_{i=1}^{N_{k,m}} a_{i,k,m} \otimes f_{i,k,m} \to \kappa_m$ in measure as $k \to \infty$, for every m we have

$$\mu(m, k, \delta_m) = \mu_{2d}\left\{\left|\sum_{i=1}^{N_{k,m}} a_{i,k,m} \otimes f_{i,k,m} - \kappa_m\right| \geq \delta_m\right\} \to 0 \text{ as } k \to \infty.$$

Now we recall that $\{B_{\boldsymbol{n},m}\}_n \xrightarrow{\text{a.c.s.}} \{A_{\boldsymbol{n}}\}_n$: for every m there exists n_m such that, for $n \geq n_m$,

$$A_{\boldsymbol{n}} = B_{\boldsymbol{n},m} + R_{\boldsymbol{n},m} + N_{\boldsymbol{n},m},$$
$$\text{rank}(R_{\boldsymbol{n},m}) \leq c(m)N(\boldsymbol{n}), \qquad \|N_{\boldsymbol{n},m}\| \leq \omega(m),$$

where

$$\lim_{m\to\infty} c(m) = \lim_{m\to\infty} \omega(m) = 0.$$

It follows that, for every m, k and every $n \geq \max(n_m, n_{k,m})$,

$$A_{\boldsymbol{n}} = \sum_{i=1}^{N_{k,m}} A_{\boldsymbol{n},m}^{(i,k)} + (R_{\boldsymbol{n},k,m} + R_{\boldsymbol{n},m}) + (N_{\boldsymbol{n},k,m} + N_{\boldsymbol{n},m}),$$
$$\text{rank}(R_{\boldsymbol{n},k,m} + R_{\boldsymbol{n},m}) \leq (c(k,m) + c(m))N(\boldsymbol{n}),$$
$$\|N_{\boldsymbol{n},k,m} + N_{\boldsymbol{n},m}\| \leq \omega(k,m) + \omega(m).$$

Choose a sequence $\{k_m\}_m$ such that $k_m \to \infty$ and

$$\lim_{m\to\infty} c(k_m, m) = \lim_{m\to\infty} \omega(k_m, m) = \lim_{m\to\infty} \mu(m, k_m, \delta_m) = 0.$$

Then, for every m and every $n \geq \max(n_m, n_{k_m,m})$,

$$A_{\boldsymbol{n}} = \sum_{i=1}^{N_{k_m,m}} A_{\boldsymbol{n},m}^{(i,k_m)} + (R_{\boldsymbol{n},k_m,m} + R_{\boldsymbol{n},m}) + (N_{\boldsymbol{n},k_m,m} + N_{\boldsymbol{n},m}),$$
$$\text{rank}(R_{\boldsymbol{n},k_m,m} + R_{\boldsymbol{n},m}) \leq (c(k_m, m) + c(m))\, N(\boldsymbol{n}),$$
$$\|N_{\boldsymbol{n},k_m,m} + N_{\boldsymbol{n},m}\| \leq \omega(k_m, m) + \omega(m).$$

It follows that

$$\left\{\sum_{i=1}^{N_{k_m,m}} A_{\boldsymbol{n},m}^{(i,k_m)}\right\}_n \xrightarrow{\text{a.c.s.}} \{A_{\boldsymbol{n}}\}_n. \tag{5.1}$$

Moreover,

$$\{A_{\boldsymbol{n},m}^{(i,k_m)}\}_n \sim_{\text{LT}} a_{i,k_m,m} \otimes f_{i,k_m,m} \tag{5.2}$$

for all m and $i = 1, \dots, N_{k_m,m}$, and

$$\sum_{i=1}^{N_{k_m,m}} a_{i,k_m,m} \otimes f_{i,k_m,m} \to \kappa \tag{5.3}$$

in measure as $m \to \infty$. Indeed, for any $\delta > 0$,

$$\begin{aligned}
&\mu_{2d}\left\{\left|\sum_{i=1}^{N_{k_m,m}} a_{i,k_m,m} \otimes f_{i,k_m,m} - \kappa\right| \geq \delta\right\} \\
&\leq \mu_{2d}\left\{\left|\sum_{i=1}^{N_{k_m,m}} a_{i,k_m,m} \otimes f_{i,k_m,m} - \kappa_m\right| \geq \delta/2\right\} + \mu_{2d}\left\{|\kappa_m - \kappa| \geq \delta/2\right\},
\end{aligned}$$

where $\mu_{2d}\{|\kappa_m - \kappa| \geq \delta/2\}$ tends to 0 by assumption (since $\kappa_m \to \kappa$ in measure) and

$$\mu_{2d}\left\{\left|\sum_{i=1}^{N_{k_m,m}} a_{i,k_m,m} \otimes f_{i,k_m,m} - \kappa_m\right| \geq \delta/2\right\} = \mu(m, k_m, \delta/2)$$

tends to 0 because it is eventually less than $\mu(m, k_m, \delta_m)$. In view of (5.1)–(5.3) and Proposition 5.1, we conclude that $\{A_{\boldsymbol{n}}\}_n \sim_{\rm GLT} \kappa$. □

Remark 5.3 (*topological closure of multilevel GLT sequences*) It is interesting to give a topological interpretation of Theorem 5.4, which is completely analogous to the topological interpretation of Corollaries 2.4 and 2.5 given in Remark 2.5. Fix a sequence of d-indices $\{\boldsymbol{n} = \boldsymbol{n}(n)\}_n \subseteq \mathbb{N}^d$ such that $\boldsymbol{n} \to \infty$ as $n \to \infty$, and let

$$\begin{aligned}
\mathscr{E} &= \{\{A_{\boldsymbol{n}}\}_n : A_{\boldsymbol{n}} \in \mathbb{C}^{N(\boldsymbol{n})\times N(\boldsymbol{n})} \text{ for every } n\}, \\
\mathfrak{M}_d &= \{\kappa : [0,1]^d \times [-\pi,\pi]^d \to \mathbb{C} : \kappa \text{ is measurable}\}.
\end{aligned}$$

We have seen in Sect. 2.7.1 and [22, Sect. 2.3.2] that $\mathscr{E}$ (resp., $\mathfrak{M}_d$) is a topological (pseudometric) space with respect to the pseudometric $d_{\text{a.c.s.}}$ (resp., d_{measure}) which induces the a.c.s. topology $\tau_{\text{a.c.s.}}$ (resp., the topology of convergence in measure τ_{measure}). Theorem 5.4 is then equivalent to saying that the set of d-level GLT pairs

$$\mathcal{G} = \left\{(\{A_{\boldsymbol{n}}\}_n, \kappa) \in \mathscr{E} \times \mathfrak{M}_d : \{A_{\boldsymbol{n}}\}_n \sim_{\rm GLT} \kappa\right\}$$

is closed in $\mathscr{E} \times \mathfrak{M}_d$ equipped with the product (pseudometrizable) topology $\tau_{\text{a.c.s.}} \times \tau_{\text{measure}}$ induced, for example, by the pseudometric

$$d_{\text{a.c.s.}\times\text{measure}}((\{A_{\boldsymbol{n}}\}_n, \kappa), (\{B_{\boldsymbol{n}}\}_n, \xi)) = d_{\text{a.c.s.}}(\{A_{\boldsymbol{n}}\}_n, \{B_{\boldsymbol{n}}\}_n) + d_{\text{measure}}(\kappa, \xi);$$

see [22, Exercise 2.2]. Indeed, Theorem 5.4 reads as follows: if a sequence of d-level GLT sequences $\{B_{\boldsymbol{n},m}\}_n \sim_{\rm GLT} \kappa_m$ converges in the a.c.s. topology to another d-level matrix-sequence $\{A_{\boldsymbol{n}}\}_n$ and if the corresponding sequence of symbols κ_m converges in measure to a measurable function κ (i.e., if a sequence of d-level GLT pairs $(\{B_{\boldsymbol{n},m}\}_n, \kappa_m)$ converges to a pair $(\{A_{\boldsymbol{n}}\}_n, \kappa)$ in $\mathscr{E} \times \mathfrak{M}_d$), then $\{A_{\boldsymbol{n}}\}_n \sim_{\rm GLT} \kappa$ (i.e., $(\{A_{\boldsymbol{n}}\}_n, \kappa)$ is a d-level GLT pair).

The approximation result for GLT sequences stated in Theorem 5.4 admits the following converse, which can be interpreted as another approximation result for multilevel GLT sequences.

Theorem 5.5 *Let $\{A_{\boldsymbol{n}}\}_n$ be a d-level matrix-sequence and let $\{\{B_{\boldsymbol{n},m}\}_n\}_m$ be a sequence of d-level matrix-sequences. Suppose that:*

1. $\{A_{\boldsymbol{n}}\}_n \sim_{\rm GLT} \kappa$;
2. $\{B_{\boldsymbol{n},m}\}_n \sim_{\rm GLT} \kappa_m$ *for every m.*

Then $\{B_{\boldsymbol{n},m}\}_n \xrightarrow{\text{a.c.s.}} \{A_{\boldsymbol{n}}\}_n$ if and only if $\kappa_m \to \kappa$ in measure.

Proof Assume that 1–2 hold. Then $\{A_{\boldsymbol{n}} - B_{\boldsymbol{n},m}\}_n \sim_{\rm GLT} \kappa - \kappa_m$ (by Lemma 5.1) and $\{A_{\boldsymbol{n}} - B_{\boldsymbol{n},m}\}_n \sim_\sigma \kappa - \kappa_m$ (by Theorem 5.1). Therefore, if $\kappa_m \to \kappa$ in measure then $\{B_{\boldsymbol{n},m}\}_n \xrightarrow{\text{a.c.s.}} \{A_{\boldsymbol{n}}\}_n$ by Theorem 2.11. Conversely, if $\{B_{\boldsymbol{n},m}\}_n \xrightarrow{\text{a.c.s.}} \{A_{\boldsymbol{n}}\}_n$ then $\kappa_m \to \kappa$ in measure by Proposition 2.5. □

Corollary 5.1 *Let $\{A_{\boldsymbol{n}}\}_n \sim_{\rm GLT} \kappa$. Then, for all functions $a_{i,m}$, $f_{i,m}$, $i = 1, \ldots, N_m$, such that*

- $a_{i,m} : [0, 1]^d \to \mathbb{C}$ *is Riemann-integrable and* $f_{i,m} \in L^1([-\pi, \pi]^d)$,
- $\sum_{i=1}^{N_m} a_{i,m} \otimes f_{i,m} \to \kappa$ *in measure,*

we have $\big\{\sum_{i=1}^{N_m} D_{\boldsymbol{n}}(a_{i,m}) T_{\boldsymbol{n}}(f_{i,m})\big\}_n \xrightarrow{\text{a.c.s.}} \{A_{\boldsymbol{n}}\}_n$. *In particular, $\{A_{\boldsymbol{n}}\}_n$ admits an a.c.s. of the form*

$$\left\{\left\{\sum_{\boldsymbol{j}=-\boldsymbol{N}_m}^{\boldsymbol{N}_m} D_{\boldsymbol{n}}(a_{\boldsymbol{j}}^{(m)})\, T_{\boldsymbol{n}}(\mathrm{e}^{\mathrm{i}\boldsymbol{j}\cdot\boldsymbol{\theta}})\right\}_n\right\}_m, \qquad a_{\boldsymbol{j}}^{(m)} \in C^\infty([0,1]^d), \qquad \boldsymbol{N}_m \in \mathbb{N}^d, \tag{5.4}$$

where $\sum_{\boldsymbol{j}=-\boldsymbol{N}_m}^{\boldsymbol{N}_m} a_{\boldsymbol{j}}^{(m)}(\mathbf{x})\, \mathrm{e}^{\mathrm{i}\boldsymbol{j}\cdot\boldsymbol{\theta}} \to \kappa(\mathbf{x}, \boldsymbol{\theta})$ *a.e.*

Proof Let $a_{i,m}$, $f_{i,m}$, $i = 1, \ldots, N_m$, be functions with the properties specified in the statement of the corollary. Then

$$\left\{\sum_{i=1}^{N_m} D_{\boldsymbol{n}}(a_{i,m}) T_{\boldsymbol{n}}(f_{i,m})\right\}_n \sim_{\rm GLT} \sum_{i=1}^{N_m} a_{i,m} \otimes f_{i,m}$$

by Theorem 4.10 and Lemma 5.1. Therefore, the convergence

$$\left\{\sum_{i=1}^{N_m} D_{\boldsymbol{n}}(a_{i,m})T_{\boldsymbol{n}}(f_{i,m})\right\}_n \xrightarrow{\text{a.c.s.}} \{A_{\boldsymbol{n}}\}_n$$

follows from Theorem 5.5 applied with

$$B_{\boldsymbol{n},m} = \sum_{i=1}^{N_m} D_{\boldsymbol{n}}(a_{i,m})T_{\boldsymbol{n}}(f_{i,m}), \qquad \kappa_m = \sum_{i=1}^{N_m} a_{i,m} \otimes f_{i,m}.$$

To obtain for $\{A_{\boldsymbol{n}}\}_n$ an a.c.s. of the form (5.4), simply use the result of this corollary in combination with Lemma 2.4. □

Remark 5.4 (*topological density in the space of multilevel GLT sequences*) With the notation of Remark 5.3, we recall that the set of d-level GLT pairs

$$\mathcal{G} = \left\{(\{A_{\boldsymbol{n}}\}_n, \kappa) \in \mathscr{E} \times \mathfrak{M}_d : \{A_{\boldsymbol{n}}\}_n \sim_{\text{GLT}} \kappa\right\}$$

is closed in $\mathscr{E} \times \mathfrak{M}_d$ equipped with the product (pseudometrizable) topology $\tau_{\text{a.c.s.}} \times \tau_{\text{measure}}$. Consider the subset of $\mathcal{G}$ consisting of the d-level GLT pairs of the form

$$\left(\sum_{i=1}^{N_m} D_{\boldsymbol{n}}(a_{i,m})T_{\boldsymbol{n}}(f_{i,m}), \sum_{i=1}^{N_m} a_{i,m} \otimes f_{i,m}\right),$$

where $a_{i,m} \in C^\infty([0,1]^d)$, $f_{i,m}$ is a d-variate trigonometric monomial in $\{e^{\mathrm{i}\boldsymbol{j}\cdot\boldsymbol{\theta}} : \boldsymbol{j} \in \mathbb{Z}^d\}$ for all $i = 1, \ldots, N_m$, and $N_m \in \mathbb{N}^d$. Then, according to Corollary 5.1, this subset is dense in $\mathcal{G}$, i.e., its closure in $\mathscr{E} \times \mathfrak{M}_d$ with respect to the topology $\tau_{\text{a.c.s.}} \times \tau_{\text{measure}}$ coincides precisely with $\mathcal{G}$.

5.3.1 *Characterizations of Multilevel GLT Sequences*

The next result is a characterization theorem for multilevel GLT sequences. All the provided characterizations have already been proved before, but it is anyway useful to collect them in a single statement.

Theorem 5.6 *Let $\{A_{\boldsymbol{n}}\}_n$ be a d-level matrix-sequence and let $\kappa : [0,1]^d \times [-\pi,\pi]^d \to \mathbb{C}$ be a measurable function. The following conditions are equivalent.*

1. $\{A_{\boldsymbol{n}}\}_n \sim_{\text{GLT}} \kappa$.
2. *For all sequences $\{\kappa_m\}_m$, $\{\{B_{\boldsymbol{n},m}\}_n\}_m$ such that*
   - $\{B_{\boldsymbol{n},m}\}_n \sim_{\text{GLT}} \kappa_m$ *for every m,*
   - $\kappa_m \to \kappa$ *in measure,*

   we have $\{B_{\boldsymbol{n},m}\}_n \xrightarrow{\text{a.c.s.}} \{A_{\boldsymbol{n}}\}_n$.

3. *There exist functions* $a_{i,m}$, $f_{i,m}$, $i = 1, \dots, N_m$, *such that*
   - $a_{i,m} : [0, 1]^d \to \mathbb{C}$ *belongs to* $C^\infty([0, 1]^d)$ *and* $f_{i,m}$ *is a* d*-varite trigonometric monomial belonging to* $\{e^{\mathrm{i}\boldsymbol{j}\cdot\boldsymbol{\theta}} : \boldsymbol{j} \in \mathbb{Z}^d\}$,
   - $\sum_{i=1}^{N_m} a_{i,m} \otimes f_{i,m} \to \kappa$ *a.e.*,
   - $\big\{\sum_{i=1}^{N_m} D_{\boldsymbol{n}}(a_{i,m})T_{\boldsymbol{n}}(f_{i,m})\big\}_n \xrightarrow{\text{a.c.s.}} \{A_{\boldsymbol{n}}\}_n$.
4. *There exist sequences* $\{\kappa_m\}_m$, $\{\{B_{\boldsymbol{n},m}\}_n\}_m$ *such that*
   - $\{B_{\boldsymbol{n},m}\}_n \sim_{\text{GLT}} \kappa_m$ *for every* m,
   - $\kappa_m \to \kappa$ *in measure*,
   - $\{B_{\boldsymbol{n},m}\}_n \xrightarrow{\text{a.c.s.}} \{A_{\boldsymbol{n}}\}_n$.

Proof The implication $1 \Longrightarrow 2$ follows from Theorem 5.5. The implication $2 \Longrightarrow 3$ follows from the observation that, on the one hand, we can find functions $a_{i,m}$, $f_{i,m}$, $i = 1, \dots, N_m$, with the first two properties specified in item 3 (by Lemma 2.4), and, on the other hand, $\{\sum_{i=1}^{N_m} D_{\boldsymbol{n}}(a_{i,m})T_{\boldsymbol{n}}(f_{i,m})\}_n \sim_{\text{GLT}} \sum_{i=1}^{N_m} a_{i,m} \otimes f_{i,m}$ (by Theorem 4.10 and Lemma 5.1). The implication $3 \Longrightarrow 4$ is obvious (it suffices to take $B_{\boldsymbol{n},m} = \sum_{i=1}^{N_m} D_{\boldsymbol{n}}(a_{i,m})T_{\boldsymbol{n}}(f_{i,m})$ and $\kappa_m = \sum_{i=1}^{N_m} a_{i,m} \otimes f_{i,m}$). Finally, the implication $4 \Longrightarrow 1$ is Theorem 5.4. □

5.3.2 *Sequences of Multilevel Diagonal Sampling Matrices*

We have seen in Sect. 4.3 the three most important examples of multilevel GLT sequences, namely multilevel zero-distributed sequences, multilevel Toeplitz sequences and sequences of multilevel diagonal sampling matrices. Concerning the latter kind of sequences, we proved that $\{D_{\boldsymbol{n}}(a)\}_n \sim_{\text{GLT}} a \otimes 1$ whenever a is Riemann-integrable. From a mathematical viewpoint, however, the GLT relation $\{D_{\boldsymbol{n}}(a)\}_n \sim_{\text{GLT}} a \otimes 1$ makes sense for all measurable functions $a : [0, 1]^d \to \mathbb{C}$, and it is therefore natural to ask whether we can drop the Riemann-integrability assumption. As an application of Theorem 5.4, in Theorem 5.7 we show that the relation $\{D_{\boldsymbol{n}}(a)\}_n \sim_{\text{GLT}} a \otimes 1$ holds for all functions $a : [0, 1]^d \to \mathbb{C}$ that are continuous a.e. in $[0, 1]^d$. Since a function $a : [0, 1]^d \to \mathbb{C}$ is Riemann-integrable if and only if a is bounded and continuous a.e. (see Sect. 2.3), Theorem 5.7 is an extension of Theorem 4.5. More precisely, in Theorem 5.7 we are dropping the boundedness assumption.

Theorem 5.7 *If* $a : [0, 1]^d \to \mathbb{C}$ *is continuous a.e. then* $\{D_{\boldsymbol{n}}(a)\}_n \sim_{\text{GLT}} a \otimes 1$ *for any sequence* $\{\boldsymbol{n} = \boldsymbol{n}(n)\}_n \subseteq \mathbb{N}^d$ *such that* $\boldsymbol{n} \to \infty$ *as* $n \to \infty$.

Proof For an arbitrary a.e. continuous function $a : [0, 1]^d \to \mathbb{C}$, we can write $a = \alpha_+ - \alpha_- + \mathrm{i}\beta_+ - \mathrm{i}\beta_-$, where $\alpha_\pm, \beta_\pm : [0, 1]^d \to \mathbb{R}$ are nonnegative a.e. continuous functions; simply take

$$\alpha_+ = \max(\Re(a), 0), \qquad \alpha_- = -\min(\Re(a), 0),$$
$$\beta_+ = \max(\Im(a), 0), \qquad \beta_- = -\min(\Im(a), 0).$$

Hence, due to Lemma 5.1 and the linearity of $D_{\boldsymbol{n}}(a)$ with respect to its argument a, it suffices to prove the relation $\{D_{\boldsymbol{n}}(a)\}_n \sim_{\rm GLT} a \otimes 1$ in the case where $a : [0, 1]^d \to \mathbb{R}$ is a nonnegative a.e. continuous function.

Let $a : [0, 1]^d \to [0, \infty)$ be a nonnegative a.e. continuous function. Denote by a_m the truncation of a at level m, i.e.,

$$a_m(\mathbf{x}) = \begin{cases} a(\mathbf{x}), & \text{if } a(\mathbf{x}) \le m, \\ m, & \text{if } a(\mathbf{x}) > m. \end{cases}$$

Since a_m is bounded and continuous a.e., a_m is Riemann-integrable, hence

$$\{D_{\boldsymbol{n}}(a_m)\}_n \sim_{\rm GLT} a_m \otimes 1$$

by Theorem 4.5. Moreover, it is clear that $a_m \to a$ pointwise, so

$$a_m \to a \text{ in measure.}$$

We show that

$$\{D_{\boldsymbol{n}}(a_m)\}_n \xrightarrow{\text{a.c.s.}} \{D_{\boldsymbol{n}}(a)\}_n,$$

after which the application of Theorem 5.4 concludes the proof. To show that $\{D_{\boldsymbol{n}}(a_m)\}_n \xrightarrow{\text{a.c.s.}} \{D_{\boldsymbol{n}}(a)\}_n$, we prove that for every m there exists n_m such that, for $n \ge n_m$,

$$\operatorname{rank}(D_{\boldsymbol{n}}(a) - D_{\boldsymbol{n}}(a_m)) = \operatorname{rank}(D_{\boldsymbol{n}}(a - a_m)) \le \varepsilon(m)N(\boldsymbol{n}), \tag{5.5}$$

where $\varepsilon(m) \to 0$ as $m \to \infty$. For any integer $k \ge 1$, consider the partition of $(0, 1]^d$ given by

$$I_{\boldsymbol{i},k} = \left(\frac{\boldsymbol{i}-\mathbf{1}}{2^k}, \frac{\boldsymbol{i}}{2^k}\right] = \left(\frac{i_1-1}{2^k}, \frac{i_1}{2^k}\right] \times \cdots \times \left(\frac{i_d-1}{2^k}, \frac{i_d}{2^k}\right], \quad \boldsymbol{i} = \mathbf{1}, \ldots, 2^k\mathbf{1}, \tag{5.6}$$

and let

$$a_{m,k} = \sum_{\boldsymbol{i}=\mathbf{1}}^{2^k\mathbf{1}} \left(\sup_{\mathbf{y}\in I_{\boldsymbol{i},k}} a_m(\mathbf{y})\right) \chi_{I_{\boldsymbol{i},k}}.$$

For every $\mathbf{x} \in (0, 1]^d$ and every m, k, we have

$$0 \le a_m(\mathbf{x}) \le a_{m,\,k+1}(\mathbf{x}) \le a_{m,k}(\mathbf{x}) \le \sup_{\mathbf{y}\in(0,1]^d} a_m(\mathbf{y}) \le m. \tag{5.7}$$

Since $a \geq 0$ on $[0, 1]^d$, for all m, n, k the rank of $D_{\boldsymbol{n}}(a - a_m)$ is equal to

$$\begin{aligned}
&\operatorname{rank}(D_{\boldsymbol{n}}(a - a_m))\\
&= \#\left\{\boldsymbol{j} \in \{\boldsymbol{1}, \ldots, \boldsymbol{n}\} : a\Big(\frac{\boldsymbol{j}}{\boldsymbol{n}}\Big) - a_m\Big(\frac{\boldsymbol{j}}{\boldsymbol{n}}\Big) \neq 0\right\}\\
&\leq \#\left\{\boldsymbol{j} \in \{\boldsymbol{1}, \ldots, \boldsymbol{n}\} : a\Big(\frac{\boldsymbol{j}}{\boldsymbol{n}}\Big) \geq m\right\}\\
&= \#\left\{\boldsymbol{j} \in \{\boldsymbol{1}, \ldots, \boldsymbol{n}\} : a_m\Big(\frac{\boldsymbol{j}}{\boldsymbol{n}}\Big) = m\right\}\\
&\leq \#\left\{\boldsymbol{j} \in \{\boldsymbol{1}, \ldots, \boldsymbol{n}\} : a_{m,k}\Big(\frac{\boldsymbol{j}}{\boldsymbol{n}}\Big) = m\right\}\\
&= \#\left\{\boldsymbol{j} \in \{\boldsymbol{1}, \ldots, \boldsymbol{n}\} : \frac{\boldsymbol{j}}{\boldsymbol{n}} \in \{a_{m,k} = m\}\right\}\\
&= \#\left\{\mathbf{x} \in \left\{\frac{\boldsymbol{j}}{\boldsymbol{n}} : \boldsymbol{j} = \boldsymbol{1}, \ldots, \boldsymbol{n}\right\} : \mathbf{x} \in \{a_{m,k} = m\}\right\}\\
&= \#\left(\left\{\frac{\boldsymbol{j}}{\boldsymbol{n}} : \boldsymbol{j} = \boldsymbol{1}, \ldots, \boldsymbol{n}\right\} \bigcap \{a_{m,k} = m\}\right)\\
&\leq \mu_d\{a_{m,k} = m\} N(\boldsymbol{n} + 2^k \boldsymbol{1}),
\end{aligned}$$

where the last inequality is justified as follows: the set $\{a_{m,k} = m\}$ is a finite union of squares from (5.6), say $r_{m,k}$ squares, and each of these squares cannot contain more than $(n_1/2^k + 1)(n_2/2^k + 1) \cdots (n_d/2^k + 1) = N(\boldsymbol{n} + 2^k\boldsymbol{1})/2^{dk}$ grid points of $\{\boldsymbol{j}/\boldsymbol{n} : \boldsymbol{j} = \boldsymbol{1}, \ldots, \boldsymbol{n}\}$, implying that $\{a_{m,k} = m\}$ contains at most $r_{m,k} N(\boldsymbol{n} + 2^k\boldsymbol{1})/2^{dk} = \mu_d\{a_{m,k} = m\} N(\boldsymbol{n} + 2^k\boldsymbol{1})$ grid points. By Lemma 2.5, $a_{m,k} \to a_m$ a.e. in $[0, 1]^d$ because a_m is Riemann-integrable. By (5.7), the convergence of $a_{m,k}$ to a_m is monotone. Thus,

$$\lim_{k\to\infty} \mu_d\{a_{m,k} = m\} = \mu_d\left(\bigcap_{k=1}^{\infty}\{a_{m,k} = m\}\right) = \mu_d\{a_m = m\} = \mu_d\{a \geq m\}.$$

For every m we choose k_m such that $\mu_d\{a_{m,k_m} = m\} \leq \mu_d\{a \geq m\} + 1/m$. Then we choose n_m such that, for $n \geq n_m$, the inequality $N(\boldsymbol{n} + 2^{k_m}\boldsymbol{1})/N(\boldsymbol{n}) \leq 2$ holds (this choice is possible because $N(\boldsymbol{n} + 2^{k_m}\boldsymbol{1})/N(\boldsymbol{n}) \to 1$ as $n \to \infty$). With these choices, we see that the inequality (5.5) is satisfied with $\varepsilon(m) = 2(\mu_d\{a \geq m\} + 1/m)$, which converges to 0 as $m \to \infty$. □

5.4 The Multilevel GLT Algebra

We investigate in this section the important algebraic properties possessed by multilevel GLT sequences, which give rise to the so-called (multilevel) GLT algebra. These properties establish that, if $\{A_{\boldsymbol{n}}^{(1)}\}_n, \ldots, \{A_{\boldsymbol{n}}^{(r)}\}_n$ are given multilevel GLT sequences with symbols $\kappa_1, \ldots, \kappa_r$, respectively, and if $A_{\boldsymbol{n}} = \mathrm{ops}(A_{\boldsymbol{n}}^{(1)}, \ldots, A_{\boldsymbol{n}}^{(r)})$ is obtained from $A_{\boldsymbol{n}}^{(1)}, \ldots, A_{\boldsymbol{n}}^{(r)}$ by means of certain operations "ops", then $\{A_{\boldsymbol{n}}\}_n$ is a multilevel GLT sequence with symbol $\kappa = \mathrm{ops}(\kappa_1, \ldots, \kappa_r)$.

Theorem 5.8 *Let* $\{A_{\boldsymbol{n}}\}_n \sim_{\mathrm{GLT}} \kappa$ *and* $\{B_{\boldsymbol{n}}\}_n \sim_{\mathrm{GLT}} \xi$. *Then*

1. $\{A_{\boldsymbol{n}}^*\}_n \sim_{\mathrm{GLT}} \overline{\kappa}$,
2. $\{\alpha A_{\boldsymbol{n}} + \beta B_{\boldsymbol{n}}\}_n \sim_{\mathrm{GLT}} \alpha\kappa + \beta\xi$ *for all* $\alpha, \beta \in \mathbb{C}$,
3. $\{A_{\boldsymbol{n}} B_{\boldsymbol{n}}\}_n \sim_{\mathrm{GLT}} \kappa\xi$.

Proof The first two items have already been settled before (see Lemma 5.1). We prove the third item. By Proposition 5.1, there exist LT sequences

$$\{A_{\boldsymbol{n}}^{(i,m)}\}_n \sim_{\mathrm{LT}} a_{i,m} \otimes f_{i,m}, \qquad i = 1, \ldots, N_m,$$
$$\{B_{\boldsymbol{n}}^{(j,m)}\}_n \sim_{\mathrm{LT}} b_{j,m} \otimes g_{j,m}, \qquad j = 1, \ldots, M_m,$$

such that

- $\sum_{i=1}^{N_m} a_{i,m} \otimes f_{i,m} \to \kappa$ in measure and $\sum_{j=1}^{M_m} b_{j,m} \otimes g_{j,m} \to \xi$ in measure,
- $\big\{\sum_{i=1}^{N_m} A_{\boldsymbol{n}}^{(i,m)}\big\}_n \xrightarrow{\text{a.c.s.}} \{A_{\boldsymbol{n}}\}_n$ and $\big\{\sum_{j=1}^{M_m} B_{\boldsymbol{n}}^{(j,m)}\big\}_n \xrightarrow{\text{a.c.s.}} \{B_{\boldsymbol{n}}\}_n$.

By Theorem 5.6, we may assume that the functions $f_{i,m}, g_{j,m}$ belong to $L^\infty([-\pi,\pi]^d)$ (actually, we might assume much more than this! We might assume that $f_{i,m}, g_{j,m}$ are d-variate trigonometric monomials, that $a_{i,m}, b_{j,m} \in C^\infty([0,1]^d)$, that $\sum_{i=1}^{N_m} a_{i,m} \otimes f_{i,m} \to \kappa$ a.e. and $\sum_{j=1}^{M_m} b_{j,m} \otimes g_{j,m} \to \xi$ a.e., and that $\{A_{\boldsymbol{n}}^{(i,m)}\}_n = \{D_{\boldsymbol{n}}(a_{i,m})T_{\boldsymbol{n}}(f_{i,m})\}_n$ and $\{B_{\boldsymbol{n}}^{(j,m)}\}_n = \{D_{\boldsymbol{n}}(b_{j,m})T_{\boldsymbol{n}}(g_{j,m})\}_n$). By Theorem 5.1 and Proposition 2.2, any multilevel GLT sequence is s.u., so in particular $\{A_{\boldsymbol{n}}\}_n$ and $\{B_{\boldsymbol{n}}\}_n$ are s.u. Thus, by Theorem 2.9,

$$\left\{\left(\sum_{i=1}^{N_m} A_{\boldsymbol{n}}^{(i,m)}\right)\left(\sum_{j=1}^{M_m} B_{\boldsymbol{n}}^{(j,m)}\right)\right\}_n = \left\{\sum_{i=1}^{N_m}\sum_{j=1}^{M_m} A_{\boldsymbol{n}}^{(i,m)} B_{\boldsymbol{n}}^{(j,m)}\right\}_n \xrightarrow{\text{a.c.s.}} \{A_{\boldsymbol{n}} B_{\boldsymbol{n}}\}_n.$$

Since $f_{i,m}, g_{j,m} \in L^\infty([-\pi,\pi]^d)$, by Theorem 4.9 we have

$$\{A_{\boldsymbol{n}}^{(i,m)} B_{\boldsymbol{n}}^{(j,m)}\}_n \sim_{\mathrm{LT}} a_{i,m}b_{j,m} \otimes f_{i,m}g_{j,m}, \qquad i = 1, \ldots, N_m, \qquad j = 1, \ldots, M_m.$$

Finally,

$$\sum_{i=1}^{N_m}\sum_{j=1}^{M_m} a_{i,m}b_{j,m} \otimes f_{i,m}g_{j,m} = \left(\sum_{i=1}^{N_m} a_{i,m} \otimes f_{i,m}\right)\left(\sum_{j=1}^{M_m} b_{j,m} \otimes g_{j,m}\right) \to \kappa\xi$$

in measure by [22, Lemma 2.3]. By Proposition 5.1 we conclude that $\{A_{\boldsymbol{n}}B_{\boldsymbol{n}}\}_n \sim_{\rm GLT} \kappa\xi$. □

Corollary 5.2 *Let* $r, q_1, \ldots, q_r \in \mathbb{N}$, $\alpha_1, \ldots, \alpha_r \in \mathbb{C}$, *and let* $\{A_{\boldsymbol{n}}^{(i,j)}\}_n \sim_{\rm GLT} \kappa_{ij}$ *for* $i = 1, \ldots, r$ *and* $j = 1, \ldots, q_i$. *Then*

$$\left\{\sum_{i=1}^{r} \alpha_i \prod_{j=1}^{q_i} A_{\boldsymbol{n}}^{(i,j)}\right\}_n \sim_{\rm GLT} \sum_{i=1}^{r} \alpha_i \prod_{j=1}^{q_i} \kappa_{ij}.$$

Remark 5.5 (*the multilevel GLT algebra*) Theorem 5.8 is enough to conclude that the set of multilevel GLT sequences is a *-algebra over the complex field $\mathbb{C}$. More precisely, fix a sequence of d-indices $\{\boldsymbol{n} = \boldsymbol{n}(n)\}_n \subseteq \mathbb{N}^d$ such that $\boldsymbol{n} \to \infty$ as $n \to \infty$, and let

$$\mathscr{E} = \left\{\{A_{\boldsymbol{n}}\}_n : A_{\boldsymbol{n}} \in \mathbb{C}^{N(\boldsymbol{n})\times N(\boldsymbol{n})} \text{ for every } n\right\},$$
$$\mathfrak{M}_d = \{\kappa : [0,1]^d \times [-\pi,\pi]^d \to \mathbb{C} : \kappa \text{ is measurable}\}.$$

The space $\mathscr{E}$ is a *-algebra over $\mathbb{C}$ with respect to the natural operations of conjugate transposition, addition, scalar-multiplication and product of matrix-sequences (see (2.44)). By Theorem 5.8, the set of d-level GLT sequences

$$\mathscr{G} = \left\{\{A_{\boldsymbol{n}}\}_n : \{A_{\boldsymbol{n}}\}_n \sim_{\rm GLT} \kappa \text{ for some } \kappa \in \mathfrak{M}_d\right\}$$

is a *-subalgebra of $\mathscr{E}$, which is referred to as the (d-level) GLT algebra. We note that, by Theorems 4.4, 4.6 and 5.7, the d-level GLT algebra contains the algebra generated by d-level zero-distributed sequences, d-level Toeplitz sequences, and sequences of d-level diagonal sampling matrices associated with a.e. continuous functions. Thus, if

$$\begin{aligned}\mathscr{B} = &\left\{\{T_{\boldsymbol{n}}(f)\}_n : f \in L^1([-\pi,\pi]^d)\right\}\\ &\cup \left\{\{D_{\boldsymbol{n}}(a)\}_n : a : [0,1]^d \to \mathbb{C} \text{ is continuous a.e.}\right\}\\ &\cup \left\{\{Z_{\boldsymbol{n}}\}_n : \{Z_{\boldsymbol{n}}\}_n \sim_\sigma 0\right\}\end{aligned}$$

and Algebra($\mathscr{B}$) denotes the subalgebra of $\mathscr{E}$ generated by $\mathscr{B}$, i.e., the smallest subalgebra of $\mathscr{E}$ containing $\mathscr{B}$, then Algebra($\mathscr{B}$) $\subseteq \mathscr{G}$. We also note that $\mathfrak{M}_d$ is a *-algebra over $\mathbb{C}$ with respect to the natural operations of complex conjugation, addition, scalar-multiplication and product of functions. Since $\mathscr{E} \times \mathfrak{M}_d$ is the product of two *-algebras over $\mathbb{C}$, it is itself a *-algebra over $\mathbb{C}$ with respect to the natural pointwise operations

$$\begin{aligned} (\{A_{\boldsymbol{n}}\}_n, \kappa)^* &= (\{A_{\boldsymbol{n}}^*\}_n, \overline{\kappa}), \\ (\{A_{\boldsymbol{n}}\}_n, \kappa) + (\{B_{\boldsymbol{n}}\}_n, \xi) &= (\{A_{\boldsymbol{n}} + B_{\boldsymbol{n}}\}_n, \kappa + \xi), \\ \alpha(\{A_{\boldsymbol{n}}\}_n, \kappa) &= (\{\alpha A_{\boldsymbol{n}}\}_n, \alpha\kappa), \\ (\{A_{\boldsymbol{n}}\}_n, \kappa)(\{B_{\boldsymbol{n}}\}_n, \xi) &= (\{A_{\boldsymbol{n}} B_{\boldsymbol{n}}\}_n, \kappa\xi). \end{aligned} \tag{5.8}$$

By Theorem 5.8, the set of d-level GLT pairs

$$\mathcal{G} = \Big\{(\{A_{\boldsymbol{n}}\}_n, \kappa) \in \mathscr{E} \times \mathfrak{M}_d : \{A_{\boldsymbol{n}}\}_n \sim_{\rm GLT} \kappa\Big\}$$

is a *-subalgebra of $\mathscr{E} \times \mathfrak{M}_d$.

We are going to see in Theorems 5.9 and 5.10 that the multilevel GLT algebra enjoys other nice properties, in addition to those of Theorem 5.8, which make it look like a "big container", closed under any type of "regular" operation.

Theorem 5.9 *If* $\{A_{\boldsymbol{n}}\}_n \sim_{\rm GLT} \kappa$ *and the* $A_{\boldsymbol{n}}$ *are Hermitian then* $\{f(A_{\boldsymbol{n}})\}_n \sim_{\rm GLT} f(\kappa)$ *for any continuous function* $f : \mathbb{C} \to \mathbb{C}$.

Proof Since every $A_{\boldsymbol{n}}$ is Hermitian and $\kappa \in \mathbb{R}$ a.e. (by Proposition 5.3), it suffices to prove the theorem for real continuous functions $f : \mathbb{R} \to \mathbb{R}$. Indeed, suppose we have proved the theorem for this kind of functions and let $f : \mathbb{C} \to \mathbb{C}$ be any continuous complex function. Denote by $\alpha, \beta : \mathbb{R} \to \mathbb{R}$ the real and imaginary parts of the restriction of f to $\mathbb{R}$. Then, α, β are continuous functions such that $f(x) = \alpha(x) + \mathrm{i}\beta(\mathrm{x})$ for all $x \in \mathbb{R}$, and since the eigenvalues of $A_{\boldsymbol{n}}$ are real we have $f(A_{\boldsymbol{n}}) = \alpha(A_{\boldsymbol{n}}) + \mathrm{i}\beta(A_{\boldsymbol{n}})$. In view of the relations $\{\alpha(A_{\boldsymbol{n}})\}_n \sim_{\rm GLT} \alpha(\kappa)$ and $\{\beta(A_{\boldsymbol{n}})\}_n \sim_{\rm GLT} \beta(\kappa)$, Theorem 5.8 yields $\{f(A_{\boldsymbol{n}})\}_n \sim_{\rm GLT} \alpha(\kappa) + \mathrm{i}\beta(\kappa) = \mathrm{f}(\kappa)$.

Let $f : \mathbb{R} \to \mathbb{R}$ be a real continuous function. For each $M > 0$, let $\{p_{m,M}\}_m$ be a sequence of polynomials that converges uniformly to f over $[-M, M]$:

$$\lim_{m\to\infty} \|f - p_{m,M}\|_{\infty,[-M,M]} = 0.$$

Note that such a sequence exists by the Weierstrass theorem; see, e.g., [31, Theorem 7.26]. For every $M > 0$ and every m, n, write

$$f(A_{\boldsymbol{n}}) = p_{m,M}(A_{\boldsymbol{n}}) + f(A_{\boldsymbol{n}}) - p_{m,M}(A_{\boldsymbol{n}}). \tag{5.9}$$

Since any multilevel GLT sequence is s.u. (by Theorem 5.1 and Proposition 2.2), the sequence $\{A_{\boldsymbol{n}}\}_n$ is s.u. Hence, by Remark 2.4, for all $M > 0$ there exists n_M such that, for $n \ge n_M$,

$$A_{\boldsymbol{n}} = \hat{A}_{\boldsymbol{n},M} + \tilde{A}_{\boldsymbol{n},M}, \qquad \operatorname{rank}(\hat{A}_{\boldsymbol{n},m}) \le r(M)N(\boldsymbol{n}), \qquad \|\tilde{A}_{\boldsymbol{n},M}\| \le M,$$

where $\lim_{M\to\infty} r(M) = 0$, the matrices $\hat{A}_{\boldsymbol{n},M}$ and $\tilde{A}_{\boldsymbol{n},M}$ are Hermitian, and for all functions $g : \mathbb{R} \to \mathbb{R}$ we have

$$g(\hat{A}_{\boldsymbol{n},M} + \tilde{A}_{\boldsymbol{n},M}) = g(\hat{A}_{\boldsymbol{n},M}) + g(\tilde{A}_{\boldsymbol{n},M}).$$

Thus, for every $M > 0$, every m and every $\boldsymbol{n} \geq \boldsymbol{n}_M$ we can write

$$\begin{aligned} f(A_{\boldsymbol{n}}) &= p_{m,M}(A_{\boldsymbol{n}}) + f(\hat{A}_{\boldsymbol{n},M}) + f(\tilde{A}_{\boldsymbol{n},M}) - p_{m,M}(\hat{A}_{\boldsymbol{n},M}) - p_{m,M}(\tilde{A}_{\boldsymbol{n},M}) \\ &= p_{m,M}(A_{\boldsymbol{n}}) + (f - p_{m,M})(\hat{A}_{\boldsymbol{n},M}) + (f - p_{m,M})(\tilde{A}_{\boldsymbol{n},M}). \end{aligned} \tag{5.10}$$

The matrix $(f - p_{m,M})(\hat{A}_{\boldsymbol{n},M})$ can be written as the sum of two terms, namely

$$(f - p_{m,M})(\hat{A}_{\boldsymbol{n},M}) = R_{\boldsymbol{n},m,M} + N'_{\boldsymbol{n},m,M},$$

where

$$\begin{aligned} R_{\boldsymbol{n},m,M} &= (f - p_{m,M})(\hat{A}_{\boldsymbol{n},M}) \cdot \chi_{S^c}((f - p_{m,M})(\hat{A}_{\boldsymbol{n},M})), \\ N'_{\boldsymbol{n},m,M} &= (f - p_{m,M})(\hat{A}_{\boldsymbol{n},M}) \cdot \chi_S((f - p_{m,M})(\hat{A}_{\boldsymbol{n},M})), \end{aligned}$$

and S is the singleton $S = \{(f - p_{m,M})(0)\}$. In other words, $R_{\boldsymbol{n},m,M}$ is the matrix obtained from $(f - p_{m,M})(\hat{A}_{\boldsymbol{n},M})$ by setting to 0 all the eigenvalues that are equal to $(f - p_{m,M})(0)$, while $N'_{\boldsymbol{n},m,M}$ is the matrix obtained from $(f - p_{m,M})(\hat{A}_{\boldsymbol{n},M})$ by setting to 0 all the eigenvalues that are different from $(f - p_{m,M})(0)$. Note that

$$\begin{aligned} \operatorname{rank}(R_{\boldsymbol{n},m,M}) &\leq \operatorname{rank}(\hat{A}_{\boldsymbol{n},M}) \leq r(M)N(\boldsymbol{n}), \\ \|N'_{\boldsymbol{n},m,M}\| &\leq |f(0) - p_{m,M}(0)|. \end{aligned}$$

Concerning the matrix $N''_{\boldsymbol{n},m,M} = (f - p_{m,M})(\tilde{A}_{\boldsymbol{n},M})$, the inequality $\|\tilde{A}_{\boldsymbol{n},M}\| \leq M$ yields

$$\|N''_{\boldsymbol{n},m,M}\| \leq \|f - p_{m,M}\|_{\infty,[-M,M]}.$$

Let

$$N_{\boldsymbol{n},m,M} = N'_{\boldsymbol{n},m,M} + N''_{\boldsymbol{n},m,M}.$$

By (5.10), for every $M > 0$, every m and every $\boldsymbol{n} \geq \boldsymbol{n}_M$ we have

$$f(A_{\boldsymbol{n}}) = p_{m,M}(A_{\boldsymbol{n}}) + R_{\boldsymbol{n},m,M} + N_{\boldsymbol{n},m,M},$$

where

$$\begin{aligned} \operatorname{rank}(R_{\boldsymbol{n},m,M}) &\leq r(M)N(\boldsymbol{n}), \\ \|N_{\boldsymbol{n},m,M}\| &\leq \|N'_{\boldsymbol{n},m,M}\| + \|N''_{\boldsymbol{n},m,M}\| \leq 2\|f - p_{m,M}\|_{\infty,[-M,M]}. \end{aligned}$$

Choose a sequence $\{M_m\}_m$ such that

$$M_m \to \infty, \qquad \|f - p_{m,M_m}\|_{\infty,[-M_m,M_m]} \to 0. \tag{5.11}$$

Then, for every m and every $n \geq n_{M_m}$,

$$f(A_{\boldsymbol{n}}) = p_{m,M_m}(A_{\boldsymbol{n}}) + R_{\boldsymbol{n},m,M_m} + N_{\boldsymbol{n},m,M_m},$$
$$\operatorname{rank}(R_{\boldsymbol{n},m,M_m}) \leq r(M_m)N(\boldsymbol{n}), \qquad \|N_{\boldsymbol{n},m,M_m}\| \leq 2\|f - p_{m,M_m}\|_{\infty,[-M_m,M_m]},$$

which implies that

$$\{p_{m,M_m}(A_{\boldsymbol{n}})\}_n \xrightarrow{\text{a.c.s.}} \{f(A_{\boldsymbol{n}})\}_n.$$

Moreover, by Theorem 5.8,

$$\{p_{m,M_m}(A_{\boldsymbol{n}})\}_n \sim_{\text{GLT}} p_{m,M_m}(\kappa).$$

Finally, by (5.11),

$$p_{m,M_m}(\kappa) \to f(\kappa) \text{ a.e.}$$

All the hypotheses of Theorem 5.4 are satisfied and $\{f(A_{\boldsymbol{n}})\}_n \sim_{\text{GLT}} f(\kappa)$. □

The last issue we are interested in is to know if $\{A_{\boldsymbol{n}}^{-1}\}_n \sim_{\text{GLT}} \kappa^{-1}$ in the case where $\{A_{\boldsymbol{n}}\}_n \sim_{\text{GLT}} \kappa$, each $A_{\boldsymbol{n}}$ is invertible, and $\kappa \neq 0$ a.e. (so that κ^{-1} is a well-defined measurable function). More in general, we may ask if $\{A_{\boldsymbol{n}}^{\dagger}\}_n \sim_{\text{GLT}} \kappa^{-1}$ when $\{A_{\boldsymbol{n}}\} \sim_{\text{GLT}} \kappa$ and $\kappa \neq 0$ a.e. The answer to both the previous questions is affirmative, as we are going to see.

Theorem 5.10 *If $\{A_{\boldsymbol{n}}\}_n \sim_{\text{GLT}} \kappa$ and $\kappa \neq 0$ a.e. then $\{A_{\boldsymbol{n}}^{\dagger}\}_n \sim_{\text{GLT}} \kappa^{-1}$.*

Proof Take a sequence of matrix-sequences $\{\{B_{\boldsymbol{n},m}\}_n\}_m$ such that

$$\{B_{\boldsymbol{n},m}\}_m \sim_{\text{GLT}} \xi_m$$

and

$$\xi_m \to \kappa^{-1} \text{ a.e.}$$

Note that a sequence $\{\{B_{\boldsymbol{n},m}\}_n\}_m$ with these properties exists. Indeed, by Lemma 2.4 there exists a sequence $\{\xi_m\}_m$, with ξ_m of the form

$$\xi_m(\mathbf{x}, \boldsymbol{\theta}) = \sum_{\boldsymbol{j}=-\boldsymbol{N}_m}^{\boldsymbol{N}_m} a_{\boldsymbol{j}}^{(m)}(\mathbf{x})\, \mathrm{e}^{\mathrm{i}\boldsymbol{j}\cdot\boldsymbol{\theta}}, \qquad a_{\boldsymbol{j}}^{(m)} \in C^\infty([0,1]^d), \qquad \boldsymbol{N}_m \in \mathbb{N}^d,$$

such that $\xi_m \to \kappa^{-1}$ a.e. Therefore, it suffices to take

$$B_{\boldsymbol{n},m} = \sum_{\boldsymbol{j}=-\boldsymbol{N}_m}^{\boldsymbol{N}_m} D_{\boldsymbol{n}}(a_{\boldsymbol{j}}^{(m)}) T_{\boldsymbol{n}}(\mathrm{e}^{\mathrm{i}\boldsymbol{j}\cdot\boldsymbol{\theta}})$$

and to observe that $\{B_{\boldsymbol{n},m}\}_n \sim_{\rm GLT} \xi_m$ by Theorems 4.10 and 5.8. We show that

$$\{B_{\boldsymbol{n},m}\}_n \xrightarrow{\text{a.c.s.}} \{A_{\boldsymbol{n}}^\dagger\}_n,$$

after which the thesis follows from Theorem 5.4.

By Theorem 5.8 we have $\{B_{\boldsymbol{n},m}A_{\boldsymbol{n}} - I_{N(\boldsymbol{n})}\}_n \sim_{\rm GLT} \xi_m\kappa - 1$, which implies that $\{B_{\boldsymbol{n},m}A_{\boldsymbol{n}} - I_{N(\boldsymbol{n})}\}_n \sim_\sigma \xi_m\kappa - 1$ by Theorem 5.1. Moreover, $\xi_m\kappa - 1 \to 0$ a.e. (and hence also in measure). Thus, by Theorem 2.11, for every m there exists n_m such that, for $n \ge n_m$,

$$\begin{aligned} B_{\boldsymbol{n},m}A_{\boldsymbol{n}} &= I_{N(\boldsymbol{n})} + R_{\boldsymbol{n},m} + N_{\boldsymbol{n},m}, \\ \operatorname{rank}(R_{\boldsymbol{n},m}) &\le c(m)N(\boldsymbol{n}), \qquad \|N_{\boldsymbol{n},m}\| \le \omega(m), \end{aligned} \tag{5.12}$$

where

$$\lim_{m\to\infty} c(m) = \lim_{m\to\infty} \omega(m) = 0.$$

Multiplying (5.12) by $A_{\boldsymbol{n}}^\dagger$, we obtain that, for every m and every $n \ge n_m$,

$$B_{\boldsymbol{n},m}\, A_{\boldsymbol{n}}A_{\boldsymbol{n}}^\dagger = A_{\boldsymbol{n}}^\dagger + (R_{\boldsymbol{n},m} + N_{\boldsymbol{n},m})A_{\boldsymbol{n}}^\dagger. \tag{5.13}$$

Since $\kappa \ne 0$ a.e. by hypothesis, $\{A_{\boldsymbol{n}}\}_n$ is s.v. by Theorem 5.1 and Proposition 2.4. It follows that $\{A_{\boldsymbol{n}}^\dagger\}_n$ is s.u. and so, by Proposition 2.1, for all $M > 0$ there is $\overline{n}_M$ such that, for $n \ge \overline{n}_M$,

$$A_{\boldsymbol{n}}^\dagger = \hat{A}_{\boldsymbol{n},M}^\dagger + \tilde{A}_{\boldsymbol{n},M}^\dagger, \qquad \operatorname{rank}(\hat{A}_{\boldsymbol{n},M}^\dagger) \le r(M)N(\boldsymbol{n}), \qquad \|\tilde{A}_{\boldsymbol{n},M}^\dagger\| \le M,$$

where $\lim_{M\to\infty} r(M) = 0$. Choosing $M_m = (\omega(m))^{-1/2}$, from (5.13) we see that, for every m and every $n \ge \max(n_m, \overline{n}_{M_m})$,

$$\begin{aligned} B_{\boldsymbol{n},m}\, A_{\boldsymbol{n}}A_{\boldsymbol{n}}^\dagger &= A_{\boldsymbol{n}}^\dagger + R'_{\boldsymbol{n},m} + N'_{\boldsymbol{n},m}, \\ \operatorname{rank}(R'_{\boldsymbol{n},m}) &\le (c(m) + r(M_m))N(\boldsymbol{n}), \qquad \|N'_{\boldsymbol{n},m}\| \le (\omega(m))^{1/2}, \end{aligned} \tag{5.14}$$

where $R'_{\boldsymbol{n},m} = R_{\boldsymbol{n},m}A_{\boldsymbol{n}}^\dagger + N_{\boldsymbol{n},m}\hat{A}_{\boldsymbol{n},M_m}^\dagger$ and $N'_{\boldsymbol{n},m} = N_{\boldsymbol{n},m}\tilde{A}_{\boldsymbol{n},M_m}^\dagger$.

If the matrices $A_{\boldsymbol{n}}$ were invertible, then $A_{\boldsymbol{n}}^\dagger = A_{\boldsymbol{n}}^{-1}$ and (5.14) would imply that $\{B_{\boldsymbol{n},m}\}_n \xrightarrow{\text{a.c.s.}} \{A_{\boldsymbol{n}}^\dagger\}_n$. In the general case where the matrices $A_{\boldsymbol{n}}$ are not invertible, the convergence $\{B_{\boldsymbol{n},m}\}_n \xrightarrow{\text{a.c.s.}} \{A_{\boldsymbol{n}}^\dagger\}_n$ will follow again from (5.14) as soon as we have proved the following: for every m there exists $\hat{n}_m$ such that, for $n \ge \hat{n}_m$,

$$A_{\boldsymbol{n}}A_{\boldsymbol{n}}^\dagger = I_{N(\boldsymbol{n})} + S_{\boldsymbol{n}}, \qquad \operatorname{rank}(S_{\boldsymbol{n}}) \le \vartheta(m)N(\boldsymbol{n}),$$

where $\lim_{m\to\infty} \vartheta(m) = 0$. This is easy, because, by definition of $A_{\boldsymbol{n}}^\dagger$, the rank of the matrix $S_{\boldsymbol{n}} = A_{\boldsymbol{n}}A_{\boldsymbol{n}}^\dagger - I_{N(\boldsymbol{n})}$ is given by $\operatorname{rank}(S_{\boldsymbol{n}}) = \#\{i \in \{1,\ldots,N(\boldsymbol{n})\} : \sigma_i(A_{\boldsymbol{n}}) = 0\}$. Hence, the previous claim is a direct consequence of the fact that $\{A_{\boldsymbol{n}}\}_n$ is s.v. □

5.5 Algebraic-Topological Definitions of Multilevel GLT Sequences

Before concluding the theory of multilevel GLT sequences, it is interesting to talk about a couple of possible abstract definitions of multilevel GLT sequences, which are based on the algebraic-topological results obtained in Sects. 5.3 and 5.4.

Fix a sequence $\{\boldsymbol{n}=\boldsymbol{n}(n)\}_n \subseteq \mathbb{N}^d$ such that $\boldsymbol{n}\to\infty$ as $n\to\infty$. Let $\mathscr{E}$ and $\mathfrak{M}_d$ be, respectively, the space of matrix-sequences corresponding to the sequence $\{\boldsymbol{n}=\boldsymbol{n}(n)\}_n$ and the space of measurable functions defined on $[0,1]^d\times[-\pi,\pi]^d$, and let $\mathscr{E}\times\mathfrak{M}_d$ be the product space:

$$\mathscr{E}=\big\{\{A_{\boldsymbol{n}}\}_n:\ A_{\boldsymbol{n}}\in\mathbb{C}^{N(\boldsymbol{n})\times N(\boldsymbol{n})}\text{ for every }n\big\},$$
$$\mathfrak{M}_d=\{\kappa:[0,1]^d\times[-\pi,\pi]^d\to\mathbb{C}:\ \kappa\text{ is measurable}\},$$
$$\mathscr{E}\times\mathfrak{M}_d=\big\{(\{A_{\boldsymbol{n}}\}_n,\kappa):\ \{A_{\boldsymbol{n}}\}_n\in\mathscr{E},\ \kappa\in\mathfrak{M}_d\big\}.$$

As we have seen in Remarks 5.3 and 5.5,

- the space $\mathscr{E}$ is a *-algebra with respect to the natural operations, and it is also a topological (pseudometric) space with respect to the topology $\tau_{\text{a.c.s.}}$ induced by the distance $d_{\text{a.c.s.}}$;
- the space $\mathfrak{M}_d$ is a *-algebra with respect to the natural operations, and it is also a topological (pseudometric) space with respect to the topology τ_{measure} induced by the distance d_{measure};
- the space $\mathscr{E}\times\mathfrak{M}_d$ is a *-algebra with respect to the natural (pointwise) operations, and it is also a topological (pseudometric) space with respect to the topology $\tau_{\text{a.c.s.}}\times\tau_{\text{measure}}$ induced by the distance $d_{\text{a.c.s.}\times\text{measure}}$.

Let $\mathcal{G}$ be the subset of $\mathscr{E}\times\mathfrak{M}_d$ consisting of the d-level GLT pairs, i.e.,

$$\mathcal{G}=\big\{(\{A_{\boldsymbol{n}}\}_n,\kappa):\ \{A_{\boldsymbol{n}}\}_n\sim_{\text{GLT}}\kappa\big\}\subseteq\mathscr{E}\times\mathfrak{M}_d.$$

By Remarks 5.3 and 5.5, $\mathcal{G}$ is a closed *-subalgebra of $\mathscr{E}\times\mathfrak{M}_d$. By Theorems 4.4, 4.6 and 5.7, $\mathcal{G}$ contains the set

$$\begin{aligned}\mathcal{B}=&\big\{(\{T_{\boldsymbol{n}}(f)\}_n,1\otimes f):\ f\in L^1([-\pi,\pi]^d)\big\}\\&\cup\big\{(\{D_{\boldsymbol{n}}(a)\}_n,a\otimes 1):\ a:[0,1]^d\to\mathbb{C}\text{ is continuous a.e.}\big\}\\&\cup\big\{(\{Z_{\boldsymbol{n}}\}_n,0):\ \{Z_{\boldsymbol{n}}\}_n\sim_\sigma 0\big\}.\end{aligned}$$

By Remark 5.4, the algebra generated by $\mathcal{B}$ is dense in $\mathcal{G}$. In conclusion,

> *the set of d-level GLT pairs $\mathcal{G}$ is the closed *-subalgebra of $\mathscr{E}\times\mathfrak{M}_d$ generated by $\mathcal{B}$, i.e., the smallest closed *-subalgebra of $\mathscr{E}\times\mathfrak{M}_d$ containing $\mathcal{B}$.*

Looking more carefully at Remark 5.4, we also note that, if we let

$$\mathcal{C} = \big\{(\{D_{\boldsymbol{n}}(a)\}_n, a \otimes 1) : a \in C^\infty([0,1]^d)\big\} \\ \cup \big\{(\{T_{\boldsymbol{n}}(e^{\mathrm{i}\boldsymbol{j}\cdot\boldsymbol{\theta}})\}_n, 1 \otimes e^{\mathrm{i}\boldsymbol{j}\cdot\boldsymbol{\theta}}) : \boldsymbol{j} \in \mathbb{Z}^d\big\},$$

then

the set of d-level GLT pairs $\mathcal{G}$ is the closure of the subalgebra of $\mathcal{E} \times \mathfrak{M}_d$ generated by $\mathcal{C}$.

One may also decide to start the theory of multilevel GLT sequences from one of these two algebraic-topological definitions instead of the traditional one (Definition 5.1). It should be said, however, that the traditional definition looks more effective to obtain the fundamental singular value and spectral distribution results expressed in Theorems 5.1 and 5.2.

Chapter 6
Summary of the Theory

We conclude the theory of multilevel GLT sequences by providing a *self-contained* summary, which contains *everything one needs to know* in order to understand the applications presented in the next chapter. As mentioned in the preface, assuming the reader possesses the necessary prerequisites, a possible way of reading this book consists in first reading this chapter and the next one, and then coming back to fill the gaps, where "fill the gaps" essentially means "read the proofs of the results reported in this chapter". The latter is substantially equivalent to reading the book. It is assumed that anyone who reads this summary is aware of the notation and terminology used throughout the book, which will be only partially repeated here for the sake of brevity. The reader can find both notation and terminology in Sect. 2.1 and/or in the index at the end.

Multi-index notation. A multi-index $\boldsymbol{i}$ of size d, also called a d-index, is a (row) vector in $\mathbb{Z}^d$; its components are denoted by $i_1, \ldots, i_d$. $\mathbf{0}$, $\mathbf{1}$, $\mathbf{2}$, ... are the vectors of all zeros, all ones, all twos, ... (their size will be clear from the context). For any d-index $\boldsymbol{m}$, we set $N(\boldsymbol{m}) = \prod_{j=1}^d m_j$ and we write $\boldsymbol{m} \to \infty$ to indicate that $\min(\boldsymbol{m}) \to \infty$. The notation $N(\boldsymbol{\alpha}) = \prod_{j=1}^d \alpha_j$ will be used for any vector $\boldsymbol{\alpha}$ with d components and not only for d-indices. If $\boldsymbol{h}, \boldsymbol{k}$ are d-indices, $\boldsymbol{h} \le \boldsymbol{k}$ means that $h_r \le k_r$ for all $r = 1, \ldots, d$. If $\boldsymbol{h}, \boldsymbol{k}$ are d-indices such that $\boldsymbol{h} \le \boldsymbol{k}$, the multi-index (or d-index) range $\boldsymbol{h}, \ldots, \boldsymbol{k}$ is the set $\{\boldsymbol{j} \in \mathbb{Z}^d : \boldsymbol{h} \le \boldsymbol{j} \le \boldsymbol{k}\}$. We assume for this set the standard lexicographic ordering:

$$\Big[\ldots \big[[(j_1, \ldots, j_d)]_{j_d=h_d,\ldots,k_d} \big]_{j_{d-1}=h_{d-1},\ldots,k_{d-1}} \ldots \Big]_{j_1=h_1,\ldots,k_1}.$$

For instance, in the case $d = 2$ the ordering is

$$(h_1, h_2),\ (h_1, h_2+1),\ \ldots,\ (h_1, k_2),\ (h_1+1, h_2),\ (h_1+1, h_2+1),\ \ldots,\ (h_1+1, k_2),$$
$$\ldots\ldots\ldots,\ (k_1, h_2),\ (k_1, h_2+1),\ \ldots,\ (k_1, k_2).$$

C. Garoni and S. Serra-Capizzano, *Generalized Locally Toeplitz Sequences: Theory and Applications*, https://doi.org/10.1007/978-3-030-02233-4_6

When a d-index $\boldsymbol{j}$ varies over a multi-index range $\boldsymbol{h},\dots,\boldsymbol{k}$ (this is often written as $\boldsymbol{j}=\boldsymbol{h},\dots,\boldsymbol{k}$), it is understood that $\boldsymbol{j}$ varies from $\boldsymbol{h}$ to $\boldsymbol{k}$ following the lexicographic ordering. For instance, if $\boldsymbol{m}\in\mathbb{N}^d$ and we write $\mathbf{x}=[x_{\boldsymbol{i}}]_{\boldsymbol{i}=\boldsymbol{1}}^{\boldsymbol{m}}$, then $\mathbf{x}$ is a vector of size $N(\boldsymbol{m})$ whose components $x_{\boldsymbol{i}}$, $\boldsymbol{i}=\boldsymbol{1},\dots,\boldsymbol{m}$, are ordered in accordance with the lexicographic ordering: the first component is $x_{\boldsymbol{1}}=x_{(1,\dots,1,1)}$, the second component is $x_{(1,\dots,1,2)}$, and so on until the last component, which is $x_{\boldsymbol{m}}=x_{(m_1,\dots,m_d)}$. Similarly, if $X=[x_{\boldsymbol{ij}}]_{\boldsymbol{i},\boldsymbol{j}=\boldsymbol{1}}^{\boldsymbol{m}}$, then X is an $N(\boldsymbol{m})\times N(\boldsymbol{m})$ matrix whose components are indexed by a pair of d-indices $\boldsymbol{i},\boldsymbol{j}$, both varying from $\boldsymbol{1}$ to $\boldsymbol{m}$ according to the lexicographic ordering. If $\boldsymbol{h},\boldsymbol{k}$ are d-indices such that $\boldsymbol{h}\le\boldsymbol{k}$, the notation $\sum_{\boldsymbol{j}=\boldsymbol{h}}^{\boldsymbol{k}}$ indicates the summation over all $\boldsymbol{j}$ in $\boldsymbol{h},\dots,\boldsymbol{k}$. If $\boldsymbol{i},\boldsymbol{j}$ are d-indices, $\boldsymbol{i}\preceq\boldsymbol{j}$ means that $\boldsymbol{i}$ precedes (or equals) $\boldsymbol{j}$ in the lexicographic ordering (which is a total ordering on $\mathbb{Z}^d$). Moreover, we define

$$\boldsymbol{i}\wedge\boldsymbol{j}=\begin{cases}\boldsymbol{i}, & \text{if } \boldsymbol{i}\preceq\boldsymbol{j},\\ \boldsymbol{j}, & \text{if } \boldsymbol{i}\succ\boldsymbol{j}.\end{cases}$$

Note that $\boldsymbol{i}\wedge\boldsymbol{j}$ is the minimum among $\boldsymbol{i}$ and $\boldsymbol{j}$ with respect to the lexicographic ordering. Operations involving d-indices that have no meaning in the vector space $\mathbb{Z}^d$ must always be interpreted in the componentwise sense. For instance, $\boldsymbol{np}=(n_1p_1,\dots,n_dp_d)$, $\alpha\boldsymbol{i}/\boldsymbol{j}=(\alpha i_1/j_1,\dots,\alpha i_d/j_d)$ for all $\alpha\in\mathbb{C}$, etc.

Matrix norms. Here is a list of important inequalities involving the p-norms and the Schatten p-norms of matrices.

- **N1.** $\|X\|\le\sqrt{|X|_1|X|_\infty}\le\max(|X|_1,|X|_\infty)$ for all $X\in\mathbb{C}^{m\times m}$.
- **N2.** $\|X\|_1\le\operatorname{rank}(X)\|X\|\le m\|X\|$ for all $X\in\mathbb{C}^{m\times m}$.
- **N3.** $\|X\|_1\le\sum_{i,j=1}^m|x_{ij}|$ for all $X\in\mathbb{C}^{m\times m}$.

We also recall that $\|X\|=\sigma_{\max}(X)=\|X\|_\infty$ for all $X\in\mathbb{C}^{m\times m}$.

Tensor products. If $X\in\mathbb{C}^{m_1\times m_2}$ and $Y\in\mathbb{C}^{\ell_1\times\ell_2}$, the tensor (Kronecker) product of X and Y is the $m_1\ell_1\times m_2\ell_2$ matrix defined by

$$X\otimes Y=\big[x_{ij}Y\big]_{\substack{i=1,\dots,m_1\\ j=1,\dots,m_2}}=\begin{bmatrix}x_{11}Y & \cdots & x_{1m_2}Y\\ \vdots & & \vdots\\ x_{m_11}Y & \cdots & x_{m_1m_2}Y\end{bmatrix}.$$

Here is a list of important properties satisfied by tensor products.

- **P1.** Associativity: $(X\otimes Y)\otimes Z=X\otimes(Y\otimes Z)$ for all matrices X,Y,Z.
- **P2.** Bilinearity: $(\alpha X+\beta Y)\otimes(\gamma W+\eta Z)=\alpha\gamma(X\otimes W)+\alpha\eta(X\otimes Z)+\beta\gamma(Y\otimes W)+\beta\eta(Y\otimes Z)$ for all $\alpha,\beta,\gamma,\eta\in\mathbb{C}$ and for all matrices X,Y,W,Z such that X,Y are summable and W,Z are summable.
- **P3.** $(X\otimes Y)^*=X^*\otimes Y^*$ and $(X\otimes Y)^T=X^T\otimes Y^T$ for all matrices X,Y.
- **P4.** $(X\otimes Y)(W\otimes Z)=(XW)\otimes(YZ)$ for all matrices X,Y,W,Z such that X,W are multipliable and Y,Z are multipliable.
- **P5.** $\|X\otimes Y\|_p=\|X\|_p\|Y\|_p$ for all square matrices X,Y and all $p\in[1,\infty]$.

P6. $\text{rank}(X \otimes Y) = \text{rank}(X)\text{rank}(Y)$ for all matrices X, Y.

P7. If $X_i \in \mathbb{C}^{m_i \times m_i}$ for $i = 1, \ldots, d$ and $\boldsymbol{m} = (m_1, \ldots, m_d)$, then

$$(X_1 \otimes \cdots \otimes X_d)_{\boldsymbol{i}\boldsymbol{j}} = (X_1)_{i_1 j_1} \cdots (X_d)_{i_d j_d}, \qquad \boldsymbol{i}, \boldsymbol{j} = \boldsymbol{1}, \ldots, \boldsymbol{m}.$$

P8. If $X_i, Y_i \in \mathbb{C}^{m_i \times m_i}$ for $i = 1, \ldots, d$ and $\boldsymbol{m} = (m_1, \ldots, m_d)$, then

$$\text{rank}(X_1 \otimes \cdots \otimes X_d - Y_1 \otimes \cdots \otimes Y_d) \le N(\boldsymbol{m}) \sum_{i=1}^{d} \frac{\text{rank}(X_i - Y_i)}{m_i}.$$

Sequences of matrices and multilevel matrix-sequences. A *sequence of matrices* is a sequence of the form $\{A_n\}_n$, where n varies in some infinite subset of $\mathbb{N}$ and A_n is a square matrix of size d_n such that $d_n \to \infty$ as $n \to \infty$. If $\{A_n\}_n$ is a sequence of matrices, with A_n of size d_n, we say that $\{A_n\}_n$ is *sparsely unbounded (s.u.)* if

$$\lim_{M\to\infty} \limsup_{n\to\infty} \frac{\#\{i \in \{1, \ldots, d_n\} : \sigma_i(A_n) > M\}}{d_n} = 0;$$

and we say that $\{A_n\}_n$ is *sparsely vanishing (s.v.)* if

$$\lim_{M\to\infty} \limsup_{n\to\infty} \frac{\#\{i \in \{1, \ldots, d_n\} : \sigma_i(A_n) < 1/M\}}{d_n} = 0.$$

A *d-level matrix-sequence* is a special sequence of matrices of the form $\{A_{\boldsymbol{n}}\}_n$, where:

- n varies in some infinite subset of $\mathbb{N}$;
- $\boldsymbol{n} = \boldsymbol{n}(n) \in \mathbb{N}^d$ and $\boldsymbol{n} \to \infty$ (i.e., $\min(\boldsymbol{n}) \to \infty$) as $n \to \infty$;
- $A_{\boldsymbol{n}}$ is a square matrix of size $N(\boldsymbol{n})$.

Singular value and eigenvalue distribution of a sequence of matrices. Let $\{A_n\}_n$ be a sequence of matrices, with A_n of size d_n, and let $f : D \subset \mathbb{R}^k \to \mathbb{C}$ be a measurable function defined on a set D with $0 < \mu_k(D) < \infty$.

- We say that $\{A_n\}_n$ has a singular value distribution described by f, and we write $\{A_n\}_n \sim_\sigma f$, if

$$\lim_{n\to\infty} \frac{1}{d_n} \sum_{i=1}^{d_n} F(\sigma_i(A_n)) = \frac{1}{\mu_k(D)} \int_D F(|f(\mathbf{x})|)\mathrm{d}\mathbf{x}, \qquad \forall F \in C_c(\mathbb{R}).$$

  In this case, f is called the *singular value symbol* of $\{A_n\}_n$.
- We say that $\{A_n\}_n$ has a spectral (or eigenvalue) distribution described by f, and we write $\{A_n\}_n \sim_\lambda f$, if

$$\lim_{n\to\infty} \frac{1}{d_n} \sum_{i=1}^{d_n} F(\lambda_i(A_n)) = \frac{1}{\mu_k(D)} \int_D F(f(\mathbf{x}))\mathrm{d}\mathbf{x}, \qquad \forall F \in C_c(\mathbb{C}).$$

In this case, f is called the *spectral (or eigenvalue) symbol* of $\{A_n\}_n$.

When we write a relation such as $\{A_n\}_n \sim_\sigma f$ or $\{A_n\}_n \sim_\lambda f$, it is understood that $\{A_n\}_n$ is a sequence of matrices and f is a measurable function defined on a subset D of some $\mathbb{R}^k$ with $0 < \mu_k(D) < \infty$. In what follows, "iff" is an abbreviation of "if and only if".

S1. If $\{A_n\}_n \sim_\sigma f$ then $\{A_n\}_n$ is s.u.
S2. If $\{A_n\}_n \sim_\sigma f$ then $\{A_n\}_n$ is s.v. iff $f \neq 0$ a.e.
S3. If $\{A_n\}_n \sim_\lambda f$ and $\Lambda(A_n) \subseteq S$ for all n then $f \in \overline{S}$ a.e.
S4. If $A_n = X_n + Y_n$ where

- each X_n is Hermitian and $\{X_n\}_n \sim_\lambda f$,
- $\|X_n\|, \|Y_n\| \leq C$ for all n, where C is a constant independent of n,
- $\lim_{n\to\infty} (d_n)^{-1}\|Y_n\|_1 = 0$,

then $\{A_n\}_n \sim_\lambda f$.

Informal meaning. Assuming that f is continuous a.e., the spectral distribution $\{A_n\}_n \sim_\lambda f$ has the following informal meaning: all the eigenvalues of A_n, except possibly for $o(d_n)$ outliers (with d_n being the size of A_n), are approximately equal to the samples of f over a uniform grid in its domain D (for n large enough). For instance, if $k = 1$ and $D = [a, b]$, then, assuming we have no outliers, the eigenvalues of A_n are approximately equal to

$$f\Big(a + i\,\frac{b-a}{d_n}\Big), \qquad i = 1, \dots, d_n,$$

for n large enough. Similarly, if $k = 2$, d_n is a perfect square and $D = [a_1, b_1] \times [a_2, b_2]$, then, assuming we have no outliers, the eigenvalues of A_n are approximately equal to

$$f\Big(a_1 + i_1\,\frac{b_1-a_1}{\sqrt{d_n}},\ a_2 + i_2\,\frac{b_2-a_2}{\sqrt{d_n}}\Big), \qquad i_1, i_2 = 1, \dots, \sqrt{d_n},$$

for n large enough. A completely analogous meaning can also be given for the singular value distribution $\{A_n\}_n \sim_\sigma f$.

Clustering and attraction.

- Let $\{A_n\}_n$ be a sequence of matrices, with A_n of size d_n, and let S be a nonempty subset of $\mathbb{C}$. We say that $\{A_n\}_n$ is weakly clustered at S if

$$\lim_{n\to\infty} \frac{\#\{j \in \{1, \dots, d_n\} : \lambda_j(A_n) \notin D(S, \epsilon)\}}{d_n} = 0, \qquad \forall \epsilon > 0.$$

- Let $\{A_n\}_n$ be a sequence of matrices, with A_n of size d_n, and let $z \in \mathbb{C}$. We say that z strongly attracts the spectrum $\Lambda(A_n)$ with infinite order if, once we have ordered the eigenvalues of A_n according to their distance from z,

$$|\lambda_1(A_n) - z| \le |\lambda_2(A_n) - z| \le \cdots \le |\lambda_{d_n}(A_n) - z|,$$

the following limit relation holds for each fixed $j \ge 1$:

$$\lim_{n\to\infty} |\lambda_j(A_n) - z| = 0.$$

CA 1. If $\{A_n\}_n \sim_\lambda f$ then $\{A_n\}_n$ is weakly clustered at $\mathcal{ER}(f)$ and each $z \in \mathcal{ER}(f)$ strongly attracts $\Lambda(A_n)$ with infinite order.

Zero-distributed sequences. A sequence of matrices $\{Z_n\}_n$ such that $\{Z_n\}_n \sim_\sigma 0$ is referred to as a zero-distributed sequence. In other words, $\{Z_n\}_n$ is zero-distributed iff

$$\lim_{n\to\infty} \frac{1}{d_n} \sum_{i=1}^{d_n} F(\sigma_i(Z_n)) = F(0), \qquad \forall\, F \in C_c(\mathbb{R}),$$

where d_n is the size of Z_n. Given a sequence of matrices $\{Z_n\}_n$, with Z_n of size d_n, the following properties hold. In what follows, we use the natural convention $C/\infty = 0$ for all numbers C.

Z 1. $\{Z_n\}_n \sim_\sigma 0$ iff $Z_n = R_n + N_n$ with $\lim_{n\to\infty} (d_n)^{-1}\mathrm{rank}(R_n) = \lim_{n\to\infty} \|N_n\| = 0$.

Z 2. $\{Z_n\}_n \sim_\sigma 0$ if there is a $p \in [1, \infty]$ such that $\lim_{n\to\infty} (d_n)^{-1/p} \|Z_n\|_p = 0$.

Sequences of multilevel diagonal sampling matrices. If $\boldsymbol{n} \in \mathbb{N}^d$ and $a : [0, 1]^d \to \mathbb{C}$, the $\boldsymbol{n}$th (d-level) diagonal sampling matrix generated by a is the $N(\boldsymbol{n}) \times N(\boldsymbol{n})$ diagonal matrix given by

$$D_{\boldsymbol{n}}(a) = \mathop{\mathrm{diag}}_{\boldsymbol{i}=\boldsymbol{1},\ldots,\boldsymbol{n}} a\Big(\frac{\boldsymbol{i}}{\boldsymbol{n}}\Big).$$

Each d-level matrix-sequence of the form $\{D_{\boldsymbol{n}}(a)\}_n$, with $\boldsymbol{n} = \boldsymbol{n}(n) \to \infty$ as $n \to \infty$, is referred to as a sequence of (d-level) diagonal sampling matrices generated by a.

Multilevel Toeplitz sequences. If $\boldsymbol{n} \in \mathbb{N}^d$ and $f : [-\pi, \pi]^d \to \mathbb{C}$ is a function in $L^1([-\pi, \pi]^d)$, the $\boldsymbol{n}$th (d-level) Toeplitz matrix generated by f is the $N(\boldsymbol{n}) \times N(\boldsymbol{n})$ matrix

$$T_{\boldsymbol{n}}(f) = [f_{\boldsymbol{i}-\boldsymbol{j}}]_{\boldsymbol{i},\boldsymbol{j}=\boldsymbol{1}}^{\boldsymbol{n}},$$

where the $f_{\boldsymbol{k}}$ are the Fourier coefficients of f,

$$f_{\boldsymbol{k}} = \frac{1}{(2\pi)^d} \int_{[-\pi,\pi]^d} f(\boldsymbol{\theta})\, \mathrm{e}^{-\mathrm{i}\boldsymbol{k}\cdot\boldsymbol{\theta}} \mathrm{d}\boldsymbol{\theta}, \qquad \boldsymbol{k} \in \mathbb{Z}^d.$$

Each d-level matrix-sequence of the form $\{T_{\boldsymbol{n}}(f)\}_n$, with $\boldsymbol{n} = \boldsymbol{n}(n) \to \infty$ as $n \to \infty$, is referred to as a (d-level) Toeplitz sequence generated by f.

T1. For every $\boldsymbol{n} \in \mathbb{N}^d$ the map $T_{\boldsymbol{n}}(\cdot) : L^1([-\pi, \pi]^d) \to \mathbb{C}^{N(\boldsymbol{n})\times N(\boldsymbol{n})}$

- is linear: $T_{\boldsymbol{n}}(\alpha f + \beta g) = \alpha T_{\boldsymbol{n}}(f) + \beta T_{\boldsymbol{n}}(g)$;
- is strictly positive: $T_{\boldsymbol{n}}(f) > O_{N(\boldsymbol{n})}$ if $f \geq 0$ and f is not a.e. equal to 0;
- satisfies $T_{\boldsymbol{n}}(1) = I_{N(\boldsymbol{n})}$ and $(T_{\boldsymbol{n}}(f))^* = T_{\boldsymbol{n}}(\overline{f})$.

T2. If f is real a.e. then $T_{\boldsymbol{n}}(f)$ is Hermitian for all $\boldsymbol{n} \in \mathbb{N}^d$.

T3. If $1 \leq p \leq \infty$ and $f \in L^p([-\pi, \pi]^d)$ then $\|T_{\boldsymbol{n}}(f)\|_p \leq \frac{N(\boldsymbol{n})^{1/p}}{(2\pi)^{d/p}} \|f\|_{L^p}$.

T4. If $f \in L^1([-\pi, \pi]^d)$ and $\{T_{\boldsymbol{n}}(f)\}_n$ is any Toeplitz sequence generated by f then $\{T_{\boldsymbol{n}}(f)\}_n \sim_\sigma f$. If in addition f is real a.e. then $\{T_{\boldsymbol{n}}(f)\}_n \sim_\lambda f$.

T5. If $f \in L^\infty([-\pi, \pi]^d)$, the interior of $\mathcal{ER}(f)$ is empty, $\mathbb{C}\backslash\mathcal{ER}(f)$ is connected and $\{T_{\boldsymbol{n}}(f)\}_n$ is any Toeplitz sequence generated by f, then $\{T_{\boldsymbol{n}}(f)\}_n \sim_\lambda f$.

T6. If f is real a.e. and m_f, M_f are its essential infimum and supremum, then

- $\Lambda(T_{\boldsymbol{n}}(f)) \subseteq [m_f, M_f]$ for all $\boldsymbol{n} \in \mathbb{N}^d$,
- $\Lambda(T_{\boldsymbol{n}}(f)) \subset (m_f, M_f)$ for all $\boldsymbol{n} \in \mathbb{N}^d$ whenever $m_f < M_f$,
- for each fixed $j \geq 1$ we have $\lambda_j(T_{\boldsymbol{n}}(f)) \to M_f$ and $\lambda_{N(\boldsymbol{n})-j+1}(T_{\boldsymbol{n}}(f)) \to m_f$ as $\boldsymbol{n} \to \infty$, where $\lambda_1(T_{\boldsymbol{n}}(f)) \geq \cdots \geq \lambda_{N(\boldsymbol{n})}(T_{\boldsymbol{n}}(f))$.

T7. If $f \in L^p([-\pi, \pi]^d)$ and $g \in L^q([-\pi, \pi]^d)$, where $1 \leq p, q \leq \infty$ are conjugate exponents, then $N(\boldsymbol{n})^{-1}\|T_{\boldsymbol{n}}(f)T_{\boldsymbol{n}}(g) - T_{\boldsymbol{n}}(fg)\|_1 \to 0$ as $\boldsymbol{n} \to \infty$.

T8. If $f_1, \ldots, f_d \in L^1([-\pi, \pi])$ and $\boldsymbol{n} \in \mathbb{N}^d$ then

$$T_{\boldsymbol{n}}(f_1 \otimes \cdots \otimes f_d) = T_{n_1}(f_1) \otimes \cdots \otimes T_{n_d}(f_d).$$

Approximating classes of sequences. Let $\{A_n\}_n$ be a sequence of matrices and $\{\{B_{n,m}\}_n\}_m$ a sequence of sequences of matrices, with A_n and $B_{n,m}$ of size d_n. We say that $\{\{B_{n,m}\}_n\}_m$ is an approximating class of sequences (a.c.s.) for $\{A_n\}_n$ if the following condition is met: for every m there exists n_m such that, for $n \geq n_m$,

$$A_n = B_{n,m} + R_{n,m} + N_{n,m}, \qquad \text{rank}(R_{n,m}) \leq c(m)d_n, \qquad \|N_{n,m}\| \leq \omega(m),$$

where n_m, $c(m)$, $\omega(m)$ depend only on m, and

$$\lim_{m\to\infty} c(m) = \lim_{m\to\infty} \omega(m) = 0.$$

We use the abbreviation "a.c.s." for both the singular "approximating class of sequences" and the plural "approximating classes of sequences". It turns out that, for each fixed sequence of positive integers d_n such that $d_n \to \infty$, the notion of a.c.s. is a notion of convergence in the space $\mathscr{E} = \{\{A_n\}_n : A_n \in \mathbb{C}^{d_n\times d_n} \text{ for every } n\}$. More precisely, for every $A \in \mathbb{C}^{\ell\times\ell}$ let

$$\begin{aligned} p(A) &= \inf\left\{\frac{\text{rank}(R)}{\ell} + \|N\| : \quad R, N \in \mathbb{C}^{\ell\times\ell}, \quad R + N = A\right\} \\ &= \min_{i=0,\ldots,\ell}\Big(\frac{i}{\ell} + \sigma_{i+1}(A)\Big), \end{aligned}$$

where $\sigma_1(A) \geq \cdots \geq \sigma_\ell(A)$ and $\sigma_{\ell+1}(A) = 0$ by convention. Set

$$p_{\text{a.c.s.}}(\{A_n\}_n) = \limsup_{n\to\infty} p(A_n), \qquad \{A_n\}_n \in \mathscr{E},$$
$$d_{\text{a.c.s.}}(\{A_n\}_n, \{B_n\}_n) = p_{\text{a.c.s.}}(\{A_n - B_n\}_n), \qquad \{A_n\}_n, \{B_n\}_n \in \mathscr{E}.$$

Then, $d_{\text{a.c.s.}}$ is a distance on $\mathscr{E}$ such that $d_{\text{a.c.s.}}(\{A_n\}_n, \{B_n\}_n) = 0$ iff $\{A_n - B_n\}_n$ is zero-distributed; moreover, $d_{\text{a.c.s.}}$ turns $\mathscr{E}$ into a pseudometric space $(\mathscr{E}, d_{\text{a.c.s.}})$ where the statement "$\{\{B_{n,m}\}_n\}_m$ converges to $\{A_n\}_n$" is equivalent to "$\{\{B_{n,m}\}_n\}_m$ is an a.c.s. for $\{A_n\}_n$". In particular, we can reformulate the definition of a.c.s. in the following way: *a sequence of sequences of matrices $\{\{B_{n,m}\}_n\}_m$ is said to be an a.c.s. for $\{A_n\}_n$ if $\{B_{n,m}\}_n$ converges to $\{A_n\}_n$ in $(\mathscr{E}, d_{\text{a.c.s.}})$ as $m \to \infty$, i.e., if $d_{\text{a.c.s.}}(\{B_{n,m}\}_n, \{A_n\}_n) \to 0$ as $m \to \infty$.* The theory of a.c.s. may then be interpreted as an approximation theory for sequences of matrices, and for this reason we will use the convergence notation $\{B_{n,m}\}_n \xrightarrow{\text{a.c.s.}} \{A_n\}_n$ to indicate that $\{\{B_{n,m}\}_n\}_m$ is an a.c.s. for $\{A_n\}_n$.

ACS 1. $\{A_n\}_n \sim_\sigma f$ iff there exist sequences of matrices $\{B_{n,m}\}_n \sim_\sigma f_m$ such that $\{B_{n,m}\}_n \xrightarrow{\text{a.c.s.}} \{A_n\}_n$ and $f_m \to f$ in measure.

ACS 2. Suppose each A_n is Hermitian. Then, $\{A_n\}_n \sim_\lambda f$ iff there exist sequences of Hermitian matrices $\{B_{n,m}\}_n \sim_\lambda f_m$ such that $\{B_{n,m}\}_n \xrightarrow{\text{a.c.s.}} \{A_n\}_n$ and $f_m \to f$ in measure.

ACS 3. If $\{B_{n,m}\}_n \xrightarrow{\text{a.c.s.}} \{A_n\}_n$ and $\{B'_{n,m}\}_n \xrightarrow{\text{a.c.s.}} \{A'_n\}_n$, with A_n and A'_n of the same size d_n, then

- $\{B^*_{n,m}\}_n \xrightarrow{\text{a.c.s.}} \{A^*_n\}_n$,
- $\{\alpha B_{n,m} + \beta B'_{n,m}\}_n \xrightarrow{\text{a.c.s.}} \{\alpha A_n + \beta A'_n\}_n$ for all $\alpha, \beta \in \mathbb{C}$,
- $\{B_{n,m} B'_{n,m}\}_n \xrightarrow{\text{a.c.s.}} \{A_n A'_n\}_n$ whenever $\{A_n\}_n, \{A'_n\}_n$ are s.u.

ACS 4. Suppose $\{A_n\}_n$ is s.v. If $\{B_{n,m}\}_n \xrightarrow{\text{a.c.s.}} \{A_n\}_n$ then $\{B^\dagger_{n,m}\}_n \xrightarrow{\text{a.c.s.}} \{A^\dagger_n\}_n$.

ACS 5. Suppose $\{A_n\}_n$ is s.u. and A_n, $B_{n,m}$ are Hermitian. If $\{B_{n,m}\}_n \xrightarrow{\text{a.c.s.}} \{A_n\}_n$ then $\{f(B_{n,m})\}_n \xrightarrow{\text{a.c.s.}} \{f(A_n)\}_n$ for every continuous function $f : \mathbb{C} \to \mathbb{C}$.

ACS 6. Let $p \in [1, \infty]$ and suppose for every m there exists n_m such that, for $n \geq n_m$, $\|A_n - B_{n,m}\|_p \leq \epsilon(m, n)(d_n)^{1/p}$, where d_n is the size of both A_n and $B_{n,m}$, and $\lim_{m\to\infty} \limsup_{n\to\infty} \epsilon(m, n) = 0$. Then $\{B_{n,m}\}_n \xrightarrow{\text{a.c.s.}} \{A_n\}_n$.

ACS 7. Suppose $\{A_n - B_{n,m}\}_n \sim_\sigma g_m$ for some g_m defined on a fixed domain (independent of m). If $g_m \to 0$ in measure then $\{B_{n,m}\}_n \xrightarrow{\text{a.c.s.}} \{A_n\}_n$.

Multilevel generalized locally Toeplitz sequences. A d-level Generalized Locally Toeplitz (GLT) sequence $\{A_{\boldsymbol{n}}\}_n$ is a special d-level matrix-sequence equipped with a measurable function $\kappa : [0, 1]^d \times [-\pi, \pi]^d \to \mathbb{C}$, the so-called *symbol* (or *kernel*). We use the notation $\{A_{\boldsymbol{n}}\}_n \sim_{\text{GLT}} \kappa$ to indicate that $\{A_{\boldsymbol{n}}\}_n$ is a d-level GLT sequence with symbol κ. The symbol of a d-level GLT sequence is unique in the sense that if

$\{A_{\boldsymbol{n}}\}_n \sim_{\rm GLT} \kappa$ and $\{A_{\boldsymbol{n}}\}_n \sim_{\rm GLT} \xi$ then $\kappa = \xi$ a.e. in $[0,1]^d \times [-\pi,\pi]^d$. Conversely, if $\{A_{\boldsymbol{n}}\}_n \sim_{\rm GLT} \kappa$ and $\kappa = \xi$ a.e. in $[0,1]^d \times [-\pi,\pi]^d$ then $\{A_{\boldsymbol{n}}\}_n \sim_{\rm GLT} \xi$.

GLT 1. If $\{A_{\boldsymbol{n}}\}_n \sim_{\rm GLT} \kappa$ then $\{A_{\boldsymbol{n}}\}_n \sim_\sigma \kappa$. If $\{A_{\boldsymbol{n}}\}_n \sim_{\rm GLT} \kappa$ and the matrices $A_{\boldsymbol{n}}$ are Hermitian then $\{A_{\boldsymbol{n}}\}_n \sim_\lambda \kappa$.

GLT 2. If $\{A_{\boldsymbol{n}}\}_n \sim_{\rm GLT} \kappa$ and $A_{\boldsymbol{n}} = X_{\boldsymbol{n}} + Y_{\boldsymbol{n}}$, where

- every $X_{\boldsymbol{n}}$ is Hermitian,
- $\|X_{\boldsymbol{n}}\|,\ \|Y_{\boldsymbol{n}}\| \le C$ for some constant C independent of n,
- $N(\boldsymbol{n})^{-1}\|Y_{\boldsymbol{n}}\|_1 \to 0$,

then $\{A_{\boldsymbol{n}}\}_n \sim_\lambda \kappa$.

GLT 3. We have

- $\{T_{\boldsymbol{n}}(f)\}_n \sim_{\rm GLT} \kappa(\mathbf{x},\boldsymbol{\theta}) = f(\boldsymbol{\theta})$ if $f \in L^1([-\pi,\pi]^d)$,
- $\{D_{\boldsymbol{n}}(a)\}_n \sim_{\rm GLT} \kappa(\mathbf{x},\boldsymbol{\theta}) = a(\mathbf{x})$ if $a : [0,1]^d \to \mathbb{C}$ is continuous a.e.,
- $\{Z_{\boldsymbol{n}}\}_n \sim_{\rm GLT} \kappa(\mathbf{x},\boldsymbol{\theta}) = 0$ iff $\{Z_{\boldsymbol{n}}\}_n \sim_\sigma 0$.

GLT 4. If $\{A_{\boldsymbol{n}}\}_n \sim_{\rm GLT} \kappa$ and $\{B_{\boldsymbol{n}}\}_n \sim_{\rm GLT} \xi$ then

- $\{A_{\boldsymbol{n}}^*\}_n \sim_{\rm GLT} \overline{\kappa}$,
- $\{\alpha A_{\boldsymbol{n}} + \beta B_{\boldsymbol{n}}\}_n \sim_{\rm GLT} \alpha\kappa + \beta\xi$ for all $\alpha, \beta \in \mathbb{C}$,
- $\{A_{\boldsymbol{n}} B_{\boldsymbol{n}}\}_n \sim_{\rm GLT} \kappa\xi$.

GLT 5. If $\{A_{\boldsymbol{n}}\}_n \sim_{\rm GLT} \kappa$ and $\kappa \ne 0$ a.e. then $\{A_{\boldsymbol{n}}^\dagger\}_n \sim_{\rm GLT} \kappa^{-1}$.

GLT 6. If $\{A_{\boldsymbol{n}}\}_n \sim_{\rm GLT} \kappa$ and each $A_{\boldsymbol{n}}$ is Hermitian, then $\{f(A_{\boldsymbol{n}})\}_n \sim_{\rm GLT} f(\kappa)$ for every continuous function $f : \mathbb{C} \to \mathbb{C}$.

GLT 7. $\{A_{\boldsymbol{n}}\}_n \sim_{\rm GLT} \kappa$ iff there exist d-level GLT sequences $\{B_{\boldsymbol{n},m}\}_n \sim_{\rm GLT} \kappa_m$ such that $\{B_{\boldsymbol{n},m}\}_n \xrightarrow{\text{a.c.s.}} \{A_{\boldsymbol{n}}\}_n$ and $\kappa_m \to \kappa$ in measure.

GLT 8. Suppose $\{A_{\boldsymbol{n}}\}_n \sim_{\rm GLT} \kappa$ and $\{B_{\boldsymbol{n},m}\}_n \sim_{\rm GLT} \kappa_m$. Then, $\{B_{\boldsymbol{n},m}\}_n \xrightarrow{\text{a.c.s.}} \{A_{\boldsymbol{n}}\}_n$ iff $\kappa_m \to \kappa$ in measure.

GLT 9. If $\{A_{\boldsymbol{n}}\}_n \sim_{\rm GLT} \kappa$ then there exist functions $a_{i,m}, f_{i,m}$, $i = 1, \ldots, N_m$, such that

- $a_{i,m} \in C^\infty([0,1]^d)$ and $f_{i,m}$ is a d-variate trigonometric polynomial,
- $\sum_{i=1}^{N_m} a_{i,m}(\mathbf{x}) f_{i,m}(\boldsymbol{\theta}) \to \kappa(\mathbf{x},\boldsymbol{\theta})$ a.e.,
- $\left\{\sum_{i=1}^{N_m} D_{\boldsymbol{n}}(a_{i,m}) T_{\boldsymbol{n}}(f_{i,m})\right\}_n \xrightarrow{\text{a.c.s.}} \{A_{\boldsymbol{n}}\}_n$.

Fix a sequence of d-indices $\{\boldsymbol{n} = \boldsymbol{n}(n)\}_n \subseteq \mathbb{N}^d$ such that $\boldsymbol{n} \to \infty$ as $n \to \infty$. Let

$$\begin{aligned}
\mathscr{E} &= \Big\{\{A_{\boldsymbol{n}}\}_n : \ A_{\boldsymbol{n}} \in \mathbb{C}^{N(\boldsymbol{n})\times N(\boldsymbol{n})} \text{ for every } n\Big\},\\
\mathfrak{M}_d &= \{\kappa : [0,1]^d \times [-\pi,\pi]^d \to \mathbb{C} : \ \kappa \text{ is measurable}\},\\
\mathscr{E} \times \mathfrak{M}_d &= \Big\{(\{A_{\boldsymbol{n}}\}_n, \kappa) : \ \{A_{\boldsymbol{n}}\}_n \in \mathscr{E}, \ \kappa \in \mathfrak{M}_d\Big\}.
\end{aligned}$$

We note the following.

- The space $\mathscr{E}$ is a *-algebra with respect to the natural operations of conjugate transposition, linear combination and product of d-level matrix-sequences:

$$\begin{aligned}\{A_{\boldsymbol{n}}\}_n^* &= \{A_{\boldsymbol{n}}^*\}_n,\\ \alpha\{A_{\boldsymbol{n}}\}_n + \beta\{B_{\boldsymbol{n}}\}_n &= \{\alpha A_{\boldsymbol{n}} + \beta B_{\boldsymbol{n}}\}_n,\\ \{A_{\boldsymbol{n}}\}_n\{B_{\boldsymbol{n}}\}_n &= \{A_{\boldsymbol{n}}B_{\boldsymbol{n}}\}_n;\end{aligned}$$

  and it is also a pseudometric space with respect to the distance $d_{\text{a.c.s.}}$ inducing the a.c.s. convergence.
- The space $\mathfrak{M}_d$ is a *-algebra with respect to the natural operations of complex conjugation, linear combination and product of functions, and it is also a pseudometric space with respect to the distance d_{measure} inducing the convergence in measure.
- The space $\mathscr{E} \times \mathfrak{M}_d$ is a *-algebra with respect to the natural (pointwise) operations:

$$\begin{aligned}(\{A_{\boldsymbol{n}}\}_n, \kappa)^* &= (\{A_{\boldsymbol{n}}^*\}_n, \overline{\kappa}),\\ \alpha(\{A_{\boldsymbol{n}}\}_n, \kappa) + \beta(\{B_{\boldsymbol{n}}\}_n, \xi) &= (\{\alpha A_{\boldsymbol{n}} + \beta B_{\boldsymbol{n}}\}_n, \alpha\kappa + \beta\xi),\\ (\{A_{\boldsymbol{n}}\}_n, \kappa)(\{B_{\boldsymbol{n}}\}_n, \xi) &= (\{A_{\boldsymbol{n}}B_{\boldsymbol{n}}\}_n, \kappa\xi);\end{aligned}$$

  and it is also a pseudometric space with respect to the product distance

$$d_{\text{a.c.s.}\times\text{measure}}((\{A_{\boldsymbol{n}}\}_n, \kappa), (\{B_{\boldsymbol{n}}\}_n, \xi)) = d_{\text{a.c.s.}}(\{A_{\boldsymbol{n}}\}_n, \{B_{\boldsymbol{n}}\}_n) + d_{\text{measure}}(\kappa, \xi).$$

Let $\mathcal{G}$ be the subset of $\mathscr{E} \times \mathfrak{M}_d$ consisting of the d-level GLT pairs, i.e.,

$$\mathcal{G} = \big\{(\{A_{\boldsymbol{n}}\}_n, \kappa) : \{A_{\boldsymbol{n}}\}_n \sim_{\text{GLT}} \kappa\big\} \subseteq \mathscr{E} \times \mathfrak{M}_d.$$

By **GLT 4** and **GLT 7**, $\mathcal{G}$ is a closed *-subalgebra of $\mathscr{E} \times \mathfrak{M}_d$. By **GLT 3**, $\mathcal{G}$ contains the set

$$\begin{aligned}\mathcal{B} = {}&\big\{(\{T_{\boldsymbol{n}}(f)\}_n, \kappa(\mathbf{x}, \boldsymbol{\theta}) = f(\boldsymbol{\theta})) : f \in L^1([-\pi, \pi]^d)\big\}\\ &\cup \big\{(\{D_{\boldsymbol{n}}(a)\}_n, \kappa(\mathbf{x}, \boldsymbol{\theta}) = a(\mathbf{x})) : a : [0, 1]^d \to \mathbb{C} \text{ is continuous a.e.}\big\}\\ &\cup \big\{(\{Z_{\boldsymbol{n}}\}_n, \kappa(\mathbf{x}, \boldsymbol{\theta}) = 0) : \{Z_{\boldsymbol{n}}\}_n \sim_\sigma 0\big\}.\end{aligned}$$

By **GLT 9**, the subalgebra of $\mathscr{E} \times \mathfrak{M}_d$ given by

$$\begin{aligned}\mathcal{C} = {}&\big\{(\{D_{\boldsymbol{n}}(a)\}_n, \kappa(\mathbf{x}, \boldsymbol{\theta}) = a(\mathbf{x})) : a \in C^\infty([0, 1]^d)\big\}\\ &\cup \big\{(\{T_{\boldsymbol{n}}(f)\}_n, \kappa(\mathbf{x}, \boldsymbol{\theta}) = f(\boldsymbol{\theta})) : f \text{ is a } d- \text{ variate trigonometric polynomial}\big\}\end{aligned}$$

is dense in $\mathcal{G}$. In conclusion:

- *the set of d-level GLT pairs* $\mathcal{G}$ *is the closed *-subalgebra of* $\mathscr{E} \times \mathfrak{M}_d$ *generated by* $\mathcal{B}$*, i.e., the smallest closed *-subalgebra of* $\mathscr{E} \times \mathfrak{M}_d$ *containing* $\mathcal{B}$;
- *the set of d-level GLT pairs* $\mathcal{G}$ *is the closure of the subalgebra of* $\mathscr{E} \times \mathfrak{M}_d$ *generated by* $\mathcal{C}$.

Both these two algebraic-topological characterizations may be taken as the definition of d-level GLT sequences.

Chapter 7
Applications

In this chapter we present several applications of the theory of multilevel GLT sequences to the computation of the singular value and eigenvalue distribution of matrix-sequences arising from the numerical discretization of PDEs. In order to understand the content of this chapter, it is enough that the reader knows the summary of Chap. 6 and possesses the necessary prerequisites, most of which have been addressed in Chap. 2 and [22]. Indeed, our arguments/derivations in this chapter will never refer to Chaps. 1–5, i.e., they will only rely on the summary of Chap. 6. For more applications than the ones presented herein, we refer the reader to [22, Sect. 1.1], where specific pointers to the available literature are provided.

7.1 Auxiliary Results

Before going into the applications of the theory of multilevel GLT sequences, we collect in this section a couple of auxiliary results. Besides simplifying the presentation of the next sections, these results are also interesting in themselves. Actually, they may be considered as further applications of the theory of multilevel GLT sequences.

7.1.1 Multilevel GLT Preconditioning

The first auxiliary result concerns the preconditioning in the context of multilevel GLT sequences. It is the multilevel version of [22, Exercise 8.4].

Theorem 7.1 *Let $\{A_{\boldsymbol{n}}\}_n$ be a sequence of Hermitian matrices such that $\{A_{\boldsymbol{n}}\}_n \sim_{\rm GLT} \kappa$, and let $\{P_{\boldsymbol{n}}\}_n$ be a sequence of HPD matrices such that $\{P_{\boldsymbol{n}}\}_n \sim_{\rm GLT} \xi$ with $\xi \neq 0$ a.e. Then, the sequence of preconditioned matrices $P_{\boldsymbol{n}}^{-1} A_{\boldsymbol{n}}$ satisfies*

C. Garoni and S. Serra-Capizzano, *Generalized Locally Toeplitz Sequences: Theory and Applications*, https://doi.org/10.1007/978-3-030-02233-4_7

$$\{P_n^{-1}A_n\}_n \sim_{\rm GLT} \xi^{-1}\kappa$$

and

$$\{P_n^{-1}A_n\}_n \sim_{\sigma,\lambda} \xi^{-1}\kappa.$$

Proof The GLT relation $\{P_n^{-1}A_n\}_n \sim_{\rm GLT} \xi^{-1}\kappa$ is a direct consequence of **GLT 4** and **GLT 5**. The singular value distribution $\{P_n^{-1}A_n\}_n \sim_\sigma \xi^{-1}\kappa$ follows immediately from **GLT 1**. The only difficult part is the spectral distribution $\{P_n^{-1}A_n\}_n \sim_\lambda \xi^{-1}\kappa$, which does not follow from **GLT 1** because $P_n^{-1}A_n$ is not Hermitian in general.

Since P_n is HPD, the eigenvalues of P_n are positive and the matrices $P_n^{1/2}$, $P_n^{-1/2}$ are well-defined. Moreover,

$$P_n^{-1}A_n \sim P_n^{-1/2}A_nP_n^{-1/2}, \tag{7.1}$$

where $X \sim Y$ means that X is similar to Y. The good news is that $P_n^{-1/2}A_nP_n^{-1/2}$ is Hermitian and, moreover, by **GLT 4**–**GLT 6** (with **GLT 6** applied to $f(z) = |z|^{1/2}$), we have

$$\{P_n^{-1/2}A_nP_n^{-1/2}\}_n \sim_{\rm GLT} |\xi|^{-1/2}\kappa|\xi|^{-1/2} = |\xi|^{-1}\kappa = \xi^{-1}\kappa;$$

note that the latter equation follows from the fact that $\xi \ge 0$ a.e. by **S 3**, since P_n is HPD and $\{P_n\}_n \sim_\lambda \xi$ by **GLT 1**. Since $P_n^{-1/2}A_nP_n^{-1/2}$ is Hermitian, **GLT 1** yields

$$\{P_n^{-1/2}A_nP_n^{-1/2}\}_n \sim_\lambda \xi^{-1}\kappa.$$

Thus, by the similarity (7.1), $\{P_n^{-1}A_n\}_n \sim_\lambda \xi^{-1}\kappa$. □

7.1.2 *Multilevel Arrow-Shaped Sampling Matrices*

If $\boldsymbol{n} \in \mathbb{N}^d$ and $a : [0,1]^d \to \mathbb{C}$, the $\boldsymbol{n}$th (d-level) arrow-shaped sampling matrix generated by a is denoted by $S_{\boldsymbol{n}}(a)$ and is defined as the following symmetric matrix of size $N(\boldsymbol{n})$:

$$(S_{\boldsymbol{n}}(a))_{i,j} = (D_{\boldsymbol{n}}(a))_{i\wedge j, i\wedge j}, \qquad i, j = 1, \dots, N(\boldsymbol{n}). \tag{7.2}$$

In multi-index notation, we have

$$S_{\boldsymbol{n}}(a) = (D_{\boldsymbol{n}}(a))_{\boldsymbol{i}\wedge \boldsymbol{j}, \boldsymbol{i}\wedge \boldsymbol{j}} = a\Big(\frac{\boldsymbol{i}\wedge\boldsymbol{j}}{\boldsymbol{n}}\Big), \qquad \boldsymbol{i}, \boldsymbol{j} = \boldsymbol{1}, \dots, \boldsymbol{n}, \tag{7.3}$$

that is,

$$S_{\boldsymbol{n}}(a) = \Big[a\Big(\frac{\boldsymbol{i}\wedge\boldsymbol{j}}{\boldsymbol{n}}\Big)\Big]_{\boldsymbol{i},\boldsymbol{j}=\boldsymbol{1}}^{\boldsymbol{n}}. \tag{7.4}$$

The motivation of the adjective "arrow-shaped" lies in the shape of the 1-level version of (7.4), as explained in [22, p. 190]. The next theorem is the multivariate version of [22, Theorem 10.4].

Theorem 7.2 *Let* $a : [0,1]^d \to \mathbb{C}$ *be continuous and let* $f(\boldsymbol{\theta}) = \sum_{\boldsymbol{j}=-\boldsymbol{r}}^{\boldsymbol{r}} f_{\boldsymbol{j}}\, \mathrm{e}^{\mathrm{i}\boldsymbol{j}\cdot\boldsymbol{\theta}}$ *be a d-variate trigonometric polynomial. Then,*

$$\|S_{\boldsymbol{n}}(a) \circ T_{\boldsymbol{n}}(f) - D_{\boldsymbol{n}}(a)T_{\boldsymbol{n}}(f)\| \le (2|\boldsymbol{r}|_\infty + 1)^d \|f\|_\infty\, \omega_a\Big(\frac{|\boldsymbol{r}|_\infty}{\min(\boldsymbol{n})}\Big) \tag{7.5}$$

for every $\boldsymbol{n} \in \mathbb{N}^d$,

$$\|S_{\boldsymbol{n}}(a) \circ T_{\boldsymbol{n}}(f)\| \le C \tag{7.6}$$

for every $\boldsymbol{n} \in \mathbb{N}^d$ *and for some constant C independent of* $\boldsymbol{n}$, *and*

$$\{S_{\boldsymbol{n}}(a) \circ T_{\boldsymbol{n}}(f)\}_n \sim_{\mathrm{GLT}} a(\mathbf{x}) f(\boldsymbol{\theta}) \tag{7.7}$$

for every sequence $\{\boldsymbol{n} = \boldsymbol{n}(n)\}_n \subseteq \mathbb{N}^d$ *such that* $\boldsymbol{n} \to \infty$ *as* $n \to \infty$.

Proof For $\boldsymbol{i}, \boldsymbol{j} = \boldsymbol{1}, \ldots, \boldsymbol{n}$, we have the following.

- If $|\boldsymbol{i} - \boldsymbol{j}|_\infty > |\boldsymbol{r}|_\infty$, then the Fourier coefficient $f_{\boldsymbol{i}-\boldsymbol{j}}$ is zero and, consequently,

$$(S_{\boldsymbol{n}}(a) \circ T_{\boldsymbol{n}}(f))_{\boldsymbol{i}\boldsymbol{j}} = (S_{\boldsymbol{n}}(a))_{\boldsymbol{i}\boldsymbol{j}}(T_{\boldsymbol{n}}(f))_{\boldsymbol{i}\boldsymbol{j}} = a\Big(\frac{\boldsymbol{i}\wedge\boldsymbol{j}}{\boldsymbol{n}}\Big) f_{\boldsymbol{i}-\boldsymbol{j}} = 0,$$
$$(D_{\boldsymbol{n}}(a)T_{\boldsymbol{n}}(f))_{\boldsymbol{i}\boldsymbol{j}} = (D_{\boldsymbol{n}}(a))_{\boldsymbol{i}\boldsymbol{i}}(T_{\boldsymbol{n}}(f))_{\boldsymbol{i}\boldsymbol{j}} = a\Big(\frac{\boldsymbol{i}}{\boldsymbol{n}}\Big) f_{\boldsymbol{i}-\boldsymbol{j}} = 0.$$

- If $|\boldsymbol{i} - \boldsymbol{j}|_\infty \le |\boldsymbol{r}|_\infty$, then, considering that $|f_{\boldsymbol{i}-\boldsymbol{j}}| \le \|f\|_\infty$, we have

$$\begin{aligned}
|(S_{\boldsymbol{n}}(a) \circ T_{\boldsymbol{n}}(f))_{\boldsymbol{i}\boldsymbol{j}} - (D_{\boldsymbol{n}}(a)T_{\boldsymbol{n}}(f))_{\boldsymbol{i}\boldsymbol{j}}| &= \Big|a\Big(\frac{\boldsymbol{i}\wedge\boldsymbol{j}}{\boldsymbol{n}}\Big) f_{\boldsymbol{i}-\boldsymbol{j}} - a\Big(\frac{\boldsymbol{i}}{\boldsymbol{n}}\Big) f_{\boldsymbol{i}-\boldsymbol{j}}\Big| \\
&\le \|f\|_\infty \Big|a\Big(\frac{\boldsymbol{i}\wedge\boldsymbol{j}}{\boldsymbol{n}}\Big) - a\Big(\frac{\boldsymbol{i}}{\boldsymbol{n}}\Big)\Big| \\
&\le \|f\|_\infty\, \omega_a\Big(\Big|\frac{\boldsymbol{j}}{\boldsymbol{n}} - \frac{\boldsymbol{i}}{\boldsymbol{n}}\Big|_\infty\Big) \\
&\le \|f\|_\infty\, \omega_a\Big(\frac{|\boldsymbol{r}|_\infty}{\min(\boldsymbol{n})}\Big).
\end{aligned}$$

It follows from the first item that the nonzero entries in each row and column of the matrix $Z_{\boldsymbol{n}} = S_{\boldsymbol{n}}(a) \circ T_{\boldsymbol{n}}(f) - D_{\boldsymbol{n}}(a)T_{\boldsymbol{n}}(f)$ are at most $(2|\boldsymbol{r}|_\infty + 1)^d$. Indeed, considering for instance the $\boldsymbol{i}$th row, we have $(Z_{\boldsymbol{n}})_{\boldsymbol{i}\boldsymbol{j}} = 0$ whenever $|\boldsymbol{i} - \boldsymbol{j}|_\infty > |\boldsymbol{r}|_\infty$, which means that the only possible nonzero entries of $Z_{\boldsymbol{n}}$ in the $\boldsymbol{i}$th row are those corresponding to the column multi-indices $\boldsymbol{j} \in \{\boldsymbol{1}, \ldots, \boldsymbol{n}\}$ such that $|\boldsymbol{i} - \boldsymbol{j}|_\infty \le |\boldsymbol{r}|_\infty$; and the number of all multi-indices $\boldsymbol{j} \in \mathbb{Z}^d$ satisfying $|\boldsymbol{i} - \boldsymbol{j}|_\infty \le |\boldsymbol{r}|_\infty$ is $(2|\boldsymbol{r}|_\infty + 1)^d$. It follows from the second item that each entry of $Z_{\boldsymbol{n}}$ is bounded in modulus by $\|f\|_\infty\,\omega_a(\frac{|\boldsymbol{r}|_\infty}{\min(\boldsymbol{n})})$. Thus, $|Z_{\boldsymbol{n}}|_1, |Z_{\boldsymbol{n}}|_\infty \le (2|\boldsymbol{r}|_\infty+1)^d\|f\|_\infty\,\omega_a(\frac{|\boldsymbol{r}|_\infty}{\min(\boldsymbol{n})})$, and the application of **N 1** yields (7.5). Using (7.5) we immediately obtain

$$\begin{aligned}\|S_{\boldsymbol{n}}(a)\circ T_{\boldsymbol{n}}(f)\| &\le \|D_{\boldsymbol{n}}(a)\|\,\|T_{\boldsymbol{n}}(f)\|+\|Z_{\boldsymbol{n}}\|\\ &\le \|a\|_\infty\|f\|_\infty+(2|\boldsymbol{r}|_\infty+1)^d\|f\|_\infty\,\omega_a\Big(\frac{|\boldsymbol{r}|_\infty}{\min(\boldsymbol{n})}\Big),\end{aligned}$$

which implies (7.6). Finally, for any sequence $\{\boldsymbol{n}=\boldsymbol{n}(n)\}_n\subseteq\mathbb{N}^d$ such that $\boldsymbol{n}\to\infty$ as $n\to\infty$, we have $\omega_a(\frac{|\boldsymbol{r}|_\infty}{\min(\boldsymbol{n})})\to 0$ as $n\to\infty$, and so $\{Z_{\boldsymbol{n}}\}_n\sim_\sigma 0$ by (7.5) and **Z 1** (or **Z 2**). Thus, (7.7) follows from the decomposition $S_{\boldsymbol{n}}(a)\circ T_{\boldsymbol{n}}(f)=D_{\boldsymbol{n}}(a)T_{\boldsymbol{n}}(f)+Z_{\boldsymbol{n}}$ in combination with **GLT 3** and **GLT 4**. □

7.2 Applications to PDE Discretizations: An Introduction

In the next sections we extend to the d-dimensional setting the GLT analysis carried out in [22, Sects. 10.5–10.7] for sequences of matrices arising from the discretization of unidimensional differential equations. More precisely, Sects. 7.3, 7.4, 7.5, 7.6, 7.7 are the d-dimensional versions of, respectively, Sects. 10.5.2, 10.6.1, 10.7.1, 10.7.2, 10.7.3 from [22]. The main observation is that no substantial differences are encountered when passing from 1 to d dimensions. In other words, *all the main "GLT ideas" have already emerged in the unidimensional setting*, and the GLT analysis of Sects. 7.3–7.7 is conceptually the same as the GLT analysis of the corresponding unidimensional subsections mentioned above. However, the d-dimensional case involves a lot of technical difficulties that are not visible in one dimension, and in order to gain familiarity with such technicalities it is necessary to see them in some detail. The most important of them is certainly the *multi-index language*, which allows one to tackle a d-dimensional GLT analysis by essentially maintaining the unidimensional notation, at the only price of turning some letters (n, p, i, j, etc.) in boldface ($\boldsymbol{n}, \boldsymbol{p}, \boldsymbol{i}, \boldsymbol{j}$, etc.).

Before going into Sects. 7.3–7.7, we outline here the main ideas of a d-dimensional GLT analysis. Consider, for example, the general d-dimensional linear second-order PDE

$$\begin{aligned}&-\sum_{\ell,k=1}^{d}a_{\ell k}\frac{\partial^2 u}{\partial x_\ell\partial x_k}+\sum_{k=1}^{d}b_k\frac{\partial u}{\partial x_k}+cu=f\\ \iff\ &-\mathbf{1}(A\circ Hu)\mathbf{1}^T+\mathbf{b}\cdot\nabla u+cu=f,\end{aligned} \tag{7.8}$$

where $a_{\ell k}, b_k, c, f$ are given functions, $A=[a_{\ell k}]_{\ell,k=1}^d$, $\mathbf{b}=[b_k]_{k=1}^d$, and Hu is the Hessian of u,

$$Hu=\left[\frac{\partial^2 u}{\partial x_\ell\partial x_k}\right]_{\ell,k=1}^d.$$

Assume we discretize (7.8) by a standard numerical method, such as, for instance, a FD scheme. The resulting discretization matrices $A_{\boldsymbol{n}}$ are parameterized by a d-index $\boldsymbol{n} = (n_1, \ldots, n_d)$, where n_i is related to the discretization step h_i in the ith direction, and $n_i \to \infty$ if and only if $h_i \to 0$ (usually, we have $h_i \approx 1/n_i$). By choosing each n_i as a function of a unique discretization parameter $n \in \mathbb{N}$, as it normally happens in practice where the most natural choice is $n_i = n$ for all $i = 1, \ldots, d$, we see that $\boldsymbol{n} = \boldsymbol{n}(n)$ and, consequently, $\{A_{\boldsymbol{n}}\}_n$ is a (d-level) matrix-sequence. The matrix $A_{\boldsymbol{n}}$ can be decomposed according to the terms of the PDE as follows:

$$A_{\boldsymbol{n}} = \sum_{\ell,k=1}^{d} K_{\boldsymbol{n},\ell k}(a_{\ell k}) + \sum_{k=1}^{d} H_{\boldsymbol{n},k}(b_k) + I_{\boldsymbol{n}}(c) + R_{\boldsymbol{n}} = K_{\boldsymbol{n}} + Z_{\boldsymbol{n}}, \tag{7.9}$$

where

$$K_{\boldsymbol{n}} = \sum_{\ell,k=1}^{d} K_{\boldsymbol{n},\ell k}(a_{\ell k}), \tag{7.10}$$

$$Z_{\boldsymbol{n}} = \sum_{k=1}^{d} H_{\boldsymbol{n},k}(b_k) + I_{\boldsymbol{n}}(c) + R_{\boldsymbol{n}}, \tag{7.11}$$

$R_{\boldsymbol{n}}$ is a small-rank perturbation due to the imposed boundary conditions,[1] and $K_{\boldsymbol{n},\ell k}(a_{\ell k})$, $H_{\boldsymbol{n},k}(b_k)$, $I_{\boldsymbol{n}}(c)$ are the matrices resulting from the discretization of the separable differential operators[2]

---

[1]Note that the boundary conditions (Dirichlet, Neumann, etc.) have not been specified precisely because they only produce a small-rank perturbation $R_{\boldsymbol{n}}$ in the resulting discretization matrix $A_{\boldsymbol{n}}$; see also the discussion in [22, p. 116] and the 2nd part of [22, Sect. 10.5.2].

[2]We say that a differential operator is separable if it is obtained by multiplying a given function with a product of partial derivatives. The general separable differential operator can be written as

$$a\frac{\partial^{r_1+\cdots+r_d} u}{\partial x_1^{r_1}\cdots\partial x_d^{r_d}}.$$

An example of a non-separable differential operator is the Laplacian, which, however, can be written (just like any other linear differential operator) as a sum of separable differential operators:

$$\Delta u = \sum_{k=1}^{d}\frac{\partial^2 u}{\partial x_k^2}.$$

As evidenced by the forthcoming discussion, *the discretization of a separable differential operator gives rise to a GLT (actually, a sLT) sequence.* For instance, after a suitable normalization that we here ignore, the matrix-sequences $\{K_{\boldsymbol{n},\ell k}(a_{\ell k})\}_n$, $\{H_{\boldsymbol{n},k}(b_k)\}_n$, $\{I_{\boldsymbol{n}}(c)\}_n$ are GLT (actually, sLT) sequences. As a consequence, the discretization of an arbitrary linear differential operator (a sum of separable differential operators) gives rise to a sum of GLT (actually, sLT) sequences, i.e., again a GLT sequence.

$$-a_{\ell k}\frac{\partial^2 u}{\partial x_\ell \partial x_k}, \qquad b_k\frac{\partial u}{\partial x_k}, \qquad cu,$$

respectively. More precisely, $K_{\boldsymbol{n},\ell k}(a_{\ell k})$, $H_{\boldsymbol{n},k}(b_k)$, $I_{\boldsymbol{n}}(c)$ are the matrices resulting from the discretization of the three left-hand side terms in the PDE

$$-a_{\ell k}\frac{\partial^2 u}{\partial x_\ell \partial x_k} + b_k\frac{\partial u}{\partial x_k} + cu = f.$$

It normally turns out that, after a suitable normalization that we ignore in this discussion,

- the matrix-sequence $\{Z_{\boldsymbol{n}}\}_n$, which results from the discretization of the lower-order differential operators of the PDE (7.8) and includes the small-rank perturbation due to boundary conditions, is zero-distributed;
- the GLT analysis of $\{A_{\boldsymbol{n}}\}_n$ reduces to the GLT analysis of the matrix-sequence $\{K_{\boldsymbol{n}}\}_n$ resulting from the discretization of the higher-order differential operator of the PDE (7.8).

In addition, every matrix-sequence $\{K_{\boldsymbol{n},\ell k}(a_{\ell k})\}_n$ appearing in the definition (7.10) of $K_{\boldsymbol{n}}$ usually turns out to be a d-level GLT sequence (actually, a d-level sLT sequence) of the form

$$K_{\boldsymbol{n},\ell k}(a_{\ell k}) = D_{\boldsymbol{n}}(a_{\ell k})K_{\boldsymbol{n},\ell k}(1) + Z_{\boldsymbol{n},\ell k}, \qquad \{Z_{\boldsymbol{n},\ell k}\}_n \sim_\sigma 0, \tag{7.12}$$

$$K_{\boldsymbol{n},\ell k}(1) = T_{\boldsymbol{n}}(H_{\ell k}) + Y_{\boldsymbol{n},\ell k}, \qquad \{Y_{\boldsymbol{n},\ell k}\}_n \sim_\sigma 0, \tag{7.13}$$

where $H_{\ell k}$ is a (separable) d-variate trigonometric polynomial. It follows immediately from (7.12)–(7.13) and **GLT 3**–**GLT 4** that

$$\{K_{\boldsymbol{n},\ell k}(1)\}_n \sim_{\rm GLT} H_{\ell k}(\boldsymbol{\theta}), \tag{7.14}$$

$$\{K_{\boldsymbol{n},\ell k}(a_{\ell k})\}_n \sim_{\rm GLT} a_{\ell k}(\mathbf{x})H_{\ell k}(\boldsymbol{\theta}). \tag{7.15}$$

As a consequence,

$$\{A_{\boldsymbol{n}}\}_n \sim_{\rm GLT} \sum_{\ell,k=1}^{d} a_{\ell k}(\mathbf{x})H_{\ell k}(\boldsymbol{\theta}) = \mathbf{1}(A(\mathbf{x}) \circ H(\boldsymbol{\theta}))\mathbf{1}^T, \tag{7.16}$$

where

$$H(\boldsymbol{\theta}) = [H_{\ell k}(\boldsymbol{\theta})]_{\ell,k=1}^d.$$

From (7.16) and **GLT 1**–**GLT 2** one often obtains the distribution relations

$$\{A_{\boldsymbol{n}}\}_n \sim_{\sigma,\lambda} \sum_{\ell,k=1}^{d} a_{\ell k}(\mathbf{x})H_{\ell k}(\boldsymbol{\theta}) = \mathbf{1}(A(\mathbf{x}) \circ H(\boldsymbol{\theta}))\mathbf{1}^T.$$

Remark 7.1 $K_{\mathbf{n},\ell k}(1)$ is the matrix resulting from the discretization of the left-hand side of the PDE

$$-\frac{\partial^2 u}{\partial x_\ell \partial x_k} = f$$

and it is therefore referred to as the matrix associated with the discretization of the second derivative $-\partial^2 u/\partial x_\ell \partial x_k$. Thus, in view of (7.14), $H_{\ell k}$ is referred to as the d-variate trigonometric polynomial associated with the discretization of $-\partial^2 u/\partial x_\ell \partial x_k$, or simply the "symbol of $-\partial^2 u/\partial x_\ell \partial x_k$". For example, if the considered discretization method is a FD scheme, then $H_{\ell k}$ is the d-variate trigonometric polynomial that represents the FD formula used to discretize $-\partial^2 u/\partial x_\ell \partial x_k$. The latter assertion will become more clear after reading Sect. 7.3.

Remark 7.2 (*formal structure of the symbol and symbol of the negative Hessian operator*) The formal analogy between the expression of the symbol $\mathbf{1}(A(\mathbf{x}) \circ H(\boldsymbol{\theta}))\mathbf{1}^T$ and the expression of the higher-order differential operator $-\mathbf{1}(A \circ Hu)\mathbf{1}^T$ in (7.8) is impressive! Because of this analogy, and especially because of (7.14) and Remark 7.1, the matrix $H(\boldsymbol{\theta})$ in the so-called "Fourier variables" $\boldsymbol{\theta} = (\theta_1, \ldots, \theta_d)$ is usually referred to as the "symbol of the negative Hessian operator", although this terminology is clearly not rigorous from a mathematical viewpoint. If we change the numerical method for the discretization of (7.8), the symbol $\mathbf{1}(A(\mathbf{x}) \circ H(\boldsymbol{\theta}))\mathbf{1}^T$ remains the same except for the matrix $H(\boldsymbol{\theta})$, which changes according to the new method. For example, if we switch from a FD scheme to another, the symbol of the negative Hessian operator switches from $H(\boldsymbol{\theta}) = [H_{\ell k}(\boldsymbol{\theta})]_{\ell,k=1}^d$ to $\tilde{H}(\boldsymbol{\theta}) = [\tilde{H}_{\ell k}(\boldsymbol{\theta})]_{\ell,k=1}^d$, where $\tilde{H}_{\ell k}$ is the (separable) d-variate trigonometric polynomial associated with the new FD formula used to discretize $-\partial^2 u/\partial x_\ell \partial x_k$.

We invite the reader to compare Remarks 7.1 and 7.2 with their univariate analog [22, Remark 10.1] and with the three paragraphs which introduce the 1st, 3rd and 4th part of [22, Sect. 10.5.2].

7.3 FD Discretization of Convection-Diffusion-Reaction PDEs

Consider the convection-diffusion-reaction problem

$$\begin{cases} -\nabla \cdot A \nabla u + \mathbf{b} \cdot \nabla u + cu = f, & \text{in } (0,1)^d, \\ u = 0, & \text{on } \partial((0,1)^d), \end{cases}$$

$$\iff \begin{cases} -\sum_{\ell,k=1}^{d} \frac{\partial}{\partial x_\ell}\Big(a_{\ell k}\frac{\partial u}{\partial x_k}\Big) + \sum_{k=1}^{d} b_k \frac{\partial u}{\partial x_k} + cu = f, & \text{in } (0,1)^d, \\ u = 0, & \text{on } \partial((0,1)^d), \end{cases} \tag{7.17}$$

where $a_{\ell k}, b_k, c, f$ are given functions, $A = [a_{\ell k}]_{\ell,k=1}^d$ and $\mathbf{b} = [b_k]_{k=1}^d$.

FD discretization. Problem (7.17) can be reformulated as follows:

$$\begin{cases} -\mathbf{1}(A \circ Hu)\mathbf{1}^T + \mathbf{s} \cdot \nabla u + cu = f, & \text{in } (0,1)^d, \\ u = 0, & \text{on } \partial((0,1)^d), \end{cases}$$

$$\iff \begin{cases} -\sum_{\ell,k=1}^{d} a_{\ell k} \frac{\partial^2 u}{\partial x_\ell \partial x_k} + \sum_{k=1}^{d} s_k \frac{\partial u}{\partial x_k} + cu = f, & \text{in } (0,1)^d, \\ u = 0, & \text{on } \partial((0,1)^d), \end{cases} \tag{7.18}$$

where Hu is the Hessian of u,

$$(Hu)_{\ell k} = \frac{\partial^2 u}{\partial x_\ell \partial x_k}, \qquad \ell, k = 1, \ldots, d,$$

and $\mathbf{s}$ collects the coefficients of the first-order derivatives,

$$s_k = b_k - \sum_{\ell=1}^{d} \frac{\partial a_{\ell k}}{\partial x_\ell}, \qquad k = 1, \ldots, d.$$

We consider the classical central FD discretization of (7.18). We choose $\boldsymbol{n} \in \mathbb{N}^d$ and we set $\boldsymbol{h} = \frac{1}{\boldsymbol{n}+\mathbf{1}}$ and $\mathbf{x}_{\boldsymbol{j}} = \boldsymbol{j}\boldsymbol{h}$ for $\boldsymbol{j} = \mathbf{0}, \ldots, \boldsymbol{n}+\mathbf{1}$.[3] Let $\mathbf{e}_k$ be the kth vector of the canonical basis of $\mathbb{R}^d$. For $\boldsymbol{j} = \mathbf{1}, \ldots, \boldsymbol{n}$, we have

$$\begin{aligned} a_{kk} \frac{\partial^2 u}{\partial x_k^2}\bigg|_{\mathbf{x}=\mathbf{x}_j} &\approx a_{kk}(\mathbf{x}_j) \frac{u(\mathbf{x}_j + h_k \mathbf{e}_k) - 2u(\mathbf{x}_j) + u(\mathbf{x}_j - h_k \mathbf{e}_k)}{h_k^2} \\ &= a_{kk}(\mathbf{x}_j) \frac{u(\mathbf{x}_{j+\mathbf{e}_k}) - 2u(\mathbf{x}_j) + u(\mathbf{x}_{j-\mathbf{e}_k})}{h_k^2} \end{aligned} \tag{7.19}$$

for $k = 1, \ldots, d$,

$$\begin{aligned} a_{\ell k} \frac{\partial^2 u}{\partial x_\ell \partial x_k}\bigg|_{\mathbf{x}=\mathbf{x}_j} &\approx a_{\ell k}(\mathbf{x}_j) \frac{\frac{\partial u}{\partial x_\ell}(\mathbf{x}_j + h_k \mathbf{e}_k) - \frac{\partial u}{\partial x_\ell}(\mathbf{x}_j - h_k \mathbf{e}_k)}{2h_k} \\ &\approx a_{\ell k}(\mathbf{x}_j) \frac{1}{2h_k} \left[\frac{u(\mathbf{x}_j + h_k \mathbf{e}_k + h_\ell \mathbf{e}_\ell) - u(\mathbf{x}_j + h_k \mathbf{e}_k - h_\ell \mathbf{e}_\ell)}{2h_\ell} \right. \\ &\qquad \left. - \frac{u(\mathbf{x}_j - h_k \mathbf{e}_k + h_\ell \mathbf{e}_\ell) - u(\mathbf{x}_j - h_k \mathbf{e}_k - h_\ell \mathbf{e}_\ell)}{2h_\ell} \right] \end{aligned}$$

[3]Recall that operations involving d-indices that have no meaning in $\mathbb{Z}^d$ must be interpreted in the componentwise sense. In the present case, given $\boldsymbol{n} = (n_1, \ldots, n_d)$ and $\boldsymbol{j} = (j_1, \ldots, j_d)$, the vector of discretization steps $\boldsymbol{h} = \frac{1}{\boldsymbol{n}+\mathbf{1}}$ and the grid point $\mathbf{x}_{\boldsymbol{j}} = \boldsymbol{j}\boldsymbol{h}$ are given by $\boldsymbol{h} = (\frac{1}{n_1+1}, \ldots, \frac{1}{n_d+1}) = (h_1, \ldots, h_d)$ and $\mathbf{x}_{\boldsymbol{j}} = (j_1 h_1, \ldots, j_d h_d)$.

$$= a_{\ell k}(\mathbf{x}_j)\frac{u(\mathbf{x}_{j+\mathbf{e}_k+\mathbf{e}_\ell}) - u(\mathbf{x}_{j+\mathbf{e}_k-\mathbf{e}_\ell}) - u(\mathbf{x}_{j-\mathbf{e}_k+\mathbf{e}_\ell}) + u(\mathbf{x}_{j-\mathbf{e}_k-\mathbf{e}_\ell})}{4h_\ell h_k} \tag{7.20}$$

for $\ell, k = 1, \ldots, d$ with $\ell \neq k$,

$$s_k \frac{\partial u}{\partial x_k}\bigg|_{\mathbf{x}=\mathbf{x}_j} \approx s_k(\mathbf{x}_j)\frac{u(\mathbf{x}_j + h_k\mathbf{e}_k) - u(\mathbf{x}_j - h_k\mathbf{e}_k)}{2h_k} = s_k(\mathbf{x}_j)\frac{u(\mathbf{x}_{j+\mathbf{e}_k}) - u(\mathbf{x}_{j-\mathbf{e}_k})}{2h_k} \tag{7.21}$$

for $k = 1, \ldots, d$,

$$cu|_{\mathbf{x}=\mathbf{x}_j} = c(\mathbf{x}_j)u(\mathbf{x}_j). \tag{7.22}$$

Thus, for every $\boldsymbol{j} = \mathbf{0}, \ldots, \boldsymbol{n} + \mathbf{1}$, we approximate the evaluation $u(\mathbf{x}_j)$ of the solution of (7.18) at the grid point $\mathbf{x}_j$ by the value u_j, where $u_j = 0$ for $\boldsymbol{j} \notin \{\mathbf{1}, \ldots, \boldsymbol{n}\}$ and the vector $\mathbf{u} = (u_\mathbf{1}, \ldots, u_\boldsymbol{n})^T$ is the solution of the linear system

$$\begin{aligned} &-\sum_{k=1}^d a_{kk}(\mathbf{x}_j)\frac{u_{j+\mathbf{e}_k} - 2u_j + u_{j-\mathbf{e}_k}}{h_k^2} \\ &-\sum_{\substack{\ell,k=1\\ \ell\neq k}}^d a_{\ell k}(\mathbf{x}_j)\frac{u_{j+\mathbf{e}_k+\mathbf{e}_\ell} - u_{j+\mathbf{e}_k-\mathbf{e}_\ell} - u_{j-\mathbf{e}_k+\mathbf{e}_\ell} + u_{j-\mathbf{e}_k-\mathbf{e}_\ell}}{4h_\ell h_k} \\ &+\sum_{k=1}^d s_k(\mathbf{x}_j)\frac{u_{j+\mathbf{e}_k} - u_{j-\mathbf{e}_k}}{2h_k} + c(\mathbf{x}_j)u_j = f(\mathbf{x}_j), \qquad \boldsymbol{j} = \mathbf{1}, \ldots, \boldsymbol{n}. \end{aligned} \tag{7.23}$$

The matrix $A_\boldsymbol{n}$ associated with this linear system admits the following natural decomposition:

$$A_\boldsymbol{n} = \sum_{\ell,k=1}^d K_{\boldsymbol{n},\ell k}(a_{\ell k}) + \sum_{k=1}^d H_{\boldsymbol{n},k}(s_k) + I_\boldsymbol{n}(c), \tag{7.24}$$

where

$$\begin{aligned} K_{\boldsymbol{n},\ell k}(a_{\ell k}) &= \frac{1}{h_\ell h_k}\Big(\operatorname*{diag}_{\boldsymbol{j}=\mathbf{1},\ldots,\boldsymbol{n}} a_{\ell k}(\mathbf{x}_j)\Big)K_{\boldsymbol{n},\ell k}, \qquad \ell, k = 1, \ldots, d, \\ H_{\boldsymbol{n},k}(s_k) &= \frac{1}{h_k}\Big(\operatorname*{diag}_{\boldsymbol{j}=\mathbf{1},\ldots,\boldsymbol{n}} s_k(\mathbf{x}_j)\Big)H_{\boldsymbol{n},k}, \qquad k = 1, \ldots, d, \\ I_\boldsymbol{n}(c) &= \Big(\operatorname*{diag}_{\boldsymbol{j}=\mathbf{1},\ldots,\boldsymbol{n}} c(\mathbf{x}_j)\Big)I_\boldsymbol{n}, \end{aligned}$$

and the matrices $K_{\boldsymbol{n},\ell k}$, $H_{\boldsymbol{n},k}$, $I_{\boldsymbol{n}}$ are defined by their actions on a generic vector $\mathbf{u} \in \mathbb{R}^{N(\boldsymbol{n})}$, as follows:

$$(K_{\boldsymbol{n},kk}\mathbf{u})_{\boldsymbol{j}} = -u_{\boldsymbol{j}-\mathbf{e}_k} + 2u_{\boldsymbol{j}} - u_{\boldsymbol{j}+\mathbf{e}_k}, \qquad \boldsymbol{j} = \mathbf{1}, \ldots, \boldsymbol{n}, \tag{7.25}$$

for $k = 1, \ldots, d$,

$$(K_{\boldsymbol{n},\ell k}\mathbf{u})_{\boldsymbol{j}} = -\frac{1}{4}(u_{\boldsymbol{j}-\mathbf{e}_\ell-\mathbf{e}_k} - u_{\boldsymbol{j}-\mathbf{e}_\ell+\mathbf{e}_k} - u_{\boldsymbol{j}+\mathbf{e}_\ell-\mathbf{e}_k} + u_{\boldsymbol{j}+\mathbf{e}_\ell+\mathbf{e}_k}), \qquad \boldsymbol{j} = \mathbf{1}, \ldots, \boldsymbol{n}, \tag{7.26}$$

for $\ell, k = 1, \ldots, d$ with $\ell \neq k$,

$$(H_{\boldsymbol{n},k}\mathbf{u})_{\boldsymbol{j}} = \frac{1}{2}(-u_{\boldsymbol{j}-\mathbf{e}_k} + u_{\boldsymbol{j}+\mathbf{e}_k}), \qquad \boldsymbol{j} = \mathbf{1}, \ldots, \boldsymbol{n}, \tag{7.27}$$

for $k = 1, \ldots, d$,

$$(I_{\boldsymbol{n}}\mathbf{u})_{\boldsymbol{j}} = u_{\boldsymbol{j}}, \qquad \boldsymbol{j} = \mathbf{1}, \ldots, \boldsymbol{n}. \tag{7.28}$$

In (7.25)–(7.28), just like in (7.23), $u_{\boldsymbol{i}} = 0$ whenever $\boldsymbol{i} \notin \{\mathbf{1}, \ldots, \boldsymbol{n}\}$.

FD discretization matrices. Thanks to the multi-index language, we are able to provide a compact and easy-to-manage expression for the matrices (7.25)–(7.28) (and hence also for the FD discretization matrix $A_{\boldsymbol{n}}$).

Lemma 7.1 *For every $\boldsymbol{n} \in \mathbb{N}^d$, we have*

$$K_{\boldsymbol{n},kk} = \left(\bigotimes_{r=1}^{k-1} I_{n_r}\right) \otimes K_{n_k} \otimes \left(\bigotimes_{r=k+1}^{d} I_{n_r}\right) \tag{7.29}$$

for $k = 1, \ldots, d$,

$$K_{\boldsymbol{n},k\ell} = K_{\boldsymbol{n},\ell k} = -\left(\bigotimes_{r=1}^{\ell-1} I_{n_r}\right) \otimes H_{n_\ell} \otimes \left(\bigotimes_{r=\ell+1}^{k-1} I_{n_r}\right) \otimes H_{n_k} \otimes \left(\bigotimes_{r=k+1}^{d} I_{n_r}\right) \tag{7.30}$$

for $1 \le \ell < k \le d$,

$$H_{\boldsymbol{n},k} = \left(\bigotimes_{r=1}^{k-1} I_{n_r}\right) \otimes H_{n_k} \otimes \left(\bigotimes_{r=k+1}^{d} I_{n_r}\right) \tag{7.31}$$

for $k = 1, \ldots, d$, and

$$I_{\boldsymbol{n}} = \bigotimes_{r=1}^{d} I_{n_r} = I_{N(\boldsymbol{n})}, \tag{7.32}$$

where the matrices K_n, H_n are defined for all n as follows:

$$K_n = \begin{bmatrix} 2 & -1 & & & \\ -1 & 2 & -1 & & \\ & \ddots & \ddots & \ddots & \\ & & -1 & 2 & -1 \\ & & & -1 & 2 \end{bmatrix} = T_n(2-2\cos\theta),$$

$$H_n = \frac{1}{2}\begin{bmatrix} 0 & 1 & & & \\ -1 & 0 & 1 & & \\ & \ddots & \ddots & \ddots & \\ & & -1 & 0 & 1 \\ & & & -1 & 0 \end{bmatrix} = -\mathrm{i}T_n(\sin\theta).$$

Proof We only prove (7.29) as the proofs of (7.30) and (7.31) are completely analogous, while (7.32) is obvious from (7.28). Let $\delta_{ij} = 1$ if $i = j$ and $\delta_{ij} = 0$ otherwise. By the crucial property **P 7**, for every $\mathbf{u} = [u_{\boldsymbol{\ell}}]_{\boldsymbol{\ell}=\mathbf{1}}^{\boldsymbol{n}} \in \mathbb{R}^{N(\boldsymbol{n})}$ and every $\boldsymbol{j} = \mathbf{1}, \dots, \boldsymbol{n}$,

$$\begin{aligned}
&\left[\left(\left(\bigotimes_{r=1}^{k-1} I_{n_r}\right)\otimes K_{n_k}\otimes\left(\bigotimes_{r=k+1}^{d} I_{n_r}\right)\right)\mathbf{u}\right]_{\boldsymbol{j}} \\
&= \sum_{\boldsymbol{\ell}=\mathbf{1}}^{\boldsymbol{n}}\left[\left(\bigotimes_{r=1}^{k-1} I_{n_r}\right)\otimes K_{n_k}\otimes\left(\bigotimes_{r=k+1}^{d} I_{n_r}\right)\right]_{\boldsymbol{j}\boldsymbol{\ell}} u_{\boldsymbol{\ell}} \\
&= \sum_{\boldsymbol{\ell}=\mathbf{1}}^{\boldsymbol{n}} u_{\boldsymbol{\ell}}\,(K_{n_k})_{j_k\ell_k}\prod_{\substack{r=1\\ r\neq k}}^{d}(I_{n_r})_{j_r\ell_r} = \sum_{\boldsymbol{\ell}=\mathbf{1}}^{\boldsymbol{n}} u_{\boldsymbol{\ell}}\,(K_{n_k})_{j_k\ell_k}\prod_{\substack{r=1\\ r\neq k}}^{d}\delta_{j_r\ell_r} \\
&= -u_{\boldsymbol{j}-\mathbf{e}_k} + 2u_{\boldsymbol{j}} - u_{\boldsymbol{j}+\mathbf{e}_k} = (K_{\boldsymbol{n},kk}\mathbf{u})_{\boldsymbol{j}},
\end{aligned}$$

where the second-to-last equality is due to the fact that, when $\boldsymbol{\ell}$ varies from $\mathbf{1}$ to $\boldsymbol{n}$,

$$(K_{n_k})_{j_k\ell_k}\prod_{\substack{r=1\\ r\neq k}}^{d}\delta_{j_r\ell_r} = \begin{cases} -1,\ 2,\ -1, & \text{for } \boldsymbol{\ell} = \boldsymbol{j}-\mathbf{e}_k,\ \boldsymbol{j},\ \boldsymbol{j}+\mathbf{e}_k, \text{ respectively,} \\ 0, & \text{otherwise.} \end{cases}$$

□

Remark 7.3 K_n, H_n, I_n are the diffusion, convection, reaction matrices resulting from the classical central FD discretization of the univariate problem

$$\begin{cases} -u''(x) + u'(x) + u(x) = f(x), & x \in (0,1), \\ u(0) = u(1) = 0. \end{cases}$$

In other words, K_n, H_n, I_n are the matrices resulting from the FD discretization of, respectively, the negative second derivative $-u''(x)$, the first derivative $u'(x)$, the identity operator $u(x)$. To see this, follow the above derivation of the FD discretization matrices in the univariate case $d = 1$ or take a look at the 3rd part of [22, Sect. 10.5.2]. Considering that $K_{\boldsymbol{n},kk} = h_k^2 K_{\boldsymbol{n},kk}(1)$ is the matrix resulting from the FD discretization of the d-variate problem

$$\begin{cases} -\dfrac{\partial^2 u}{\partial x_k^2} = f, & \text{in } (0,1)^d, \\ u = 0, & \text{on } \partial((0,1)^d), \end{cases}$$

that is, the matrix associated with the FD discretization of the negative second derivative $-\partial^2 u/\partial x_k^2$, it is immediately clear from (7.29) the relationship that exists between K_{n}, I_n and $K_{\boldsymbol{n},kk}$(!) Similar considerations also apply to $K_{\boldsymbol{n},\ell k}$ with $\ell \neq k$, $H_{\boldsymbol{n},k}$ and $I_{\boldsymbol{n}}$.

Remark 7.4 It follows from Lemma 7.1 and **T 8** that

$$K_{\boldsymbol{n},kk} = T_{\boldsymbol{n}}(2 - 2\cos\theta_k), \qquad k = 1, \ldots, d, \tag{7.33}$$

$$K_{\boldsymbol{n},\ell k} = T_{\boldsymbol{n}}(\sin\theta_\ell \sin\theta_k), \qquad \ell, k = 1, \ldots, d, \qquad \ell \neq k, \tag{7.34}$$

$$H_{\boldsymbol{n},k} = -\mathrm{i}\, T_{\boldsymbol{n}}(\sin\theta_k), \qquad k = 1, \ldots, d. \tag{7.35}$$

In particular,

$$K_{\boldsymbol{n},\ell k} = T_{\boldsymbol{n}}(H_{\ell k}), \qquad \ell, k = 1, \ldots, d, \tag{7.36}$$

where $H(\boldsymbol{\theta})$ is the $d \times d$ symmetric matrix defined as follows:

$$H_{\ell k}(\boldsymbol{\theta}) = \begin{cases} 2 - 2\cos\theta_k, & \text{if } \ell = k, \\ \sin\theta_\ell \sin\theta_k, & \text{if } \ell \neq k. \end{cases} \tag{7.37}$$

Since $K_{\boldsymbol{n},\ell k}(1) = (h_\ell h_k)^{-1} K_{\boldsymbol{n},\ell k}$, according to (7.36) and the discussion in Sect. 7.2 (see in particular (7.14)), we may predict that $H(\boldsymbol{\theta})$ is, up to some normalization, the symbol of the negative Hessian operator. For instance, assuming $\boldsymbol{n} + \boldsymbol{1} = \boldsymbol{\nu} n$ for some fixed vector $\boldsymbol{\nu} \in \mathbb{Q}^d$ with positive components, from (7.36), **GLT 3** and **GLT 4** we infer that

$$\{n^{-2} K_{\boldsymbol{n},\ell k}(1) = \nu_\ell \nu_k\, T_{\boldsymbol{n}}(H_{\ell k})\}_n \sim_{\mathrm{GLT}} \nu_\ell \nu_k\, H_{\ell k}(\boldsymbol{\theta}) = H_{\ell k}^{(\boldsymbol{\nu})}(\boldsymbol{\theta}), \tag{7.38}$$

where $H^{(\boldsymbol{\nu})}(\boldsymbol{\theta}) = \mathrm{diag}(\boldsymbol{\nu}) H(\boldsymbol{\theta}) \mathrm{diag}(\boldsymbol{\nu})$. This means that, assuming $\boldsymbol{n} + \boldsymbol{1} = \boldsymbol{\nu} n$ and after normalization by n^{-2}, the symbol of the negative Hessian operator is $H^{(\boldsymbol{\nu})}(\boldsymbol{\theta})$, which coincides with $H(\boldsymbol{\theta})$ up to a trivial transformation by $\mathrm{diag}(\boldsymbol{\nu})$. With the same argument as in the proof of [17, Theorem 2.2], one can show that the matrix $H(\boldsymbol{\theta})$ is SPSD for all $\boldsymbol{\theta} \in [-\pi, \pi]^d$, and it is SPD for all $\boldsymbol{\theta} \in [-\pi, \pi]^d$ such that $\theta_1 \cdots \theta_d \neq 0$.

GLT analysis of the FD discretization matrices. Using the theory of multilevel GLT sequences, we now derive the spectral and singular value distribution of the sequence of normalized FD discretization matrices $\{n^{-2} A_{\boldsymbol{n}}\}_n$ under the assumption that $\boldsymbol{n}+\mathbf{1}=\boldsymbol{\nu}n$ for some fixed vector $\boldsymbol{\nu}$. This assumption essentially says that each stepsize $h_i = \frac{1}{n_i+1}$ tends to 0 with the same asymptotic speed as the others.

Theorem 7.3 *Suppose that the following conditions on the PDE coefficients are satisfied:*

- *for every $\ell, k = 1,\ldots,d$, the function $a_{\ell k}:[0,1]^d \to \mathbb{R}$ belongs to $C([0,1]^d)$ and its partial derivatives $\partial a_{\ell k}/\partial x_1,\ldots,\partial a_{\ell k}/\partial x_d:[0,1]^d\to\mathbb{R}$ are bounded;*
- *for every $k=1,\ldots,d$, the function $b_k:[0,1]^d\to\mathbb{R}$ is bounded;*
- *$c:[0,1]^d\to\mathbb{R}$ is bounded.*

Let $\boldsymbol{\nu}\in\mathbb{Q}^d$ be a vector with positive components and assume that $\boldsymbol{n}+\mathbf{1}=\boldsymbol{\nu}n$ (it is understood that n varies in the infinite subset of $\mathbb{N}$ such that $\boldsymbol{n}+\mathbf{1}=\boldsymbol{\nu}n\in\mathbb{N}^d$). Then

$$\{n^{-2}A_{\boldsymbol{n}}\}_n \sim_{\rm GLT} f^{(\boldsymbol{\nu})}(\mathbf{x},\boldsymbol{\theta}) \tag{7.39}$$

and

$$\{n^{-2}A_{\boldsymbol{n}}\}_n \sim_{\sigma,\lambda} f^{(\boldsymbol{\nu})}(\mathbf{x},\boldsymbol{\theta}), \tag{7.40}$$

where

$$f^{(\boldsymbol{\nu})}(\mathbf{x},\boldsymbol{\theta}) = \sum_{\ell,k=1}^{d} a_{\ell k}(\mathbf{x})H_{\ell k}^{(\boldsymbol{\nu})}(\boldsymbol{\theta}) = \mathbf{1}(A(\mathbf{x})\circ H^{(\boldsymbol{\nu})}(\boldsymbol{\theta}))\mathbf{1}^T = \boldsymbol{\nu}(A(\mathbf{x})\circ H(\boldsymbol{\theta}))\boldsymbol{\nu}^T,$$
$$H^{(\boldsymbol{\nu})}(\boldsymbol{\theta}) = \mathrm{diag}(\boldsymbol{\nu})H(\boldsymbol{\theta})\mathrm{diag}(\boldsymbol{\nu}),$$

and $H(\boldsymbol{\theta})$ is defined in (7.37).

Proof The proof consists of the following steps. In what follows, the letter C denotes a generic constant independent of n. While reading this proof, the reader should keep in mind the relation $\boldsymbol{n}+\mathbf{1}=\boldsymbol{\nu}n$.

Step 1. In view of (7.24), we decompose $n^{-2}A_{\boldsymbol{n}}$ as follows:

$$n^{-2}A_{\boldsymbol{n}} = n^{-2}K_{\boldsymbol{n}} + n^{-2}Z_{\boldsymbol{n}}, \tag{7.41}$$

where

$$\begin{aligned} n^{-2}K_{\boldsymbol{n}} &= n^{-2}\sum_{\ell,k=1}^{d}\frac{1}{h_\ell h_k}\Big(\operatorname*{diag}_{\boldsymbol{j}=\mathbf{1},\ldots,\boldsymbol{n}} a_{\ell k}(\mathbf{x}_{\boldsymbol{j}})\Big)K_{\boldsymbol{n},\ell k} \\ &= \sum_{\ell,k=1}^{d}\nu_\ell\nu_k\Big(\operatorname*{diag}_{\boldsymbol{j}=\mathbf{1},\ldots,\boldsymbol{n}} a_{\ell k}(\mathbf{x}_{\boldsymbol{j}})\Big)K_{\boldsymbol{n},\ell k} \end{aligned} \tag{7.42}$$

is the diffusion matrix, resulting from the FD discretization of the higher-order (diffusion) term in (7.18), while

$$n^{-2}Z_n = n^{-2}\sum_{k=1}^{d}\frac{1}{h_k}\Big(\operatorname*{diag}_{j=1,\ldots,n} s_k(\mathbf{x}_j)\Big)H_{n,k} + n^{-2}\Big(\operatorname*{diag}_{j=1,\ldots,n} c(\mathbf{x}_j)\Big)I_n$$

$$= \sum_{k=1}^{d} n^{-1}\nu_k\Big(\operatorname*{diag}_{j=1,\ldots,n} s_k(\mathbf{x}_j)\Big)H_{n,k} + n^{-2}\Big(\operatorname*{diag}_{j=1,\ldots,n} c(\mathbf{x}_j)\Big) \tag{7.43}$$

is the matrix resulting from the FD discretization of the lower-order terms (the convection and reaction terms). We show that

$$\|n^{-2}K_n\| \le C, \tag{7.44}$$

$$\|n^{-2}Z_n\| \le Cn^{-1}. \tag{7.45}$$

We have

$$\|n^{-2}K_n\| = \left\|\sum_{\ell,k=1}^{d}\nu_\ell\nu_k\Big(\operatorname*{diag}_{j=1,\ldots,n} a_{\ell k}(\mathbf{x}_j)\Big)K_{n,\ell k}\right\|$$

$$\le \sum_{\ell,k=1}^{d}\nu_\ell\nu_k\left\|\operatorname*{diag}_{j=1,\ldots,n} a_{\ell k}(\mathbf{x}_j)\right\|\,\|K_{n,\ell k}\|$$

$$\le \sum_{\ell,k=1}^{d}\nu_\ell\nu_k\|a_{\ell k}\|_\infty\,\|K_{n,\ell k}\| \le 4\sum_{\ell,k=1}^{d}\nu_\ell\nu_k\|a_{\ell k}\|_\infty = C,$$

where in the last inequality we used the fact that $\|K_{n,\ell k}\| \le 4$, which follows from either Lemma 7.1 and **P 5** (taking into account that $\|K_n\| \le 4$ and $\|H_n\| \le 1$ for all n by **N 1**) or (7.36) and **T 3** (taking into account that $\|H_{\ell k}\|_\infty \le 4$ for all $\ell, k = 1, \ldots, d$). Note that $\|a_{\ell k}\|_\infty$ is finite because of the assumption that $a_{\ell k} \in C([0, 1]^d)$. This completes the proof of (7.44). The proof of (7.45) is analogous:

$$\|n^{-2}Z_n\| = \left\|\sum_{k=1}^{d} n^{-1}\nu_k\Big(\operatorname*{diag}_{j=1,\ldots,n} s_k(\mathbf{x}_j)\Big)H_{n,k} + n^{-2}\Big(\operatorname*{diag}_{j=1,\ldots,n} c(\mathbf{x}_j)\Big)\right\|$$

$$\le \sum_{k=1}^{d} n^{-1}\nu_k\left\|\operatorname*{diag}_{j=1,\ldots,n} s_k(\mathbf{x}_j)\right\|\,\|H_{n,k}\| + n^{-2}\left\|\operatorname*{diag}_{j=1,\ldots,n} c(\mathbf{x}_j)\right\|$$

$$\le \sum_{k=1}^{d} n^{-1}\nu_k\|s_k\|_\infty\,\|H_{n,k}\| + n^{-2}\|c\|_\infty$$

$$\le \sum_{k=1}^{d} n^{-1}\nu_k\|s_k\|_\infty + n^{-2}\|c\|_\infty \le Cn^{-1},$$

where in the second-to-last inequality we used the fact that $\|H_{n,k}\| \le 1$, which follows from either Lemma 7.1 and **P 5** (taking into account that $\|H_n\| \le 1$ for all n) or (7.35) and **T 3** (taking into account that $\| \sin \theta_k \|_\infty \le 1$ for all $k = 1, \ldots, d$). Note that $\|c\|_\infty$ and $\|s_k\|_\infty$ are finite because of the assumption that c, b_k and the partial derivatives $\partial a_{\ell k}/\partial x_r$ are bounded.

Step 2. Define the symmetric matrix

$$n^{-2}\tilde{K}_{\boldsymbol{n}} = n^{-2} \sum_{\ell,k=1}^{d} \frac{1}{h_\ell h_k} S_{\boldsymbol{n}}(a_{\ell k}) \circ T_{\boldsymbol{n}}(H_{\ell k}) = \sum_{\ell,k=1}^{d} S_{\boldsymbol{n}}(a_{\ell k}) \circ T_{\boldsymbol{n}}(H_{\ell k}^{(\boldsymbol{\nu})}) \tag{7.46}$$

and consider the following decomposition of $n^{-2}A_{\boldsymbol{n}}$:

$$n^{-2}A_{\boldsymbol{n}} = n^{-2}\tilde{K}_{\boldsymbol{n}} + (n^{-2}K_{\boldsymbol{n}} - n^{-2}\tilde{K}_{\boldsymbol{n}}) + n^{-2}Z_{\boldsymbol{n}}. \tag{7.47}$$

By Theorem 7.2 and **GLT 4**, $\|n^{-2}\tilde{K}_{\boldsymbol{n}}\| \le C$ and $\{n^{-2}\tilde{K}_{\boldsymbol{n}}\}_n \sim_{\rm GLT} f^{(\boldsymbol{\nu})}(\mathbf{x}, \boldsymbol{\theta})$. In the next step we show that $n^{-2}\tilde{K}_{\boldsymbol{n}}$ is a symmetric approximation of $n^{-2}K_{\boldsymbol{n}}$, in the sense that

$$\|n^{-2}K_{\boldsymbol{n}} - n^{-2}\tilde{K}_{\boldsymbol{n}}\| \to 0. \tag{7.48}$$

Once this is done, the theorem is proved. Indeed, from (7.48) and Step 1 we have $\|n^{-2}K_{\boldsymbol{n}} - n^{-2}\tilde{K}_{\boldsymbol{n}} + n^{-2}Z_{\boldsymbol{n}}\| \to 0$. Hence, the GLT relation (7.39) follows from (7.47), **Z 1** (or **Z 2**) and **GLT 3**–**GLT 4**; the singular value distribution in (7.40) follows from (7.39) and **GLT 1**; and the spectral distribution in (7.40) follows from (7.39) and **GLT 2** applied to the decomposition (7.47), taking into account **N 2**.

Step 3. To prove (7.48), we use (7.36) and the fact that

$$\begin{aligned}\Big\| \operatorname*{diag}_{\boldsymbol{j}=\boldsymbol{1},\ldots,\boldsymbol{n}} a(\mathbf{x}_{\boldsymbol{j}}) - D_{\boldsymbol{n}}(a) \Big\| &= \max_{\boldsymbol{j}=\boldsymbol{1},\ldots,\boldsymbol{n}} \Big| a(\mathbf{x}_{\boldsymbol{j}}) - a\Big(\frac{\boldsymbol{j}}{\boldsymbol{n}}\Big) \Big| \\ &\le \max_{\boldsymbol{j}=\boldsymbol{1},\ldots,\boldsymbol{n}} \omega_a\Big(\Big| \mathbf{x}_{\boldsymbol{j}} - \frac{\boldsymbol{j}}{\boldsymbol{n}} \Big|_\infty \Big) \le \omega_a(\max(\boldsymbol{h}))\end{aligned} \tag{7.49}$$

for all functions $a \in C([0, 1]^d)$. We have

$$\begin{aligned}&\|n^{-2}K_{\boldsymbol{n}} - n^{-2}\tilde{K}_{\boldsymbol{n}}\| \\ &= \Big\| \sum_{\ell,k=1}^{d} \nu_\ell \nu_k \Big(\operatorname*{diag}_{\boldsymbol{j}=\boldsymbol{1},\ldots,\boldsymbol{n}} a_{\ell k}(\mathbf{x}_{\boldsymbol{j}}) \Big) K_{\boldsymbol{n},\ell k} - \sum_{\ell,k=1}^{d} S_{\boldsymbol{n}}(a_{\ell k}) \circ T_{\boldsymbol{n}}(H_{\ell k}^{(\boldsymbol{\nu})}) \Big\| \\ &= \Big\| \sum_{\ell,k=1}^{d} \Big[\Big(\operatorname*{diag}_{\boldsymbol{j}=\boldsymbol{1},\ldots,\boldsymbol{n}} a_{\ell k}(\mathbf{x}_{\boldsymbol{j}}) \Big) T_{\boldsymbol{n}}(H_{\ell k}^{(\boldsymbol{\nu})}) - D_{\boldsymbol{n}}(a_{\ell k}) T_{\boldsymbol{n}}(H_{\ell k}^{(\boldsymbol{\nu})}) \Big] \\ &\quad + \sum_{\ell,k=1}^{d} \Big[D_{\boldsymbol{n}}(a_{\ell k}) T_{\boldsymbol{n}}(H_{\ell k}^{(\boldsymbol{\nu})}) - S_{\boldsymbol{n}}(a_{\ell k}) \circ T_{\boldsymbol{n}}(H_{\ell k}^{(\boldsymbol{\nu})}) \Big] \Big\|\end{aligned}$$

$$\leq \sum_{\ell,k=1}^{d} \left\| \operatorname*{diag}_{j=1,\dots,n} a_{\ell k}(\mathbf{x}_j) - D_{\boldsymbol{n}}(a_{\ell k}) \right\| \|T_{\boldsymbol{n}}(H_{\ell k}^{(\boldsymbol{\nu})})\|$$

$$+ \sum_{\ell,k=1}^{d} \|D_{\boldsymbol{n}}(a_{\ell k})T_{\boldsymbol{n}}(H_{\ell k}^{(\boldsymbol{\nu})}) - S_{\boldsymbol{n}}(a_{\ell k}) \circ T_{\boldsymbol{n}}(H_{\ell k}^{(\boldsymbol{\nu})})\|,$$

which tends to 0 by Theorem 7.2 and (7.49), taking into account that $a_{\ell k} \in C([0,1]^d)$ by assumption and $\|T_{\boldsymbol{n}}(H_{\ell k}^{(\boldsymbol{\nu})})\| \leq \|H_{\ell k}^{(\boldsymbol{\nu})}\|_\infty$ by **T3**. □

We remark the the proof of Theorem 7.3 is conceptually analogous to the proof of its univariate version [22, Theorem 10.8] in the form suggested by [22, Remark 10.4].

7.4 FE Discretization of Convection-Diffusion-Reaction PDEs

Consider the convection-diffusion-reaction problem

$$\begin{cases} -\nabla \cdot A\nabla u + \mathbf{b} \cdot \nabla u + cu = f, & \text{in } (0,1)^d, \\ u = 0, & \text{on } \partial((0,1)^d), \end{cases}$$

$$\iff \begin{cases} -\displaystyle\sum_{\ell,k=1}^{d} \frac{\partial}{\partial x_\ell}\Big(a_{\ell k}\frac{\partial u}{\partial x_k}\Big) + \sum_{k=1}^{d} b_k \frac{\partial u}{\partial x_k} + cu = f, & \text{in } (0,1)^d, \\ u = 0, & \text{on } \partial((0,1)^d), \end{cases} \tag{7.50}$$

where $a_{\ell k}, b_k, c, f$ are given functions, $A = [a_{\ell k}]_{\ell,k=1}^d$ and $\mathbf{b} = [b_k]_{k=1}^d$.

FE discretization. The weak form of (7.50) reads as follows [10, Chap. 9]: find $u \in H_0^1([0,1]^d)$ such that

$$\mathrm{a}(u,w) = \mathrm{f}(w), \qquad \forall w \in H_0^1([0,1]^d),$$

where

$$\mathrm{a}(u,w) = \int_{(0,1)^d} ((\nabla w)^T A \nabla u + (\nabla u)^T \mathbf{b}\, w + cuw), \qquad \mathrm{f}(w) = \int_{(0,1)^d} fw.$$

Let $\boldsymbol{n} \in \mathbb{N}^d$, set $\boldsymbol{h} = \frac{\mathbf{1}}{\boldsymbol{n}+\mathbf{1}}$ and $\mathbf{x}_{\boldsymbol{j}} = \boldsymbol{j}\boldsymbol{h}$ for $\boldsymbol{j} = \mathbf{0},\dots,\boldsymbol{n}+\mathbf{1}$. In what follows, we use the notation $x_{j_i} = j_i h_i$ for $j_i = 0,\dots,n_i+1$ and $i = 1,\dots,d$, so that we can write $\mathbf{x}_{\boldsymbol{j}} = (x_{j_1},\dots,x_{j_d})$ for all $\boldsymbol{j} = \mathbf{0},\dots,\boldsymbol{n}+\mathbf{1}$. Pay attention to the fact that this notation is used to simplify the presentation but is not rigorous from a mathematical viewpoint. Fix the subspace $\mathscr{W}_{\boldsymbol{n}} = \mathrm{span}(\varphi_{\mathbf{1}},\dots,\varphi_{\boldsymbol{n}})$, where $\varphi_{\mathbf{1}},\dots,\varphi_{\boldsymbol{n}} : [0,1]^d \to \mathbb{R}$ are the so-called tensor-product hat-functions. They are defined as follows:

$$\varphi_{\boldsymbol{j}} = \varphi_{j_1} \otimes \cdots \otimes \varphi_{j_d}, \qquad \boldsymbol{j} = \mathbf{1}, \ldots, \boldsymbol{n}, \tag{7.51}$$

where the functions $\varphi_{j_i} : [0, 1] \to \mathbb{R}$, $j_i = 1, \ldots, n_i$, are the hat-functions corresponding to the discretization step $h_i = \frac{1}{n_i+1}$. We recall from [22, Sect. 10.6.1] that the hat-functions $\varphi_j : [0, 1] \to \mathbb{R}$, $j = 1, \ldots, n$, corresponding to the discretization step $h = \frac{1}{n+1}$ are defined as

$$\varphi_j(x) = \frac{x - x_{j-1}}{x_j - x_{j-1}} \chi_{[x_{j-1}, x_j)}(x) + \frac{x_{j+1} - x}{x_{j+1} - x_j} \chi_{[x_j, x_{j+1})}(x), \qquad j = 1, \ldots, n; \tag{7.52}$$

see also [22, Fig. 10.4]. It can be shown that $\mathscr{W}_{\boldsymbol{n}} \subset H_0^1([0, 1]^d)$ and the tensor-product hat-functions (7.51) form a basis for $\mathscr{W}_{\boldsymbol{n}}$ (i.e., they are linearly independent). Moreover, any partial Sobolev derivative of each tensor-product hat-function $\varphi_{\boldsymbol{j}}$ coincides with the classical partial derivative (which exists a.e. in $[0, 1]^d$). A few properties of tensor-product hat-functions, which can be easily established on the basis of (7.52), are reported below.

- Local support property:

$$\mathrm{supp}(\varphi_{\boldsymbol{i}}) = [\mathbf{x}_{\boldsymbol{i}-\mathbf{1}}, \mathbf{x}_{\boldsymbol{i}+\mathbf{1}}] = [x_{i_1-1}, x_{i_1+1}] \times \cdots \times [x_{i_d-1}, x_{i_d+1}] \tag{7.53}$$

for all $\boldsymbol{i} = \mathbf{1}, \ldots, \boldsymbol{n}$. In particular, $\mu_d(\mathrm{supp}(\varphi_{\boldsymbol{i}})) = 2^d / N(\boldsymbol{n} + \mathbf{1})$.

- Bound for partial derivatives:

$$\left|\frac{\partial \varphi_{\boldsymbol{i}}}{\partial x_k}(\mathbf{x})\right| = \left|\varphi'_{i_k}(x_k) \prod_{\substack{r=1 \\ r \neq k}}^{d} \varphi_{i_r}(x_r)\right| \le n_k + 1 \tag{7.54}$$

for all $\boldsymbol{i} = \mathbf{1}, \ldots, \boldsymbol{n}$ and for a.e. $\mathbf{x} \in [0, 1]^d$.

- Bound for the sum of partial derivatives:

$$\sum_{\boldsymbol{i}=\mathbf{1}}^{\boldsymbol{n}} \left|\frac{\partial \varphi_{\boldsymbol{i}}}{\partial x_k}(\mathbf{x})\right| \le 2(n_k + 1) \tag{7.55}$$

for a.e. $\mathbf{x} \in [0, 1]^d$.

In the linear FE approach, we look for an approximation $u_{\mathscr{W}_{\boldsymbol{n}}}$ of u by solving the following (Galerkin) problem: find $u_{\mathscr{W}_{\boldsymbol{n}}} \in \mathscr{W}_{\boldsymbol{n}}$ such that

$$\mathrm{a}(u_{\mathscr{W}_{\boldsymbol{n}}}, w) = \mathrm{f}(w), \qquad \forall\, w \in \mathscr{W}_{\boldsymbol{n}}.$$

Since $\{\varphi_{\mathbf{1}}, \ldots, \varphi_{\boldsymbol{n}}\}$ is a basis for $\mathscr{W}_{\boldsymbol{n}}$, we can write $u_{\mathscr{W}_{\boldsymbol{n}}} = \sum_{\boldsymbol{j}=\mathbf{1}}^{\boldsymbol{n}} u_{\boldsymbol{j}} \varphi_{\boldsymbol{j}}$ for a unique vector $\mathbf{u} = (u_{\mathbf{1}}, \ldots, u_{\boldsymbol{n}})^T$. By linearity, the computation of $u_{\mathscr{W}_{\boldsymbol{n}}}$ (i.e., of $\mathbf{u}$) reduces to solving the linear system

$$A_n\mathbf{u} = \mathbf{f},$$

where $\mathbf{f} = (\mathrm{f}(\varphi_1), \ldots, \mathrm{f}(\varphi_n))^T$ and A_n is the stiffness matrix,

$$A_n = [\mathrm{a}(\varphi_j, \varphi_i)]_{i,j=1}^n.$$

Note that A_n admits the following decomposition:

$$A_n = K_n + Z_n, \tag{7.56}$$

where

$$K_n = \left[\int_{(0,1)^d} (\nabla\varphi_i)^T A \nabla\varphi_j\right]_{i,j=1}^n \tag{7.57}$$

is the diffusion matrix and

$$Z_n = \left[\int_{(0,1)^d} (\nabla\varphi_j)^T \mathbf{b}\, \varphi_i\right]_{i,j=1}^n + \left[\int_{(0,1)^d} c\varphi_j\varphi_i\right]_{i,j=1}^n \tag{7.58}$$

is the sum of the convection and reaction matrices.

FE discretization matrices. As explained in Sect. 7.2, the crucial object for determining the symbol of a sequence of matrices arising from the discretization of the PDE (7.50) is the "symbol of the negative Hessian operator". Let us then investigate the structure of the FE discretization matrices so as to "guess" this crucial object. Let $E_{\ell k}$ be the $d \times d$ matrix with 1 in position (ℓ, k) and 0 elsewhere, $1 \le \ell, k \le d$. By looking at the expression of K_n, we see that

$$K_n = \sum_{\ell,k=1}^{d} K_{n,\ell k}(a_{\ell k}), \tag{7.59}$$

where

$$K_{n,\ell k}(a_{\ell k}) = \left[\int_{(0,1)^d} a_{\ell k} \frac{\partial\varphi_i}{\partial x_\ell}\frac{\partial\varphi_j}{\partial x_k}\right]_{i,j=1}^n \tag{7.60}$$

is the matrix obtained from the FE discretization of (7.50) with A replaced by $a_{\ell k}E_{\ell k}$, that is,

$$\begin{cases} -\dfrac{\partial}{\partial x_\ell}\left(a_{\ell k}\dfrac{\partial u}{\partial x_k}\right) = f, & \text{in } (0,1)^d, \\ u = 0, & \text{on } \partial((0,1)^d). \end{cases} \tag{7.61}$$

According to the discussion in Sect. 7.2, in order to guess the symbol of the negative Hessian operator, we should understand which is the symbol of (a properly normalized version of) the matrix-sequence $\{K_{n,\ell k}(1)\}_n$, once we have fixed a suitable relation $\boldsymbol{n} = \boldsymbol{n}(n)$. The matrix $K_{\boldsymbol{n},\ell k}(1)$ is in fact the matrix resulting from the FE discretization of (7.61) with $a_{\ell k} = 1$ identically, and it is therefore the matrix resulting from the FE discretization of the second derivative $-\partial^2 u/\partial x_\ell \partial x_k$. To sum up, we have to study the matrix

$$K_{\boldsymbol{n},\ell k} = K_{\boldsymbol{n},\ell k}(1) = \left[\int_{(0,1)^d} \frac{\partial \varphi_{\boldsymbol{i}}}{\partial x_\ell} \frac{\partial \varphi_{\boldsymbol{j}}}{\partial x_k}\right]_{\boldsymbol{i},\boldsymbol{j}=\boldsymbol{1}}^{\boldsymbol{n}}. \tag{7.62}$$

Lemma 7.2 *For every $\boldsymbol{n} \in \mathbb{N}^d$, we have*

$$K_{\boldsymbol{n},kk} = \left(\bigotimes_{r=1}^{k-1} M_{n_r}\right) \otimes K_{n_k} \otimes \left(\bigotimes_{r=k+1}^{d} M_{n_r}\right) \tag{7.63}$$

for $k = 1, \ldots, d$, and

$$K_{\boldsymbol{n},k\ell} = K_{\boldsymbol{n},\ell k} = -\left(\bigotimes_{r=1}^{\ell-1} M_{n_r}\right) \otimes H_{n_\ell} \otimes \left(\bigotimes_{r=\ell+1}^{k-1} M_{n_r}\right) \otimes H_{n_k} \otimes \left(\bigotimes_{r=k+1}^{d} M_{n_r}\right) \tag{7.64}$$

for $1 \leq \ell < k \leq d$, where the matrices K_n, H_n, M_n are defined in terms of the hat-functions (7.52) *as follows*:

$$K_n = \left[\int_0^1 \varphi_j'(x)\varphi_i'(x)\mathrm{d}x\right]_{i,j=1}^n = (n+1)T_n(2 - 2\cos\theta), \tag{7.65}$$

$$H_n = \left[\int_0^1 \varphi_j'(x)\varphi_i(x)\mathrm{d}x\right]_{i,j=1}^n = \left[-\int_0^1 \varphi_j(x)\varphi_i'(x)\mathrm{d}x\right]_{i,j=1}^n = -\mathrm{i}T_n(\sin\theta), \tag{7.66}$$

$$M_n = \left[\int_0^1 \varphi_j(x)\varphi_i(x)\mathrm{d}x\right]_{i,j=1}^n = \frac{1}{n+1}T_n\left(\frac{2}{3} + \frac{1}{3}\cos\theta\right), \tag{7.67}$$

with the last equalities in (7.65)–(7.67) *following from direct computations based on* (7.52).

Proof We only prove (7.63) because (7.64) is proved in the same way. For every $\boldsymbol{i}, \boldsymbol{j} = \boldsymbol{1}, \ldots, \boldsymbol{n}$,

$$
\begin{aligned}
(K_{\boldsymbol{n},kk})_{\boldsymbol{ij}} &= \int_{(0,1)^d} \frac{\partial \varphi_{\boldsymbol{i}}}{\partial x_k}(\mathbf{x}) \frac{\partial \varphi_{\boldsymbol{j}}}{\partial x_k}(\mathbf{x})\mathrm{d}\mathbf{x} = \int_{(0,1)^d} \varphi'_{i_k}(x_k)\varphi'_{j_k}(x_k) \prod_{\substack{r=1\\ r\neq k}}^{d} \varphi_{i_r}(x_r)\varphi_{j_r}(x_r)\mathrm{d}\mathbf{x} \\
&= \int_0^1 \varphi'_{j_k}(x_k)\varphi'_{i_k}(x_k)\mathrm{d}x_k \prod_{\substack{r=1\\ r\neq k}}^{d} \int_0^1 \varphi_{j_r}(x_r)\varphi_{i_r}(x_r)\mathrm{d}x_r = (K_{n_k})_{i_k j_k} \prod_{\substack{r=1\\ r\neq k}}^{d} (M_{n_r})_{i_r j_r} \\
&= (M_{n_1} \otimes \cdots \otimes M_{n_{k-1}} \otimes K_{n_k} \otimes M_{n_{k+1}} \otimes \cdots \otimes M_{n_d})_{\boldsymbol{ij}},
\end{aligned}
$$

where the last equality follows from the crucial property **P 7**. □

Remark 7.5 As in the case of FDs (see Remark 7.3), K_n, H_n, M_n are the diffusion, convection, reaction matrices resulting from the FE discretization of the univariate problem

$$
\begin{cases} -u''(x) + u'(x) + u(x) = f(x), & x \in (0,1), \\ u(0) = u(1) = 0. \end{cases}
$$

In other words, K_n, H_n, M_n are the matrices resulting from the FE discretization of, respectively, the negative second derivative $-u''(x)$, the first derivative $u'(x)$, the identity operator $u(x)$. To see this, follow the above derivation of the FE discretization matrices in the univariate case $d = 1$ or take a look at [22, Sect. 10.6.1].

Remark 7.6 It follows from Lemma 7.2 and **T 8** that, for all $\ell, k = 1, \dots, d$,

$$
K_{\boldsymbol{n},\ell k} = \frac{(n_\ell + 1)(n_k + 1)}{N(\boldsymbol{n} + \mathbf{1})} T_{\boldsymbol{n}}(H_{\ell k}), \tag{7.68}
$$

where $H(\boldsymbol{\theta})$ is the $d \times d$ symmetric matrix defined as follows:

$$
H_{\ell k}(\boldsymbol{\theta}) = \begin{cases} (2 - 2\cos\theta_k) \prod_{\substack{r=1\\ r\neq k}}^{d} (\frac{2}{3} + \frac{1}{3}\cos\theta_r), & \text{if } \ell = k, \\ \sin\theta_\ell \sin\theta_k \prod_{\substack{r=1\\ r\neq \ell,k}}^{d} (\frac{2}{3} + \frac{1}{3}\cos\theta_r), & \text{if } \ell \neq k. \end{cases} \tag{7.69}
$$

The matrix $H(\boldsymbol{\theta})$ is what we may predict to be, up to some normalization, the symbol of the negative Hessian operator. For instance, assuming $\boldsymbol{n} + \mathbf{1} = \boldsymbol{\nu} n$ for some fixed vector $\boldsymbol{\nu} \in \mathbb{Q}^d$ with positive components, from (7.68), **GLT 3** and **GLT 4** we infer that

$$
\Big\{n^{d-2} K_{\boldsymbol{n},\ell k} = \frac{\nu_\ell \nu_k}{N(\boldsymbol{\nu})} T_{\boldsymbol{n}}(H_{\ell k})\Big\}_n \sim_{\mathrm{GLT}} \frac{\nu_\ell \nu_k}{N(\boldsymbol{\nu})} H_{\ell k}(\boldsymbol{\theta}) = H^{(\boldsymbol{\nu})}_{\ell k}(\boldsymbol{\theta}), \tag{7.70}
$$

where $H^{(\boldsymbol{\nu})}(\boldsymbol{\theta}) = \mathrm{diag}(\boldsymbol{\nu}) H(\boldsymbol{\theta}) \mathrm{diag}(\boldsymbol{\nu}) / N(\boldsymbol{\nu})$. This means that, assuming $\boldsymbol{n} + \mathbf{1} = \boldsymbol{\nu} n$ and after normalization by n^{d-2}, the symbol of the negative Hessian operator is $H^{(\boldsymbol{\nu})}(\boldsymbol{\theta})$, which coincides with $H(\boldsymbol{\theta})$ up to a trivial transformation. With the same argument as in the proof of [17, Theorem 2.2], one can show that the matrix $H(\boldsymbol{\theta})$ is SPSD for all $\boldsymbol{\theta} \in [-\pi, \pi]^d$, and it is SPD for all $\boldsymbol{\theta} \in [-\pi, \pi]^d$ such that $\theta_1 \cdots \theta_d \neq 0$.

GLT analysis of the FE discretization matrices. Using the theory of multilevel GLT sequences, we now derive the spectral and singular value distribution of the sequence of normalized stiffness matrices $\{n^{d-2}A_{\boldsymbol{n}}\}_n$ under the assumptions that $\boldsymbol{n}+\mathbf{1}=\boldsymbol{\nu}n$ for some fixed vector $\boldsymbol{\nu}$ and that the matrix of diffusion coefficients $A(\mathbf{x})$ is symmetric.

Theorem 7.4 *Suppose that the following conditions on the PDE coefficients are satisfied:*

- $a_{\ell k}\in L^\infty((0,1)^d)$ *for every* $\ell,k=1,\ldots,d$;
- $b_k\in L^\infty((0,1)^d)$ *for every* $k=1,\ldots,d$;
- $c\in L^\infty((0,1)^d)$;
- $A(\mathbf{x})=[a_{\ell k}(\mathbf{x})]_{\ell,k=1}^d$ *is symmetric for every* $\mathbf{x}\in(0,1)^d$.

Let $\boldsymbol{\nu}\in\mathbb{Q}^d$ *be a vector with positive components and assume that* $\boldsymbol{n}+\mathbf{1}=\boldsymbol{\nu}n$ *(it is understood that* n *varies in the infinite subset of* $\mathbb{N}$ *such that* $\boldsymbol{n}+\mathbf{1}=\boldsymbol{\nu}n\in\mathbb{N}^d$*). Then*

$$\{n^{d-2}A_{\boldsymbol{n}}\}_n\sim_{\mathrm{GLT}} f^{(\boldsymbol{\nu})}(\mathbf{x},\boldsymbol{\theta}) \tag{7.71}$$

and

$$\{n^{d-2}A_{\boldsymbol{n}}\}_n\sim_{\sigma,\lambda} f^{(\boldsymbol{\nu})}(\mathbf{x},\boldsymbol{\theta}), \tag{7.72}$$

where

$$f^{(\boldsymbol{\nu})}(\mathbf{x},\boldsymbol{\theta})=\sum_{\ell,k=1}^d a_{\ell k}(\mathbf{x})H_{\ell k}^{(\boldsymbol{\nu})}(\boldsymbol{\theta})=\mathbf{1}(A(\mathbf{x})\circ H^{(\boldsymbol{\nu})}(\boldsymbol{\theta}))\mathbf{1}^T=\frac{\boldsymbol{\nu}(A(\mathbf{x})\circ H(\boldsymbol{\theta}))\boldsymbol{\nu}^T}{N(\boldsymbol{\nu})},$$
$$H^{(\boldsymbol{\nu})}(\boldsymbol{\theta})=\frac{\mathrm{diag}(\boldsymbol{\nu})H(\boldsymbol{\theta})\mathrm{diag}(\boldsymbol{\nu})}{N(\boldsymbol{\nu})},$$

and $H(\boldsymbol{\theta})$ *is defined in* (7.69).

Proof The proof consists of the following steps. Throughout the proof, the letter C denotes a generic constant independent of n. While reading the proof, the reader should keep in mind the relation $\boldsymbol{n}+\mathbf{1}=\boldsymbol{\nu}n$ and the notation $E_{\ell k}$ for the $d\times d$ matrix having 1 in position (ℓ,k) and 0 elsewhere $(1\le\ell,k\le d)$.

Step 1. We show that

$$\|n^{d-2}K_{\boldsymbol{n}}\|\le C, \tag{7.73}$$
$$\|n^{d-2}Z_{\boldsymbol{n}}\|\le Cn^{-1}. \tag{7.74}$$

To prove (7.73) we note that, for all $\boldsymbol{i},\boldsymbol{j}=\mathbf{1},\ldots,\boldsymbol{n}$, we have the following.

- If $|\boldsymbol{i}-\boldsymbol{j}|_\infty > 1$ then there exists $k \in \{1,\ldots,d\}$ such that $|i_k - j_k| > 1$, which implies by (7.53) that $\mathrm{supp}(\varphi_{\boldsymbol{i}})$ and $\mathrm{supp}(\varphi_{\boldsymbol{j}})$ intersect at most on a set of zero measure. Thus,
$$(K_{\boldsymbol{n}})_{\boldsymbol{ij}} = \int_{(0,1)^d} (\nabla\varphi_{\boldsymbol{i}})^T A \nabla\varphi_{\boldsymbol{j}} = 0.$$
- By (7.53) and (7.54),

$$\begin{aligned}
|(K_{\boldsymbol{n}})_{\boldsymbol{ij}}| &\le \int_{(0,1)^d} |(\nabla\varphi_{\boldsymbol{i}})^T A \nabla\varphi_{\boldsymbol{j}}| \le \int_{(0,1)^d} \sum_{\ell,k=1}^d |a_{\ell k}| \left|\frac{\partial\varphi_{\boldsymbol{i}}}{\partial x_\ell}\right| \left|\frac{\partial\varphi_{\boldsymbol{j}}}{\partial x_k}\right| \\
&\le \max_{\ell,k=1,\ldots,d} \|a_{\ell k}\|_{L^\infty} \sum_{\ell,k=1}^d \int_{\mathrm{supp}(\varphi_{\boldsymbol{i}})} (n_\ell+1)(n_k+1) \\
&\le \max_{\ell,k=1,\ldots,d} \|a_{\ell k}\|_{L^\infty} \sum_{\ell,k=1}^d (n_\ell+1)(n_k+1) \frac{2^d}{N(\boldsymbol{n}+\boldsymbol{1})} \le Cn^{2-d}.
\end{aligned}$$

Thus, each row and column of $K_{\boldsymbol{n}}$ has at most 3^d nonzero entries whose moduli are bounded by Cn^{2-d}, which implies (7.73) by **N 1**. The proof of (7.74) is conceptually identical. For all $\boldsymbol{i},\boldsymbol{j} = \boldsymbol{1},\ldots,\boldsymbol{n}$, we have the following.

- If $|\boldsymbol{i}-\boldsymbol{j}|_\infty > 1$ then $(Z_{\boldsymbol{n}})_{\boldsymbol{ij}} = 0$ for the same reason for which $(K_{\boldsymbol{n}})_{\boldsymbol{ij}} = 0$.
- By (7.53), (7.54) and the obvious inequality $|\varphi_{\boldsymbol{i}}(\mathbf{x})| \le 1$,

$$\begin{aligned}
|(Z_{\boldsymbol{n}})_{\boldsymbol{ij}}| &\le \int_{(0,1)^d} |(\nabla\varphi_{\boldsymbol{j}})^T \mathbf{b}\,\varphi_{\boldsymbol{i}}| + \int_{(0,1)^d} |c\varphi_{\boldsymbol{j}}\varphi_{\boldsymbol{i}}| \\
&\le \int_{\mathrm{supp}(\varphi_{\boldsymbol{i}})} \sum_{k=1}^d |b_k| \left|\frac{\partial\varphi_{\boldsymbol{j}}}{\partial x_k}\right| + \int_{\mathrm{supp}(\varphi_{\boldsymbol{i}})} \|c\|_{L^\infty} \\
&\le \max_{k=1,\ldots,d} \|b_k\|_{L^\infty} \sum_{k=1}^d \int_{\mathrm{supp}(\varphi_{\boldsymbol{i}})} (n_k+1) + \frac{2^d\|c\|_{L^\infty}}{N(\boldsymbol{n}+\boldsymbol{1})} \\
&\le \max_{k=1,\ldots,d} \|b_k\|_{L^\infty} \frac{2^d\sum_{k=1}^d (n_k+1)}{N(\boldsymbol{n}+\boldsymbol{1})} + \frac{2^d\|c\|_{L^\infty}}{N(\boldsymbol{n}+\boldsymbol{1})} \\
&\le Cn^{1-d} + Cn^{-d} \le Cn^{1-d}.
\end{aligned}$$

Thus, each row and column of $Z_{\boldsymbol{n}}$ has at most 3^d nonzero entries whose moduli are bounded by Cn^{1-d}, which implies (7.74) by **N 1**.

Step 2. Let $L^1([0,1]^d, \mathbb{R}^{d\times d})$ be the space of functions $L : [0,1]^d \to \mathbb{R}^{d\times d}$ such that $L_{ij} \in L^1([0,1]^d)$ for all $i,j = 1,\ldots,d$. Consider the linear operator $K_{\boldsymbol{n}}(\cdot) : L^1([0,1]^d, \mathbb{R}^{d\times d}) \to \mathbb{R}^{N(\boldsymbol{n})\times N(\boldsymbol{n})}$,

$$K_{\boldsymbol{n}}(L) = \left[\int_{(0,1)^d} (\nabla\varphi_{\boldsymbol{i}})^T L \nabla\varphi_{\boldsymbol{j}}\right]_{\boldsymbol{i},\boldsymbol{j}=\boldsymbol{1}}^{\boldsymbol{n}}. \tag{7.75}$$

The next four steps are devoted to showing that

$$\{n^{d-2}K_{\boldsymbol{n}}(L)\}_n \sim_{\rm GLT} \mathbf{1}(L(\mathbf{x}) \circ H^{(\boldsymbol{\nu})}(\boldsymbol{\theta}))\mathbf{1}^T, \qquad \forall\, L \in L^1([0,1]^d, \mathbb{R}^{d\times d}). \tag{7.76}$$

Once this is done, the theorem is proved. Indeed, by applying (7.76) with $L = A$, we get $\{n^{d-2}K_{\boldsymbol{n}}\}_n \sim_{\rm GLT} f^{(\boldsymbol{\nu})}(\mathbf{x},\boldsymbol{\theta})$. Moreover, $\{n^{d-2}Z_{\boldsymbol{n}}\}_n$ is zero-distributed by Step 1 and **Z 1** (or **Z 2**). Hence, the GLT relation (7.71) follows from the decomposition $n^{d-2}A_{\boldsymbol{n}} = n^{d-2}K_{\boldsymbol{n}} + n^{d-2}Z_{\boldsymbol{n}}$ (see (7.56)) and **GLT 3**−**GLT 4**; the singular value distribution in (7.72) follows from (7.71) and **GLT 1**; and the spectral distribution in (7.72) follows from (7.71) and **GLT 2** applied to the decomposition $n^{d-2}A_{\boldsymbol{n}} = n^{d-2}K_{\boldsymbol{n}} + n^{d-2}Z_{\boldsymbol{n}}$, taking into account what we have seen in Step 1, the inequality **N 2**, and the fact that $K_{\boldsymbol{n}}$ is symmetric (because $A(\mathbf{x})$ is symmetric for all $\mathbf{x} \in (0,1)^d$ by assumption).

Step 3. We first prove (7.76) in the constant-coefficient case where $L(\mathbf{x}) = E_{\ell k}$ identically. In this case, we have $K_{\boldsymbol{n}}(E_{\ell k}) = K_{\boldsymbol{n},\ell k}$ and (7.76) is nothing else than (7.70).

Step 4. We now prove (7.76) in the case where $L(\mathbf{x}) = a(\mathbf{x})E_{\ell k}$ with $a \in C([0,1]^d)$. To this end, we show that

$$\|Y_{\boldsymbol{n}}\| \to 0 \tag{7.77}$$

where $Y_{\boldsymbol{n}} = n^{d-2}K_{\boldsymbol{n}}(aE_{\ell k}) - n^{d-2}D_{\boldsymbol{n}}(a)K_{\boldsymbol{n}}(E_{\ell k})$. Once this is done, from **Z 1** (or **Z 2**) we have $\{Y_{\boldsymbol{n}}\}_n \sim_\sigma 0$. Hence, from Step 3 and **GLT 3**−**GLT 4** applied to the obvious decomposition $n^{d-2}K_{\boldsymbol{n}}(aE_{\ell k}) = n^{d-2}D_{\boldsymbol{n}}(a)K_{\boldsymbol{n}}(E_{\ell k}) + Y_{\boldsymbol{n}}$, we obtain

$$\{n^{d-2}K_{\boldsymbol{n}}(aE_{\ell k})\}_n \sim_{\rm GLT} a(\mathbf{x})H_{\ell k}^{(\boldsymbol{\nu})}(\boldsymbol{\theta}),$$

as required. For every $\boldsymbol{i},\boldsymbol{j} = \mathbf{1},\ldots,\boldsymbol{n}$, we have the following.

- If $|\boldsymbol{i}-\boldsymbol{j}|_\infty > 1$ then $(K_{\boldsymbol{n}}(L))_{\boldsymbol{ij}} = 0$ for all $L \in L^1([0,1]^d, \mathbb{R}^{d\times d})$, because of the local support property of the tensor-product hat-functions already exploited in Step 1. Thus, $(Y_{\boldsymbol{n}})_{\boldsymbol{ij}} = 0$.
- If $|\boldsymbol{i}-\boldsymbol{j}|_\infty \le 1$ then, using (7.53)–(7.54) and taking into account that $\boldsymbol{i}/\boldsymbol{n} \in \mathrm{supp}(\varphi_{\boldsymbol{i}})$, we obtain

$$\begin{aligned}
|(Y_{\boldsymbol{n}})_{\boldsymbol{ij}}| &= n^{d-2}\left|\int_{(0,1)^d}\left[a(\mathbf{x}) - a\Big(\frac{\boldsymbol{i}}{\boldsymbol{n}}\Big)\right]\frac{\partial\varphi_{\boldsymbol{i}}}{\partial x_\ell}(\mathbf{x})\frac{\partial\varphi_{\boldsymbol{j}}}{\partial x_k}(\mathbf{x})\mathrm{d}\mathbf{x}\right| \\
&\le n^{d-2}\int_{\mathrm{supp}(\varphi_{\boldsymbol{i}})}\left|a(\mathbf{x}) - a\Big(\frac{\boldsymbol{i}}{\boldsymbol{n}}\Big)\right|(n_\ell+1)(n_k+1)\mathrm{d}\mathbf{x} \\
&\le n^{d-2}Cn^2\omega_a\Big(\max_{\mathbf{x}\in\mathrm{supp}(\varphi_{\boldsymbol{i}})}\Big|\mathbf{x} - \frac{\boldsymbol{i}}{\boldsymbol{n}}\Big|_\infty\Big)\int_{\mathrm{supp}(\varphi_{\boldsymbol{i}})}\mathrm{d}\mathbf{x} \\
&\le Cn^d\omega_a\Big(\frac{2}{\min(\boldsymbol{n})+1}\Big)\frac{2^d}{N(\boldsymbol{n}+\mathbf{1})} \le C\omega_a\Big(\frac{2}{\min(\boldsymbol{n})+1}\Big).
\end{aligned}$$

Thus, each row and column of $Y_{\boldsymbol{n}}$ has at most 3^d nonzero entries whose moduli are bounded by $C\omega_a(\frac{2}{\min(\boldsymbol{n})+1})$, which implies (7.77) by **N 1**.

Step 5. We now prove (7.76) in the case where $L(\mathbf{x}) = a(\mathbf{x})E_{\ell k}$ with $a \in L^1([0,1]^d)$. Take $a_m \in C([0,1]^d)$ such that $a_m \to a$ in $L^1([0,1]^d)$. We have

$$\{n^{d-2}K_{\boldsymbol{n}}(a_m E_{\ell k})\}_n \sim_{\rm GLT} a_m(\mathbf{x})H_{\ell k}^{(\boldsymbol{\nu})}(\boldsymbol{\theta})$$

by Step 4, and

$$a_m(\mathbf{x})H_{\ell k}^{(\boldsymbol{\nu})}(\boldsymbol{\theta}) \to a(\mathbf{x})H_{\ell k}^{(\boldsymbol{\nu})}(\boldsymbol{\theta})$$

in measure as $m \to \infty$. We prove that

$$\{n^{d-2}K_{\boldsymbol{n}}(a_m E_{\ell k})\}_m \xrightarrow{\text{a.c.s.}} \{n^{d-2}K_{\boldsymbol{n}}(aE_{\ell k})\}_n, \tag{7.78}$$

after which the desired result follows from **GLT 7**. By **N 3** and the bounds for the sum of partial derivatives of tensor-product hat-functions (7.55), we have

$$\begin{aligned}
\|K_{\boldsymbol{n}}(aE_{\ell k}) - K_{\boldsymbol{n}}(a_m E_{\ell k})\|_1 = \|K_{\boldsymbol{n}}((a-a_m)E_{\ell k})\|_1 &\le \sum_{\boldsymbol{i},\boldsymbol{j}=\mathbf{1}}^{\boldsymbol{n}} |(K_{\boldsymbol{n}}((a-a_m)E_{\ell k}))_{\boldsymbol{i}\boldsymbol{j}}| \\
&\le \sum_{\boldsymbol{i},\boldsymbol{j}=\mathbf{1}}^{\boldsymbol{n}} \int_{(0,1)^d} |a(\mathbf{x})-a_m(\mathbf{x})| \left|\frac{\partial\varphi_{\boldsymbol{i}}}{\partial x_\ell}(\mathbf{x})\right| \left|\frac{\partial\varphi_{\boldsymbol{j}}}{\partial x_k}(\mathbf{x})\right| \mathrm{d}\mathbf{x} \\
&\le \int_{(0,1)^d} |a(\mathbf{x})-a_m(\mathbf{x})| \sum_{\boldsymbol{i}=\mathbf{1}}^{\boldsymbol{n}} \left|\frac{\partial\varphi_{\boldsymbol{i}}}{\partial x_\ell}(\mathbf{x})\right| \sum_{\boldsymbol{j}=\mathbf{1}}^{\boldsymbol{n}} \left|\frac{\partial\varphi_{\boldsymbol{j}}}{\partial x_k}(\mathbf{x})\right| \mathrm{d}\mathbf{x} \\
&\le 4(n_\ell+1)(n_k+1)\|a-a_m\|_{L^1} \le Cn^2\|a-a_m\|_{L^1},
\end{aligned}$$

hence

$$\|n^{d-2}K_{\boldsymbol{n}}(aE_{\ell k}) - n^{d-2}K_{\boldsymbol{n}}(a_m E_{\ell k})\|_1 \le Cn^d\|a-a_m\|_{L^1} \le CN(\boldsymbol{n})\|a-a_m\|_{L^1},$$

and the a.c.s. convergence (7.78) follows from **ACS 6**.

Step 6. Finally, we prove (7.76) for an arbitrary $L \in L^1([0,1]^d, \mathbb{R}^{d\times d})$. Write

$$L(\mathbf{x}) = \sum_{\ell,k=1}^{d} L_{\ell k}(\mathbf{x})E_{\ell k}$$

and note that, by linearity,

$$K_{\boldsymbol{n}}(L) = \sum_{\ell,k=1}^{d} K_{\boldsymbol{n}}(L_{\ell k}E_{\ell k}).$$

Hence, by Step 5 and **GLT 4**,

$$\{n^{d-2}K_{\boldsymbol{n}}(L)\}_n \sim_{\mathrm{GLT}} \sum_{\ell,k=1}^{d} L_{\ell k}(\mathbf{x})H_{\ell k}^{(\boldsymbol{\nu})}(\boldsymbol{\theta}) = \mathbf{1}(L(\mathbf{x}) \circ H^{(\boldsymbol{\nu})}(\boldsymbol{\theta}))\mathbf{1}^T,$$

which concludes the proof. □

We invite the reader to compare the proof of Theorem 7.4 with the proof of its univariate version [22, Theorem 10.12]: they are essentially the same!

We remark that the GLT relation (7.71) and the singular value distribution in (7.72) remain true even without the hypothesis that the matrix of diffusion coefficients $A(\mathbf{x})$ is symmetric for all $\mathbf{x} \in (0, 1)^d$. Indeed, as it is clear from the proof of Theorem 7.4 and especially from Step 2, the symmetry of A is used only in the proof of the eigenvalue distribution in (7.72). Actually, also the eigenvalue distribution remains true without the symmetry assumption on A, as long as we assume that the entries of A are continuous. This result, which has never been observed before in the literature, is proved in the next theorem.

Theorem 7.5 *Suppose that the following conditions on the PDE coefficients are satisfied:*

- $a_{\ell k} \in C([0, 1]^d)$ *for every* $\ell, k = 1, \dots, d$;
- $b_k \in L^\infty((0, 1)^d)$ *for every* $k = 1, \dots, d$;
- $c \in L^\infty((0, 1)^d)$.

Let $\boldsymbol{\nu} \in \mathbb{Q}^d$ *be a vector with positive components and assume that* $\boldsymbol{n} + \mathbf{1} = \boldsymbol{\nu} n$ *(it is understood that* n *varies in the infinite subset of* $\mathbb{N}$ *such that* $\boldsymbol{n} + \mathbf{1} = \boldsymbol{\nu} n \in \mathbb{N}^d$*). Then, both* (7.71) *and* (7.72) *are satisfied.*

Proof As noted before the statement of the theorem, we only have to prove the eigenvalue distribution in (7.72). The underlying idea is that, even in the case where A is not symmetric, the matrix $K_{\boldsymbol{n}}$ is "almost" symmetric (as long as the entries of A are continuous). We can then derive the eigenvalue distribution in (7.72) from **GLT 2** applied to a suitable decomposition of $A_{\boldsymbol{n}}$ obtained from (7.56) by replacing $K_{\boldsymbol{n}}$ with one of its symmetric approximations. Let us work out the details. By (7.59)–(7.60),

$$K_{\boldsymbol{n}} = \sum_{\ell,k=1}^{d} K_{\boldsymbol{n},\ell k}(a_{\ell k}), \qquad K_{\boldsymbol{n},\ell k}(a) = \left[\int_{(0,1)^d} a \frac{\partial \varphi_{\boldsymbol{i}}}{\partial x_\ell} \frac{\partial \varphi_{\boldsymbol{j}}}{\partial x_k}\right]_{\boldsymbol{i},\boldsymbol{j}=\mathbf{1}}^{\boldsymbol{n}}.$$

Let $\tilde{K}_{\boldsymbol{n}}$ be the symmetric approximation of $K_{\boldsymbol{n}}$ given by

$$\tilde{K}_{\boldsymbol{n}} = \sum_{\ell,k=1}^{d} \tilde{K}_{\boldsymbol{n},\ell k}(a_{\ell k}), \qquad \tilde{K}_{\boldsymbol{n},\ell k}(a) = S_{\boldsymbol{n}}(a) \circ K_{\boldsymbol{n},\ell k}(1)$$
$$= S_{\boldsymbol{n}}(a) \circ \frac{(n_\ell + 1)(n_k + 1)}{N(\boldsymbol{n} + \mathbf{1})} T_{\boldsymbol{n}}(H_{\ell k})$$
$$= S_{\boldsymbol{n}}(a) \circ n^{2-d} T_{\boldsymbol{n}}(H_{\ell k}^{(\boldsymbol{\nu})}).$$

We prove that

$$\|n^{d-2} K_{\boldsymbol{n}} - n^{d-2} \tilde{K}_{\boldsymbol{n}}\| \to 0. \tag{7.79}$$

Once this is done, the eigenvalue distribution in (7.72) follows from (7.71) and **GLT 2** applied to the decomposition

$$n^{d-2} A_{\boldsymbol{n}} = n^{d-2} \tilde{K}_{\boldsymbol{n}} + (n^{d-2} K_{\boldsymbol{n}} - n^{d-2} \tilde{K}_{\boldsymbol{n}}) + n^{d-2} Z_{\boldsymbol{n}},$$

taking into account what we have seen in Step 1 of the proof of Theorem 7.4, the inequality **N 2**, and the symmetry of $\tilde{K}_{\boldsymbol{n}}$. To prove (7.79), it is enough to show that

$$\|n^{d-2} K_{\boldsymbol{n},\ell k}(a) - n^{d-2} \tilde{K}_{\boldsymbol{n},\ell k}(a)\| \to 0$$

for every $a \in C([0,1]^d)$ and every $\ell, k = 1, \ldots, d$. With the notation used in the proof of Theorem 7.4 (see in particular (7.75)), we have $K_{\boldsymbol{n},\ell k}(a) = K_{\boldsymbol{n}}(a E_{\ell k})$. Thus, keeping in mind that $K_{\boldsymbol{n},\ell k}(1) = K_{\boldsymbol{n}}(E_{\ell k}) = n^{2-d} T_{\boldsymbol{n}}(H_{\ell k}^{(\boldsymbol{\nu})})$,

$$\|n^{d-2} K_{\boldsymbol{n},\ell k}(a) - n^{d-2} \tilde{K}_{\boldsymbol{n},\ell k}(a)\|$$
$$\le \|n^{d-2} K_{\boldsymbol{n}}(a E_{\ell k}) - n^{d-2} D_{\boldsymbol{n}}(a) K_{\boldsymbol{n}}(E_{\ell k})\| + \|D_{\boldsymbol{n}}(a) T_{\boldsymbol{n}}(H_{\ell k}^{(\boldsymbol{\nu})}) - S_{\boldsymbol{n}}(a) \circ T_{\boldsymbol{n}}(H_{\ell k}^{(\boldsymbol{\nu})})\|,$$

which tends to 0 by Theorem 7.2 and what we have seen in Step 4 of the proof of Theorem 7.4. □

7.5 B-Spline IgA Collocation Discretization of Convection-Diffusion-Reaction PDEs

Consider the convection-diffusion-reaction problem

$$\begin{cases} -\nabla \cdot A \nabla u + \mathbf{b} \cdot \nabla u + cu = f, & \text{in } \Omega, \\ u = 0, & \text{on } \partial\Omega, \end{cases}$$
$$\iff \begin{cases} -\displaystyle\sum_{\ell,k=1}^{d} \frac{\partial}{\partial x_\ell}\Big(a_{\ell k} \frac{\partial u}{\partial x_k}\Big) + \sum_{k=1}^{d} b_k \frac{\partial u}{\partial x_k} + cu = f, & \text{in } \Omega, \\ u = 0, & \text{on } \partial\Omega, \end{cases} \tag{7.80}$$

where $a_{\ell k}$, b_k, c, f are given functions, $A = [a_{\ell k}]_{\ell,k=1}^d$, $\mathbf{b} = [b_k]_{k=1}^d$, and Ω is a bounded open domain in $\mathbb{R}^d$.

Isogeometric collocation approximation. Problem (7.80) can be reformulated as follows:

$$\begin{cases} -\mathbf{1}(A \circ Hu)\mathbf{1}^T + \mathbf{s} \cdot \nabla u + cu = f, & \text{in } \Omega, \\ u = 0, & \text{on } \partial\Omega, \end{cases}$$
$$\iff \begin{cases} -\displaystyle\sum_{\ell,k=1}^d a_{\ell k} \frac{\partial^2 u}{\partial x_\ell \partial x_k} + \sum_{k=1}^d s_k \frac{\partial u}{\partial x_k} + cu = f, & \text{in } \Omega, \\ u = 0, & \text{on } \partial\Omega, \end{cases} \tag{7.81}$$

where Hu is the Hessian of u,

$$(Hu)_{\ell k} = \frac{\partial^2 u}{\partial x_\ell \partial x_k}, \qquad \ell, k = 1, \ldots, d,$$

and $\mathbf{s}$ collects the coefficients of the first-order derivatives,

$$s_k = b_k - \sum_{\ell=1}^d \frac{\partial a_{\ell k}}{\partial x_\ell}, \qquad k = 1, \ldots, d.$$

In the standard collocation method, we choose a finite dimensional vector space $\mathcal{W}$, consisting of sufficiently smooth functions defined on $\overline{\Omega}$ and vanishing on $\partial\Omega$; we call $\mathcal{W}$ the approximation space. Then, we introduce a set of $N = \dim \mathcal{W}$ collocation points $\{\boldsymbol{\tau}_1, \ldots, \boldsymbol{\tau}_N\} \subset \Omega$, and we look for a function $u_{\mathcal{W}} \in \mathcal{W}$ satisfying the PDE (7.81) at the points $\boldsymbol{\tau}_i$, i.e.,

$$\begin{aligned} &-\mathbf{1}(A(\boldsymbol{\tau}_i) \circ Hu_{\mathcal{W}}(\boldsymbol{\tau}_i))\mathbf{1}^T + \mathbf{s}(\boldsymbol{\tau}_i) \cdot \nabla u_{\mathcal{W}}(\boldsymbol{\tau}_i) + c(\boldsymbol{\tau}_i) u_{\mathcal{W}}(\boldsymbol{\tau}_i) = f(\boldsymbol{\tau}_i), \\ &i = 1, \ldots, N. \end{aligned} \tag{7.82}$$

The function $u_{\mathcal{W}}$ is taken as an approximation to the solution u of (7.81). If $\{\varphi_1, \ldots, \varphi_N\}$ is a basis of $\mathcal{W}$, then we have $u_{\mathcal{W}} = \sum_{j=1}^N u_j \varphi_j$ for a unique vector $\mathbf{u} = (u_1, \ldots, u_N)^T$, and, by linearity, the computation of $u_{\mathcal{W}}$ (i.e., of $\mathbf{u}$) reduces to solving the linear system

$$A^C \mathbf{u} = \mathbf{f}, \tag{7.83}$$

where $\mathbf{f} = [f(\boldsymbol{\tau}_i)]_{i=1}^N$ and A^C is the collocation matrix,

$$
\begin{aligned}
A^C &= \left[-\mathbf{1}(A(\boldsymbol{\tau}_i) \circ H\varphi_j(\boldsymbol{\tau}_i))\mathbf{1}^T + \mathbf{s}(\boldsymbol{\tau}_i) \cdot \nabla\varphi_j(\boldsymbol{\tau}_i) + c(\boldsymbol{\tau}_i)\varphi_j(\boldsymbol{\tau}_i)\right]_{i,j=1}^N \\
&= \left[-\sum_{\ell,k=1}^d a_{\ell k}(\boldsymbol{\tau}_i)\frac{\partial^2\varphi_j}{\partial x_\ell \partial x_k}(\boldsymbol{\tau}_i) + \sum_{k=1}^d s_k(\boldsymbol{\tau}_i)\frac{\partial\varphi_j}{\partial x_k}(\boldsymbol{\tau}_i) + c(\boldsymbol{\tau}_i)\varphi_j(\boldsymbol{\tau}_i)\right]_{i,j=1}^N \\
&= -\sum_{\ell,k=1}^d \left(\operatorname*{diag}_{i=1,\dots,N} a_{\ell k}(\boldsymbol{\tau}_i)\right)\left[\frac{\partial^2\varphi_j}{\partial x_\ell \partial x_k}(\boldsymbol{\tau}_i)\right]_{i,j=1}^N \\
&\quad + \sum_{k=1}^d \left(\operatorname*{diag}_{i=1,\dots,N} s_k(\boldsymbol{\tau}_i)\right)\left[\frac{\partial\varphi_j}{\partial x_k}(\boldsymbol{\tau}_i)\right]_{i,j=1}^N + \left(\operatorname*{diag}_{i=1,\dots,N} c(\boldsymbol{\tau}_i)\right)\left[\varphi_j(\boldsymbol{\tau}_i)\right]_{i,j=1}^N.
\end{aligned}
\tag{7.84}
$$

Now, suppose that the physical domain Ω can be described by a global geometry function $\mathbf{G} : [0, 1]^d \to \overline{\Omega}$, which is invertible and satisfies $\mathbf{G}(\partial([0, 1]^d)) = \partial\overline{\Omega}$. Let

$$\{\hat{\varphi}_1, \dots, \hat{\varphi}_N\} \tag{7.85}$$

be a set of basis functions defined on the parametric (or reference) domain $[0, 1]^d$ and vanishing on the boundary $\partial([0, 1]^d)$. Let

$$\{\hat{\boldsymbol{\tau}}_1, \dots, \hat{\boldsymbol{\tau}}_N\} \tag{7.86}$$

be a set of N collocation points in $(0, 1)^d$. In the isogeometric collocation approach, we find an approximation $u_{\mathscr{W}}$ of u by using the standard collocation method described above, in which

- the approximation space is chosen as $\mathscr{W} = \text{span}(\varphi_1, \dots, \varphi_N)$, with

$$\varphi_i(\mathbf{x}) = \hat{\varphi}_i(\mathbf{G}^{-1}(\mathbf{x})) = \hat{\varphi}_i(\hat{\mathbf{x}}), \qquad \mathbf{x} = \mathbf{G}(\hat{\mathbf{x}}), \qquad i = 1, \dots, N, \tag{7.87}$$

- the collocation points in the physical domain Ω are defined as

$$\boldsymbol{\tau}_i = \mathbf{G}(\hat{\boldsymbol{\tau}}_i), \qquad i = 1, \dots, N. \tag{7.88}$$

The resulting collocation matrix A^C is given by (7.84), with the basis functions φ_i and the collocation points $\boldsymbol{\tau}_i$ defined as in (7.87) and (7.88).

Assuming that $\mathbf{G}$ and $\hat{\varphi}_i$, $i = 1, \dots, N$, are sufficiently regular, we can apply standard differential calculus to express A^C in terms of $\mathbf{G}$ and $\hat{\varphi}_i$, $\hat{\boldsymbol{\tau}}_i$, $i = 1, \dots, N$. Let us work out this expression. For any $u : \overline{\Omega} \to \mathbb{R}$ consider the corresponding function $\hat{u} : [0, 1]^d \to \mathbb{R}$, which is defined on the parametric domain by

$$\hat{u}(\hat{\mathbf{x}}) = u(\mathbf{x}), \qquad \mathbf{x} = \mathbf{G}(\hat{\mathbf{x}}). \tag{7.89}$$

In other words, $\hat{u} = u(\mathbf{G})$. Then, u satisfies (7.81) if and only if $\hat{u}$ satisfies the corresponding transformed problem

$$\begin{cases} -\mathbf{1}(A_{\mathbf{G}} \circ H\hat{u})\mathbf{1}^T + \mathbf{s}_{\mathbf{G}} \cdot \nabla\hat{u} + c_{\mathbf{G}}\hat{u} = f(\mathbf{G}), & \text{in } (0,1)^d, \\ \hat{u} = 0, & \text{on } \partial((0,1)^d), \end{cases}$$

$$\iff \begin{cases} -\displaystyle\sum_{\ell,k=1}^{d} a_{\mathbf{G},\ell k}\frac{\partial^2\hat{u}}{\partial\hat{x}_\ell\partial\hat{x}_k} + \sum_{k=1}^{d} s_{\mathbf{G},k}\frac{\partial\hat{u}}{\partial\hat{x}_k} + c_{\mathbf{G}}\hat{u} = f(\mathbf{G}), & \text{in } (0,1)^d, \\ \hat{u} = 0 & \text{on } \partial((0,1)^d). \end{cases} \tag{7.90}$$

In (7.90), $H\hat{u}$ is the Hessian of $\hat{u}$,

$$(H\hat{u})_{\ell k} = \frac{\partial^2\hat{u}}{\partial\hat{x}_\ell\partial\hat{x}_k}, \qquad \ell, k = 1, \ldots, d,$$

while $A_{\mathbf{G}} = [a_{\mathbf{G},\ell k}]_{\ell,k=1}^d$, $\mathbf{s}_{\mathbf{G}} = [s_{\mathbf{G},k}]_{k=1}^d$, $c_{\mathbf{G}}$ are the transformed diffusion, convection and reaction coefficients. Through standard differential calculus, one can show that

$$A_{\mathbf{G}} = (J_{\mathbf{G}})^{-1}A(\mathbf{G})(J_{\mathbf{G}})^{-T}, \qquad c_{\mathbf{G}} = c(\mathbf{G}), \tag{7.91}$$

where $J_{\mathbf{G}}$ is the Jacobian matrix of $\mathbf{G}$,

$$J_{\mathbf{G}} = \left[\frac{\partial G_i}{\partial\hat{x}_j}\right]_{i,j=1}^d = \left[\frac{\partial x_i}{\partial\hat{x}_j}\right]_{i,j=1}^d.$$

The expression of $\mathbf{s}_{\mathbf{G}}$ in terms of A, $\mathbf{s}$, $\mathbf{G}$ is complicated and hence not reported here. The collocation matrix A^C in (7.84) can be expressed in terms of $\mathbf{G}$ and $\hat{\varphi}_i$, $\hat{\boldsymbol{\tau}}_i$, $i = 1, \ldots, N$, as follows:

$$\begin{aligned} A^C &= \left[-\mathbf{1}(A_{\mathbf{G}}(\hat{\boldsymbol{\tau}}_i) \circ H\hat{\varphi}_j(\hat{\boldsymbol{\tau}}_i))\mathbf{1}^T + \mathbf{s}_{\mathbf{G}}(\hat{\boldsymbol{\tau}}_i) \cdot \nabla\hat{\varphi}_j(\hat{\boldsymbol{\tau}}_i) + c_{\mathbf{G}}(\hat{\boldsymbol{\tau}}_i)\hat{\varphi}_j(\hat{\boldsymbol{\tau}}_i)\right]_{i,j=1}^N \\ &= -\sum_{\ell,k=1}^{d}\left(\operatorname*{diag}_{i=1,\ldots,N} a_{\mathbf{G},\ell k}(\hat{\boldsymbol{\tau}}_i)\right)\left[\frac{\partial^2\hat{\varphi}_j}{\partial\hat{x}_\ell\partial\hat{x}_k}(\hat{\boldsymbol{\tau}}_i)\right]_{i,j=1}^N \\ &\quad + \sum_{k=1}^{d}\left(\operatorname*{diag}_{i=1,\ldots,N} s_{\mathbf{G},k}(\hat{\boldsymbol{\tau}}_i)\right)\left[\frac{\partial\hat{\varphi}_j}{\partial\hat{x}_k}(\hat{\boldsymbol{\tau}}_i)\right]_{i,j=1}^N + \left(\operatorname*{diag}_{i=1,\ldots,N} c_{\mathbf{G}}(\hat{\boldsymbol{\tau}}_i)\right)\left[\hat{\varphi}_j(\hat{\boldsymbol{\tau}}_i)\right]_{i,j=1}^N. \end{aligned} \tag{7.92}$$

In the IgA context, the geometry map $\mathbf{G}$ is expressed in terms of the functions $\hat{\varphi}_i$, in accordance with theisoparametric approach [12, Sect. 3.1]. Moreover, the functions $\hat{\varphi}_i$ are usually tensor-product B-splines or their rational versions, the so-called NURBS. In this section, the role of the $\hat{\varphi}_i$ will be played by tensor-product B-splines over uniform knot sequences. Furthermore, we do not limit ourselves to the isoparametric approach, but we allow the geometry map $\mathbf{G}$ to be any sufficiently regular function from $[0,1]^d$ to $\overline{\Omega}$, not necessarily expressed in terms of tensor-product B-splines. Finally, following [1], the collocation points $\hat{\boldsymbol{\tau}}_i$ will be chosen as the (tensor-product) Greville abscissae corresponding to the tensor-product B-splines $\hat{\varphi}_i$. For the same analysis as in this section, but with tensor-product B-splines replaced by NURBS, we refer the reader to [18].

Remark 7.7 For later purposes, we point out that the functions $s_{\mathbf{G},k}$, $k = 1, \ldots, d$, are bounded over Ω if the following conditions are satisfied:

- for every $\ell, k = 1, \ldots, d$, the function $a_{\ell k} : \overline{\Omega} \to \mathbb{R}$ is bounded and its partial derivatives $\partial a_{\ell k}/\partial x_1, \ldots, \partial a_{\ell k}/\partial x_d : \Omega \to \mathbb{R}$ are bounded;
- for every $k = 1, \ldots, d$, the function $b_k : \overline{\Omega} \to \mathbb{R}$ is bounded;
- $\mathbf{G} \in C^2([0, 1]^d)$ and $\det(J_{\mathbf{G}}) \neq 0$ in $[0, 1]^d$.

To understand why this is true (without computing the complicated expression of $\mathbf{s}_{\mathbf{G}}$), we suggest that the reader have a look at the unidimensional case [22, Sect. 10.7.1], especially [22, Eq. (10.115)], where $\mathbf{s}_{\mathbf{G}} = s_G$ is explicitly given in terms of $A = a$, $\mathbf{s} = s$, $\mathbf{G} = G$.

Tensor-product B-splines and Greville abscissae. For $\boldsymbol{p}, \boldsymbol{n} \in \mathbb{N}^d$ and $k = 1, \ldots, d$, let $N_{i_k,[p_k]}$, $i_k = 1, \ldots, n_k + p_k$, be the B-splines of degree p_k defined on the knot sequence

$$t_1 = \cdots = t_{p_k+1} = 0 < t_{p_k+2} < \cdots < t_{p_k+n_k} < 1 = t_{p_k+n_k+1} = \cdots = t_{2p_k+n_k+1}, \tag{7.93}$$

where

$$t_{i_k+p_k+1} = \frac{i_k}{n_k}, \qquad i_k = 0, \ldots, n_k. \tag{7.94}$$

For $i_k = 1, \ldots, n_k + p_k$, let $\xi_{i_k,[p_k]}$ be the Greville abscissa associated with the B-spline $N_{i_k,[p_k]}$. Note that the B-splines $N_{i,[p]}$, $i = 1, \ldots, n + p$, and the associated Greville abscissae $\xi_{i,[p]}$, $i = 1, \ldots, n + p$, have been defined in [22, p. 232] for all $p, n \geq 1$; see also [22, Fig. 10.8]. We define the tensor-product B-splines $N_{\boldsymbol{i},[\boldsymbol{p}]} : [0, 1]^d \to \mathbb{R}$ as follows:

$$N_{\boldsymbol{i},[\boldsymbol{p}]} = N_{i_1,[p_1]} \otimes \cdots \otimes N_{i_d,[p_d]}, \qquad \boldsymbol{i} = \mathbf{1}, \ldots, \boldsymbol{n} + \boldsymbol{p}. \tag{7.95}$$

The (tensor-product) Greville abscissa $\boldsymbol{\xi}_{\boldsymbol{i},[\boldsymbol{p}]}$ associated with the tensor-product B-spline $N_{\boldsymbol{i},[\boldsymbol{p}]}$ is defined by

$$\boldsymbol{\xi}_{\boldsymbol{i},[\boldsymbol{p}]} = (\xi_{i_1,[p_1]}, \ldots, \xi_{i_d,[p_d]}), \qquad \boldsymbol{i} = \mathbf{1}, \ldots, \boldsymbol{n} + \boldsymbol{p}. \tag{7.96}$$

The properties of B-splines and Greville abscissae reported in [22, pp. 232–235] imply the following analogous properties of tensor-product B-splines and Greville abscissae.

- Local support property:

$$\operatorname{supp}(N_{\boldsymbol{i},[\boldsymbol{p}]}) = [\mathbf{t}_{\boldsymbol{i}}, \mathbf{t}_{\boldsymbol{i}+\boldsymbol{p}+\mathbf{1}}], \qquad \boldsymbol{i} = \mathbf{1}, \ldots, \boldsymbol{n} + \boldsymbol{p}, \tag{7.97}$$

where $[\mathbf{t}_{\boldsymbol{i}}, \mathbf{t}_{\boldsymbol{i}+\boldsymbol{p}+\mathbf{1}}] = [t_{i_1}, t_{i_1+p_1+1}] \times \cdots \times [t_{i_d}, t_{i_d+p_d+1}]$. In particular, for the measure of the support we have $\mu_d(\mathrm{supp}(N_{\boldsymbol{i},[\boldsymbol{p}]})) \le N(\boldsymbol{p}+\mathbf{1})/N(\boldsymbol{n})$.

- Vanishing property on the boundary:

$$N_{\boldsymbol{i},[\boldsymbol{p}]}(\mathbf{t}) = 0, \qquad \mathbf{t} \in \partial([0,1]^d), \qquad \boldsymbol{i} = \mathbf{2}, \dots, \boldsymbol{n}+\boldsymbol{p}-\mathbf{1}. \tag{7.98}$$

- Nonnegative partition of unity:

$$N_{\boldsymbol{i},[\boldsymbol{p}]}(\mathbf{t}) \ge 0, \qquad \mathbf{t} \in [0,1]^d, \qquad \boldsymbol{i} = \mathbf{1}, \dots, \boldsymbol{n}+\boldsymbol{p}, \tag{7.99}$$

$$\sum_{\boldsymbol{i}=\mathbf{1}}^{\boldsymbol{n}+\boldsymbol{p}} N_{\boldsymbol{i},[\boldsymbol{p}]}(\mathbf{t}) = 1, \qquad \mathbf{t} \in [0,1]^d. \tag{7.100}$$

- Bounds for derivatives:

$$\sum_{\boldsymbol{i}=\mathbf{1}}^{\boldsymbol{n}+\boldsymbol{p}} \left| \frac{\partial N_{\boldsymbol{i},[\boldsymbol{p}]}}{\partial t_k}(\mathbf{t}) \right| \le 2p_k n_k, \qquad \mathbf{t} \in [0,1]^d, \qquad k = 1, \dots, d, \tag{7.101}$$

$$\sum_{\boldsymbol{i}=\mathbf{1}}^{\boldsymbol{n}+\boldsymbol{p}} \left| \frac{\partial^2 N_{\boldsymbol{i},[\boldsymbol{p}]}}{\partial t_\ell \partial t_k}(\mathbf{t}) \right| \le 4p_\ell p_k n_\ell n_k, \qquad \mathbf{t} \in [0,1]^d, \qquad \ell, k = 1, \dots, d. \tag{7.102}$$

In (7.101) and (7.102), just as in their univariate analogs [22, Eqs. (10.128) and (10.129)], it is understood that the undefined values are counted as 0 in the summations.

- $\boldsymbol{\xi}_{\boldsymbol{i},[\boldsymbol{p}]}$ lies in the support of $N_{\boldsymbol{i},[\boldsymbol{p}]}$,

$$\boldsymbol{\xi}_{\boldsymbol{i},[\boldsymbol{p}]} \in \mathrm{supp}(N_{\boldsymbol{i},[\boldsymbol{p}]}) = [\mathbf{t}_{\boldsymbol{i}}, \mathbf{t}_{\boldsymbol{i}+\boldsymbol{p}+\mathbf{1}}], \qquad \boldsymbol{i} = \mathbf{1}, \dots, \boldsymbol{n}+\boldsymbol{p}. \tag{7.103}$$

- The Greville abscissae are somehow equivalent, in an asymptotic sense, to the uniform knots in $[0,1]^d$. More precisely,

$$\left| \boldsymbol{\xi}_{\boldsymbol{i},[\boldsymbol{p}]} - \frac{\boldsymbol{i}}{\boldsymbol{n}+\boldsymbol{p}} \right|_\infty \le \frac{C_{\boldsymbol{p}}}{\min(\boldsymbol{n})}, \qquad \boldsymbol{i} = \mathbf{1}, \dots, \boldsymbol{n}+\boldsymbol{p}, \tag{7.104}$$

where $C_{\boldsymbol{p}}$ depends only on $\boldsymbol{p}$.

B-spline IgA collocation matrices. In the IgA collocation approach based on (uniform) tensor-product B-splines, the basis functions $\hat{\varphi}_1, \dots, \hat{\varphi}_N$ in (7.85) are chosen as the tensor-product B-splines

$$N_{\boldsymbol{i}+\mathbf{1},[\boldsymbol{p}]}, \qquad \boldsymbol{i} = \mathbf{1}, \dots, \boldsymbol{n}+\boldsymbol{p}-\mathbf{2}, \tag{7.105}$$

and the collocation points $\hat{\boldsymbol{\tau}}_1, \dots, \hat{\boldsymbol{\tau}}_N$ in (7.86) are chosen as the Greville abscissae

$$\boldsymbol{\xi}_{\boldsymbol{i}+\mathbf{1},[\boldsymbol{p}]}, \qquad \boldsymbol{i} = \mathbf{1}, \dots, \boldsymbol{n}+\boldsymbol{p}-\mathbf{2}. \tag{7.106}$$

In this d-dimensional setting, we have $N = N(\boldsymbol{n} + \boldsymbol{p} - \mathbf{2})$. Of course, the basis functions (7.105) and the collocation points (7.106) are ordered in accordance with the standard lexicographic ordering. Throughout this section, we assume $\boldsymbol{p} \geq \mathbf{2}$, so as to ensure that the second-order partial derivative $\partial^2 N_{\boldsymbol{j}+\mathbf{1},[\boldsymbol{p}]}/\partial t_\ell \partial t_k$ is defined at the Greville abscissa $\boldsymbol{\xi}_{\boldsymbol{i}+\mathbf{1},[\boldsymbol{p}]}$ for every $\boldsymbol{i}, \boldsymbol{j} = \mathbf{1}, \ldots, \boldsymbol{n} + \boldsymbol{p} - \mathbf{2}$ and every $\ell, k = 1, \ldots, d$.

The collocation matrix (7.92) resulting from the choices of $\hat{\varphi}_i$, $\hat{\boldsymbol{\tau}}_i$ as in (7.105) and (7.106) will be denoted by $A^{[\boldsymbol{p}]}_{\mathbf{G},\boldsymbol{n}}$ so as to emphasize its dependence on $\mathbf{G}, \boldsymbol{n}, \boldsymbol{p}$:

$$\begin{aligned} A^{[\boldsymbol{p}]}_{\mathbf{G},\boldsymbol{n}} &= \Big[-\mathbf{1}(A_{\mathbf{G}}(\boldsymbol{\xi}_{\boldsymbol{i}+\mathbf{1},[\boldsymbol{p}]}) \circ H N_{\boldsymbol{j}+\mathbf{1},[\boldsymbol{p}]}(\boldsymbol{\xi}_{\boldsymbol{i}+\mathbf{1},[\boldsymbol{p}]}))\mathbf{1}^T \\ &\quad + \mathbf{s}_{\mathbf{G}}(\boldsymbol{\xi}_{\boldsymbol{i}+\mathbf{1},[\boldsymbol{p}]}) \cdot \nabla N_{\boldsymbol{j}+\mathbf{1},[\boldsymbol{p}]}(\boldsymbol{\xi}_{\boldsymbol{i}+\mathbf{1},[\boldsymbol{p}]}) \\ &\quad + c_{\mathbf{G}}(\boldsymbol{\xi}_{\boldsymbol{i}+\mathbf{1},[\boldsymbol{p}]}) N_{\boldsymbol{j}+\mathbf{1},[\boldsymbol{p}]}(\boldsymbol{\xi}_{\boldsymbol{i}+\mathbf{1},[\boldsymbol{p}]})\Big]_{\boldsymbol{i},\boldsymbol{j}=\mathbf{1}}^{\boldsymbol{n}+\boldsymbol{p}-\mathbf{2}} \\ &= \sum_{\ell,k=1}^{d} K^{[\boldsymbol{p}]}_{\boldsymbol{n},\ell k}(a_{\mathbf{G},\ell k}) + \sum_{k=1}^{d} H^{[\boldsymbol{p}]}_{\boldsymbol{n},k}(s_{\mathbf{G},k}) + M^{[\boldsymbol{p}]}_{\boldsymbol{n}}(c_{\mathbf{G}}), \end{aligned} \tag{7.107}$$

where

$$\begin{aligned} K^{[\boldsymbol{p}]}_{\boldsymbol{n},\ell k}(a_{\mathbf{G},\ell k}) &= D^{[\boldsymbol{p}]}_{\boldsymbol{n}}(a_{\mathbf{G},\ell k}) K^{[\boldsymbol{p}]}_{\boldsymbol{n},\ell k}, \qquad \ell, k = 1, \ldots, d, \\ H^{[\boldsymbol{p}]}_{\boldsymbol{n},k}(s_{\mathbf{G},k}) &= D^{[\boldsymbol{p}]}_{\boldsymbol{n}}(s_{\mathbf{G},k}) H^{[\boldsymbol{p}]}_{\boldsymbol{n},k}, \qquad k = 1, \ldots, d, \\ M^{[\boldsymbol{p}]}_{\boldsymbol{n}}(c_{\mathbf{G}}) &= D^{[\boldsymbol{p}]}_{\boldsymbol{n}}(c_{\mathbf{G}}) M^{[\boldsymbol{p}]}_{\boldsymbol{n}}, \end{aligned}$$

with

$$D^{[\boldsymbol{p}]}_{\boldsymbol{n}}(y) = \operatorname*{diag}_{\boldsymbol{i}=\mathbf{1},\ldots,\boldsymbol{n}+\boldsymbol{p}-\mathbf{2}} y(\boldsymbol{\xi}_{\boldsymbol{i}+\mathbf{1},[\boldsymbol{p}]})$$

being the d-level diagonal sampling matrix containing the samples of the function $y : [0, 1]^d \to \mathbb{R}$ at the Greville abscissae (7.106), and

$$K^{[\boldsymbol{p}]}_{\boldsymbol{n},\ell k} = \left[-\frac{\partial^2 N_{\boldsymbol{j}+\mathbf{1},[\boldsymbol{p}]}}{\partial \hat{x}_\ell \partial \hat{x}_k}(\boldsymbol{\xi}_{\boldsymbol{i}+\mathbf{1},[\boldsymbol{p}]})\right]_{\boldsymbol{i},\boldsymbol{j}=\mathbf{1}}^{\boldsymbol{n}+\boldsymbol{p}-\mathbf{2}}, \qquad \ell, k = 1, \ldots, d, \tag{7.108}$$

$$H^{[\boldsymbol{p}]}_{\boldsymbol{n},k} = \left[\frac{\partial N_{\boldsymbol{j}+\mathbf{1},[\boldsymbol{p}]}}{\partial \hat{x}_k}(\boldsymbol{\xi}_{\boldsymbol{i}+\mathbf{1},[\boldsymbol{p}]})\right]_{\boldsymbol{i},\boldsymbol{j}=\mathbf{1}}^{\boldsymbol{n}+\boldsymbol{p}-\mathbf{2}}, \qquad k = 1, \ldots, d, \tag{7.109}$$

$$M^{[\boldsymbol{p}]}_{\boldsymbol{n}} = \left[N_{\boldsymbol{j}+\mathbf{1},[\boldsymbol{p}]}(\boldsymbol{\xi}_{\boldsymbol{i}+\mathbf{1},[\boldsymbol{p}]})\right]_{\boldsymbol{i},\boldsymbol{j}=\mathbf{1}}^{\boldsymbol{n}+\boldsymbol{p}-\mathbf{2}}. \tag{7.110}$$

Note that $A^{[\boldsymbol{p}]}_{\mathbf{G},\boldsymbol{n}}$ can be decomposed as follows:

$$A^{[\boldsymbol{p}]}_{\mathbf{G},\boldsymbol{n}} = K^{[\boldsymbol{p}]}_{\mathbf{G},\boldsymbol{n}} + Z^{[\boldsymbol{p}]}_{\mathbf{G},\boldsymbol{n}}, \tag{7.111}$$

where

$$K_{\mathbf{G},\boldsymbol{n}}^{[\boldsymbol{p}]} = \sum_{\ell,k=1}^{d} K_{\boldsymbol{n},\ell k}^{[\boldsymbol{p}]}(a_{\mathbf{G},\ell k}) = \sum_{\ell,k=1}^{d} D_{\boldsymbol{n}}^{[\boldsymbol{p}]}(a_{\mathbf{G},\ell k}) K_{\boldsymbol{n},\ell k}^{[\boldsymbol{p}]} \tag{7.112}$$

is the collocation diffusion matrix, resulting from the collocation discretization of the higher-order (diffusion) term in (7.81), and

$$\begin{aligned} Z_{\mathbf{G},\boldsymbol{n}}^{[\boldsymbol{p}]} &= \sum_{k=1}^{d} H_{\boldsymbol{n},k}^{[\boldsymbol{p}]}(s_{\mathbf{G},k}) + M_{\boldsymbol{n}}^{[\boldsymbol{p}]}(c_{\mathbf{G}}) \\ &= \sum_{k=1}^{d} D_{\boldsymbol{n}}^{[\boldsymbol{p}]}(s_{\mathbf{G},k}) H_{\boldsymbol{n},k}^{[\boldsymbol{p}]} + D_{\boldsymbol{n}}^{[\boldsymbol{p}]}(c_{\mathbf{G}}) M_{\boldsymbol{n}}^{[\boldsymbol{p}]} \end{aligned} \tag{7.113}$$

is the matrix resulting from the discretization of the lower-order terms (the convection and reaction terms). The next lemma highlights the structure of the matrices $K_{\boldsymbol{n},\ell k}^{[\boldsymbol{p}]} = K_{\boldsymbol{n},\ell k}^{[\boldsymbol{p}]}(1)$, $H_{\boldsymbol{n},k}^{[\boldsymbol{p}]} = H_{\boldsymbol{n},k}^{[\boldsymbol{p}]}(1)$, $M_{\boldsymbol{n}}^{[\boldsymbol{p}]} = M_{\boldsymbol{n}}^{[\boldsymbol{p}]}(1)$.

Lemma 7.3 *For $\boldsymbol{p}, \boldsymbol{n} \in \mathbb{N}^d$ with $\boldsymbol{p} \geq \mathbf{2}$, we have*

$$K_{\boldsymbol{n},kk}^{[\boldsymbol{p}]} = \left(\bigotimes_{r=1}^{k-1} M_{n_r}^{[p_r]} \right) \otimes K_{n_k}^{[p_k]} \otimes \left(\bigotimes_{r=k+1}^{d} M_{n_r}^{[p_r]} \right) \tag{7.114}$$

for $k = 1, \ldots, d$,

$$\begin{aligned} K_{\boldsymbol{n},k\ell}^{[\boldsymbol{p}]} &= K_{\boldsymbol{n},\ell k}^{[\boldsymbol{p}]} \\ &= -\left(\bigotimes_{r=1}^{\ell-1} M_{n_r}^{[p_r]} \right) \otimes H_{n_\ell}^{[p_\ell]} \otimes \left(\bigotimes_{r=\ell+1}^{k-1} M_{n_r}^{[p_r]} \right) \otimes H_{n_k}^{[p_k]} \otimes \left(\bigotimes_{r=k+1}^{d} M_{n_r}^{[p_r]} \right) \end{aligned} \tag{7.115}$$

for $1 \leq \ell < k \leq d$,

$$H_{\boldsymbol{n},k}^{[\boldsymbol{p}]} = \left(\bigotimes_{r=1}^{k-1} M_{n_r}^{[p_r]} \right) \otimes H_{n_k}^{[p_k]} \otimes \left(\bigotimes_{r=k+1}^{d} M_{n_r}^{[p_r]} \right) \tag{7.116}$$

for $k = 1, \ldots, d$, and

$$M_{\boldsymbol{n}}^{[\boldsymbol{p}]} = \bigotimes_{r=1}^{d} M_{n_r}^{[p_r]}, \tag{7.117}$$

where the matrices $K_n^{[p]}$, $H_n^{[p]}$, $M_n^{[p]}$ are defined for all $p, n \in \mathbb{N}$ with $p \geq 2$ as follows:

$$K_n^{[p]} = \left[-N''_{j+1,[p]}(\xi_{i+1,[p]})\right]_{i,j=1}^{n+p-2}, \tag{7.118}$$

$$H_n^{[p]} = \left[N'_{j+1,[p]}(\xi_{i+1,[p]})\right]_{i,j=1}^{n+p-2}, \tag{7.119}$$

$$M_n^{[p]} = \left[N_{j+1,[p]}(\xi_{i+1,[p]})\right]_{i,j=1}^{n+p-2}, \tag{7.120}$$

with $N_{i,[p]}$, $i = 1, \dots, n+p$, and $\xi_{i,[p]}$, $i = 1, \dots, n+p$, being, respectively, the B-splines of degree p defined on the knot sequence

$$\Big\{\underbrace{0,\dots,0}_{p+1},\frac{1}{n},\frac{2}{n},\dots,\frac{n-1}{n},\underbrace{1,\dots,1}_{p+1}\Big\}$$

and the associated Greville abscissae; see [22, pp. 232–235] *for the corresponding definitions and properties.*

Proof We only prove (7.114) as the proof of the other equations is the same. By the crucial property **P7**, for all $\boldsymbol{i}, \boldsymbol{j} = \boldsymbol{1}, \dots, \boldsymbol{n}+\boldsymbol{p}-\boldsymbol{2}$ we have

$$\left[\left(\bigotimes_{r=1}^{k-1} M_{n_r}^{[p_r]}\right) \otimes K_{n_k}^{[p_k]} \otimes \left(\bigotimes_{r=k+1}^{d} M_{n_r}^{[p_r]}\right)\right]_{\boldsymbol{i}\boldsymbol{j}} = (K_{n_k}^{[p_k]})_{i_k j_k} \prod_{\substack{r=1\\ r\neq k}}^{d} (M_{n_r}^{[p_r]})_{i_r j_r}$$

$$= -N''_{j_k+1,[p_k]}(\xi_{i_k+1,[p_k]}) \prod_{\substack{r=1\\ r\neq k}}^{d} N_{j_r+1,[p_r]}(\xi_{i_r+1,[p_r]}) = -\frac{\partial^2 N_{\boldsymbol{j}+\boldsymbol{1},[\boldsymbol{p}]}}{\partial \hat{x}_k^2}(\boldsymbol{\xi}_{\boldsymbol{i}+\boldsymbol{1},[\boldsymbol{p}]})$$

$$= (K_{\boldsymbol{n},kk}^{[\boldsymbol{p}]})_{\boldsymbol{i}\boldsymbol{j}}.$$ □

Remark 7.8 As in the case of FDs and FEs (see Remarks 7.3 and 7.5), $K_n^{[p]}$, $H_n^{[p]}$, $M_n^{[p]}$ are the diffusion, convection, reaction matrices resulting from the B-spline IgA collocation discretization of the univariate problem

$$\begin{cases} -u''(x) + u'(x) + u(x) = f(x), & x \in (0,1),\\ u(0) = u(1) = 0. \end{cases}$$

In other words, $K_n^{[p]}$, $H_n^{[p]}$, $M_n^{[p]}$ are the matrices resulting from the B-spline IgA collocation discretization of, respectively, the negative second derivative $-u''(x)$, the first derivative $u'(x)$, the identity operator $u(x)$. To see this, follow the above derivation of the IgA collocation matrices in the univariate case $d = 1$ or take a look at [22, Sect. 10.7.1]. In view of what follows, we recall from [22, Eqs. (10.150)–(10.152)] that

$$n^{-2} K_n^{[p]} = T_{n+p-2}(f_p) + R_n^{[p]}, \qquad \mathrm{rank}(R_n^{[p]}) \le 4(p-1), \tag{7.121}$$

$$-\mathrm{i}\, n^{-1} H_n^{[p]} = T_{n+p-2}(g_p) + S_n^{[p]}, \qquad \mathrm{rank}(S_n^{[p]}) \le 4(p-1), \tag{7.122}$$

$$M_n^{[p]} = T_{n+p-2}(h_p) + V_n^{[p]}, \qquad \mathrm{rank}(V_n^{[p]}) \le 4(p-1), \tag{7.123}$$

where f_p, g_p, h_p are real trigonometric polynomials whose definitions are given in [22, Eqs. (10.147)–(10.149)]. Moreover, we know from [22, Eq. (10.153)] that

$$\|n^{-2}K_n^{[p]}\|,\ \|n^{-1}H_n^{[p]}\|,\ \|M_n^{[p]}\| \le C^{[p]} \tag{7.124}$$

for some constant $C^{[p]}$ depending only on p.

Remark 7.9 It follows from Lemma 7.3, Eqs. (7.121)–(7.123), **T 8** and **P 8** that

$$K_{\boldsymbol{n},\ell k}^{[\boldsymbol{p}]} = n_\ell n_k\, T_{\boldsymbol{n}+\boldsymbol{p}-\mathbf{2}}((H_{\boldsymbol{p}})_{\ell k}) + R_{\boldsymbol{n},\ell k}^{[\boldsymbol{p}]}, \qquad \ell, k = 1, \dots, d, \tag{7.125}$$

where

$$\operatorname{rank}(R_{\boldsymbol{n},\ell k}^{[\boldsymbol{p}]}) \le N(\boldsymbol{n}+\boldsymbol{p}-\mathbf{2}) \sum_{i=1}^{d} \frac{4(p_i-1)}{n_i+p_i-2}, \qquad \ell, k = 1, \dots, d, \tag{7.126}$$

and $H_{\boldsymbol{p}}(\boldsymbol{\theta})$ is the $d \times d$ symmetric matrix defined as follows:

$$(H_{\boldsymbol{p}})_{\ell k}(\boldsymbol{\theta}) = \begin{cases} f_{p_k}(\theta_k) \prod_{\substack{r=1\\ r\ne k}}^{d} h_{p_r}(\theta_r), & \text{if } \ell = k, \\ g_{p_\ell}(\theta_\ell) g_{p_k}(\theta_k) \prod_{\substack{r=1\\ r\ne \ell,k}}^{d} h_{p_r}(\theta_r), & \text{if } \ell \ne k. \end{cases} \tag{7.127}$$

Since $K_{\boldsymbol{n},\ell k}^{[\boldsymbol{p}]}(1) = K_{\boldsymbol{n},\ell k}^{[\boldsymbol{p}]}$, according to (7.125) and the discussion in Sect. 7.2 (see in particular (7.14)), we may predict that $H_{\boldsymbol{p}}(\boldsymbol{\theta})$ is, up to some normalization, the symbol of the negative Hessian operator. More precisely, we should call it the symbol of the *negative Hessian operator in the parametric variables* $\hat{x}_i$, because $K_{\boldsymbol{n},\ell k}^{[\boldsymbol{p}]}(1)$ is the matrix resulting from the B-spline IgA collocation discretization of the d-variate problem

$$\begin{cases} -\dfrac{\partial^2 \hat{u}}{\partial \hat{x}_\ell \partial \hat{x}_k} = f, & \text{in } (0,1)^d, \\ \hat{u} = 0, & \text{on } \partial((0,1)^d), \end{cases}$$

in which the physical domain Ω and the physical variables x_i are replaced by the parametric domain $(0,1)^d$ and the parametric variables $\hat{x}_i$. For instance, assuming $\boldsymbol{n} = \boldsymbol{\nu} n$ for some fixed vector $\boldsymbol{\nu} \in \mathbb{Q}^d$ with positive components, from (7.125), (7.126), **Z 1**, **GLT 3** and **GLT 4** we infer that

$$\left\{n^{-2}K_{\boldsymbol{n},\ell k}^{[\boldsymbol{p}]}(1) = \nu_\ell \nu_k\, T_{\boldsymbol{n}+\boldsymbol{p}-\mathbf{2}}((H_{\boldsymbol{p}})_{\ell k}) + n^{-2}R_{\boldsymbol{n},\ell k}^{[\boldsymbol{p}]}\right\}_n$$
$$\sim_{\rm GLT} \nu_\ell \nu_k\, (H_{\boldsymbol{p}})_{\ell k}(\boldsymbol{\theta}) = (H_{\boldsymbol{p}}^{(\boldsymbol{\nu})})_{\ell k}(\boldsymbol{\theta}), \tag{7.128}$$

where $H_{\boldsymbol{p}}^{(\boldsymbol{\nu})}(\boldsymbol{\theta}) = \operatorname{diag}(\boldsymbol{\nu}) H_{\boldsymbol{p}}(\boldsymbol{\theta}) \operatorname{diag}(\boldsymbol{\nu})$. This means that, assuming $\boldsymbol{n} = \boldsymbol{\nu} n$ and after normalization by n^{-2}, the symbol of the negative Hessian operator (in the parametric variables) is $H_{\boldsymbol{p}}^{(\boldsymbol{\nu})}(\boldsymbol{\theta})$, which coincides with $H_{\boldsymbol{p}}(\boldsymbol{\theta})$ up to a trivial transformation. It was proved in [17, Theorem 6.1] that, for all $p \geq 2$, the matrix $H_p(\theta)$ is SPSD for all $\theta \in [-\pi, \pi]^d$ and SPD for all $\theta \in [-\pi, \pi]^d$ such that $\theta_1 \cdots \theta_d \neq 0$.

GLT analysis of the B-spline IgA collocation matrices. Using the theory of multilevel GLT sequences, we now derive the spectral and singular value distribution of the sequence of normalized IgA collocation matrices $\{n^{-2}A^{[p]}_{\mathbf{G},\boldsymbol{n}}\}_n$ under the assumptions that $\boldsymbol{n} = \boldsymbol{\nu} n$ for some fixed vector $\boldsymbol{\nu}$ and that the geometry map $\mathbf{G}$ is sufficiently smooth.

Theorem 7.6 *Let Ω be a bounded open domain in $\mathbb{R}^d$ and suppose that the following conditions on the PDE coefficients and the geometry map are satisfied:*

- *for every $\ell, k = 1, \ldots, d$, the function $a_{\ell k} : \overline{\Omega} \to \mathbb{R}$ belongs to $C(\overline{\Omega})$ and its partial derivatives $\partial a_{\ell k}/\partial x_1, \ldots, \partial a_{\ell k}/\partial x_d : \Omega \to \mathbb{R}$ are bounded;*
- *for every $k = 1, \ldots, d$, the function $b_k : \overline{\Omega} \to \mathbb{R}$ is bounded;*
- *$c : \overline{\Omega} \to \mathbb{R}$ is bounded;*
- *$\mathbf{G} \in C^2([0,1]^d)$ and $\det(J_\mathbf{G}) \neq 0$ over $[0,1]^d$.*

Let $p \geq 2$, let $\boldsymbol{\nu} \in \mathbb{Q}^d$ be a vector with positive components, and assume that $\boldsymbol{n} = \boldsymbol{\nu} n$ (it is understood that n varies in the infinite subset of $\mathbb{N}$ such that $\boldsymbol{n} = \boldsymbol{\nu} n \in \mathbb{N}^d$). Then

$$\{n^{-2}A^{[p]}_{\mathbf{G},\boldsymbol{n}}\}_n \sim_{\rm GLT} f^{(\boldsymbol{\nu})}_{\mathbf{G},p} \tag{7.129}$$

and

$$\{n^{-2}A^{[p]}_{\mathbf{G},\boldsymbol{n}}\}_n \sim_{\sigma,\lambda} f^{(\boldsymbol{\nu})}_{\mathbf{G},p}, \tag{7.130}$$

where

$$\begin{aligned} f^{(\boldsymbol{\nu})}_{\mathbf{G},p}(\hat{\mathbf{x}}, \boldsymbol{\theta}) &= \sum_{\ell,k=1}^{d} a_{\mathbf{G},\ell k}(\hat{\mathbf{x}})(H^{(\boldsymbol{\nu})}_p)_{\ell k}(\boldsymbol{\theta}) \\ &= \mathbf{1}(A_\mathbf{G}(\hat{\mathbf{x}}) \circ H^{(\boldsymbol{\nu})}_p(\boldsymbol{\theta}))\mathbf{1}^T = \boldsymbol{\nu}\big(A_\mathbf{G}(\hat{\mathbf{x}}) \circ H_p(\boldsymbol{\theta})\big)\boldsymbol{\nu}^T, \qquad (7.131) \\ H^{(\boldsymbol{\nu})}_p(\boldsymbol{\theta}) &= \mathrm{diag}(\boldsymbol{\nu})H_p(\boldsymbol{\theta})\mathrm{diag}(\boldsymbol{\nu}), \end{aligned}$$

and $A_\mathbf{G}(\hat{\mathbf{x}})$ and $H_p(\boldsymbol{\theta})$ are defined, respectively, in (7.91) *and* (7.127).

Proof The proof consists of the following steps. Throughout the proof, the letter C denotes a generic constant independent of n. While reading the proof, the reader should keep in mind the relation $\boldsymbol{n} = \boldsymbol{\nu} n$.

Step 1. We show that

$$\|n^{-2}K^{[p]}_{\mathbf{G},\boldsymbol{n}}\| \leq C, \tag{7.132}$$

$$\|n^{-2}Z^{[p]}_{\mathbf{G},\boldsymbol{n}}\| \leq Cn^{-1}. \tag{7.133}$$

By Lemma 7.3, the property **P 5** and the inequalities (7.124), we have

$$\|K_{n,\ell k}^{[p]}\| \le C^{[p]} n_\ell n_k, \qquad \ell, k = 1, \ldots, d, \tag{7.134}$$

$$\|H_{n,k}^{[p]}\| \le C^{[p]} n_k, \qquad k = 1, \ldots, d, \tag{7.135}$$

$$\|M_n^{[p]}\| \le C^{[p]}, \tag{7.136}$$

where $C^{[p]} = C^{[p_1]} \cdots C^{[p_d]}$. Moreover, due to the assumptions on the geometry map $\mathbf{G}$ and the PDE coefficients $a_{\ell k}, b_k, c$ (see also Remark 7.7), we have

$$\|D_n^{[p]}(a_{\mathbf{G},\ell k})\| \le \|a_{\mathbf{G},\ell k}\|_\infty < \infty, \qquad \ell, k = 1, \ldots, d, \tag{7.137}$$

$$\|D_n^{[p]}(s_{\mathbf{G},k})\| \le \|s_{\mathbf{G},k}\|_\infty < \infty, \qquad k = 1, \ldots, d, \tag{7.138}$$

$$\|D_n^{[p]}(c_{\mathbf{G}})\| \le \|c_{\mathbf{G}}\|_\infty < \infty. \tag{7.139}$$

Thus, from (7.112) and (7.113) we obtain

$$\|K_{\mathbf{G},n}^{[p]}\| \le \sum_{\ell,k=1}^{d} \|D_n^{[p]}(a_{\mathbf{G},\ell k})\| \, \|K_{n,\ell k}^{[p]}\| \le \sum_{\ell,k=1}^{d} \|a_{\mathbf{G},\ell k}\|_\infty C^{[p]} n_\ell n_k \le C n^2,$$

$$\|Z_{\mathbf{G},n}^{[p]}\| \le \sum_{k=1}^{d} \|D_n^{[p]}(s_{\mathbf{G},k})\| \, \|H_{n,k}^{[p]}\| + \|D_n^{[p]}(c_{\mathbf{G}})\| \, \|M_n^{[p]}\|$$

$$\le \sum_{k=1}^{d} \|s_{\mathbf{G},k}\|_\infty C^{[p]} n_k + \|c_{\mathbf{G}}\|_\infty C^{[p]} \le C n,$$

which imply (7.132) and (7.133).

Step 2. Define the symmetric matrix

$$n^{-2} \tilde{K}_{\mathbf{G},n}^{[p]} = n^{-2} \sum_{\ell,k=1}^{d} S_{n+p-2}(a_{\mathbf{G},\ell k}) \circ n_\ell n_k T_{n+p-2}((H_p)_{\ell k})$$

$$= \sum_{\ell,k=1}^{d} S_{n+p-2}(a_{\mathbf{G},\ell k}) \circ T_{n+p-2}((H_p^{(\nu)})_{\ell k}) \tag{7.140}$$

and consider the following decomposition of $n^{-2} A_{\mathbf{G},n}^{[p]}$:

$$n^{-2} A_{\mathbf{G},n}^{[p]} = n^{-2} \tilde{K}_{\mathbf{G},n}^{[p]} + (n^{-2} K_{\mathbf{G},n}^{[p]} - n^{-2} \tilde{K}_{\mathbf{G},n}^{[p]}) + n^{-2} Z_{\mathbf{G},n}^{[p]}. \tag{7.141}$$

By Theorem 7.2 and **GLT 4**, $\|n^{-2} \tilde{K}_{\mathbf{G},n}^{[p]}\| \le C$ and $\{n^{-2} \tilde{K}_{\mathbf{G},n}^{[p]}\}_n \sim_{\rm GLT} f_{\mathbf{G},p}^{(\nu)}(\hat{\mathbf{x}}, \boldsymbol{\theta})$. In the next step we show that $n^{-2} \tilde{K}_{\mathbf{G},n}^{[p]}$ is a symmetric approximation of $n^{-2} K_{\mathbf{G},n}^{[p]}$, in the sense that

$$\|n^{-2} K_{\mathbf{G},n}^{[p]} - n^{-2} \tilde{K}_{\mathbf{G},n}^{[p]}\|_1 = o(n^d). \tag{7.142}$$

Once this is done, the theorem is proved. Indeed, from (7.142), **N 2** and Step 1 we have

$$\|n^{-2}K_{\mathbf{G},\boldsymbol{n}}^{[p]} - n^{-2}\tilde{K}_{\mathbf{G},\boldsymbol{n}}^{[p]} + n^{-2}Z_{\mathbf{G},\boldsymbol{n}}^{[p]}\|_1 = o(n^d),$$
$$\|n^{-2}K_{\mathbf{G},\boldsymbol{n}}^{[p]} - n^{-2}\tilde{K}_{\mathbf{G},\boldsymbol{n}}^{[p]} + n^{-2}Z_{\mathbf{G},\boldsymbol{n}}^{[p]}\| \le C,$$

and so (7.129)–(7.130) follow from **Z 2** and **GLT 1**–**GLT 4**, with **GLT 2** applied to the decomposition (7.141).

Step 3. To prove (7.142), we use the fact that, by (7.104),

$$\begin{aligned}\|D_{\boldsymbol{n}}^{[p]}(a) - D_{\boldsymbol{n}+\boldsymbol{p}-\boldsymbol{2}}(a)\| &= \max_{\boldsymbol{j}=\boldsymbol{1},\ldots,\boldsymbol{n}+\boldsymbol{p}-\boldsymbol{2}} \left|a(\boldsymbol{\xi}_{\boldsymbol{j}+\boldsymbol{1},[\boldsymbol{p}]}) - a\Big(\frac{\boldsymbol{j}}{\boldsymbol{n}+\boldsymbol{p}-\boldsymbol{2}}\Big)\right| \\ &\le \max_{\boldsymbol{j}=\boldsymbol{1},\ldots,\boldsymbol{n}+\boldsymbol{p}-\boldsymbol{2}} \omega_a\Big(\Big|\boldsymbol{\xi}_{\boldsymbol{j}+\boldsymbol{1},[\boldsymbol{p}]} - \frac{\boldsymbol{j}}{\boldsymbol{n}+\boldsymbol{p}-\boldsymbol{2}}\Big|_\infty\Big) \\ &\le \omega_a\Big(\frac{C_p}{\min(\boldsymbol{n})}\Big) \le \omega_a(Cn^{-1}) \end{aligned} \tag{7.143}$$

for all functions $a \in C([0,1]^d)$. We have

$$n^{-2}K_{\mathbf{G},\boldsymbol{n}}^{[p]} - n^{-2}\tilde{K}_{\mathbf{G},\boldsymbol{n}}^{[p]}$$
$$= n^{-2}\sum_{\ell,k=1}^{d} D_{\boldsymbol{n}}^{[p]}(a_{\mathbf{G},\ell k})K_{\boldsymbol{n},\ell k}^{[p]} - \sum_{\ell,k=1}^{d} S_{\boldsymbol{n}+\boldsymbol{p}-\boldsymbol{2}}(a_{\mathbf{G},\ell k}) \circ T_{\boldsymbol{n}+\boldsymbol{p}-\boldsymbol{2}}((H_{\boldsymbol{p}}^{(\boldsymbol{\nu})})_{\ell k})$$
$$= \sum_{\ell,k=1}^{d}\Big[D_{\boldsymbol{n}}^{[p]}(a_{\mathbf{G},\ell k})n^{-2}K_{\boldsymbol{n},\ell k}^{[p]} - D_{\boldsymbol{n}}^{[p]}(a_{\mathbf{G},\ell k})T_{\boldsymbol{n}+\boldsymbol{p}-\boldsymbol{2}}((H_{\boldsymbol{p}}^{(\boldsymbol{\nu})})_{\ell k})\Big] \tag{7.144}$$
$$+ \sum_{\ell,k=1}^{d}\Big[D_{\boldsymbol{n}}^{[p]}(a_{\mathbf{G},\ell k})T_{\boldsymbol{n}+\boldsymbol{p}-\boldsymbol{2}}((H_{\boldsymbol{p}}^{(\boldsymbol{\nu})})_{\ell k})$$
$$- D_{\boldsymbol{n}+\boldsymbol{p}-\boldsymbol{2}}(a_{\mathbf{G},\ell k})T_{\boldsymbol{n}+\boldsymbol{p}-\boldsymbol{2}}((H_{\boldsymbol{p}}^{(\boldsymbol{\nu})})_{\ell k})\Big] \tag{7.145}$$
$$+ \sum_{\ell,k=1}^{d}\Big[D_{\boldsymbol{n}+\boldsymbol{p}-\boldsymbol{2}}(a_{\mathbf{G},\ell k})T_{\boldsymbol{n}+\boldsymbol{p}-\boldsymbol{2}}((H_{\boldsymbol{p}}^{(\boldsymbol{\nu})})_{\ell k})$$
$$- S_{\boldsymbol{n}+\boldsymbol{p}-\boldsymbol{2}}(a_{\mathbf{G},\ell k}) \circ T_{\boldsymbol{n}+\boldsymbol{p}-\boldsymbol{2}}((H_{\boldsymbol{p}}^{(\boldsymbol{\nu})})_{\ell k})\Big]. \tag{7.146}$$

We consider separately the three summations in (7.144)–(7.146) and we show that their trace-norms are $o(n^d)$. Once this is done, the proof of the theorem is complete.

- Consider the first summation (7.144). For $\ell, k = 1, \ldots, d$, let

$$X_{\boldsymbol{n},\ell k}^{[p]} = D_{\boldsymbol{n}}^{[p]}(a_{\mathbf{G},\ell k})n^{-2}K_{\boldsymbol{n},\ell k}^{[p]} - D_{\boldsymbol{n}}^{[p]}(a_{\mathbf{G},\ell k})T_{\boldsymbol{n}+\boldsymbol{p}-\boldsymbol{2}}((H_{\boldsymbol{p}}^{(\boldsymbol{\nu})})_{\ell k}).$$

By (7.125),

$$\operatorname{rank}(X_{n,\ell k}^{[p]}) \le Cn^{d-1}, \qquad \ell, k = 1, \dots, d.$$

Moreover, by (7.134), (7.137) and **T 3**,

$$\|X_{n,\ell k}^{[p]}\| \le C, \qquad \ell, k = 1, \dots, d.$$

Thus, by **N 2**,

$$\|X_{n,\ell k}^{[p]}\|_1 \le \operatorname{rank}(X_{n,\ell k}^{[p]})\|X_{n,\ell k}^{[p]}\| \le Cn^{d-1}.$$

This proves that the trace-norm of the first summation (7.144) is $o(n^d)$ (actually, $O(n^{d-1})$).

- Consider now the second summation (7.145). For $\ell, k = 1, \dots, d$, let

$$Y_{n,\ell k}^{[p]} = D_n^{[p]}(a_{\mathbf{G},\ell k})T_{n+p-2}((H_p^{(\nu)})_{\ell k}) - D_{n+p-2}(a_{\mathbf{G},\ell k})T_{n+p-2}((H_p^{(\nu)})_{\ell k}).$$

Taking into account that the $a_{\mathbf{G},\ell k}$ are continuous on $[0, 1]^d$ due to the assumptions on $\mathbf{G}$ and the $a_{\ell k}$, by (7.143) and **T 3** we have

$$\|Y_{n,\ell k}^{[p]}\| \le C\omega_{a_{\mathbf{G},\ell k}}(Cn^{-1}), \qquad \ell, k = 1, \dots, d,$$

and so, by **N 2**,

$$\|Y_{n,\ell k}^{[p]}\|_1 \le Cn^d\omega_{a_{\mathbf{G},\ell k}}(Cn^{-1}), \qquad \ell, k = 1, \dots, d.$$

This proves that the trace-norm of the second summation (7.145) is $o(n^d)$.

- Finally, consider the third summation (7.146). For $\ell, k = 1, \dots, d$, let

$$V_{n,\ell k}^{[p]} = D_{n+p-2}(a_{\mathbf{G},\ell k})T_{n+p-2}((H_p^{(\nu)})_{\ell k}) - S_{n+p-2}(a_{\mathbf{G},\ell k}) \circ T_{n+p-2}((H_p^{(\nu)})_{\ell k}).$$

By Theorem 7.2,

$$\|V_{n,\ell k}^{[p]}\| \le C\omega_{a_{\mathbf{G},\ell k}}(Cn^{-1}), \qquad \ell, k = 1, \dots, d,$$

and so, by **N 2**,

$$\|V_{n,\ell k}^{[p]}\|_1 \le Cn^d\omega_{a_{\mathbf{G},\ell k}}(n^{-1}), \qquad \ell, k = 1, \dots, d.$$

This proves that the trace-norm of the third summation (7.146) is $o(n^d)$. □

We remark that the proof of Theorem 7.6 is essentially the same(!) as the proof of its univariate version [22, Theorem 10.14].

Remark 7.10 (formal structure of the symbol) It is important to note the impressive analogy between the expression of the symbol $\mathbf{1}(A_{\mathbf{G}}(\hat{\mathbf{x}}) \circ H_p^{(\nu)}(\boldsymbol{\theta}))\mathbf{1}^T$ in (7.131) and the expression of the higher-order differential operator $-\mathbf{1}(A_{\mathbf{G}}(\hat{\mathbf{x}}) \circ H\hat{u}(\hat{\mathbf{x}}))\mathbf{1}^T$ in the transformed problem (7.90). We invite the reader to think about this analogy in the light of Remark 7.9 and the introductory discussion in Sect. 7.2.

7.6 Galerkin B-Spline IgA Discretization of Convection-Diffusion-Reaction PDEs

Consider the convection-diffusion-reaction problem

$$\begin{cases} -\nabla \cdot A\nabla u + \mathbf{b} \cdot \nabla u + cu = f, & \text{in } \Omega, \\ u = 0, & \text{on } \partial\Omega, \end{cases}$$
$$\iff \begin{cases} -\sum_{\ell,k=1}^{d} \frac{\partial}{\partial x_\ell}\left(a_{\ell k}\frac{\partial u}{\partial x_k}\right) + \sum_{k=1}^{d} b_k \frac{\partial u}{\partial x_k} + cu = f, & \text{in } \Omega, \\ u = 0, & \text{on } \partial\Omega, \end{cases} \tag{7.147}$$

where $a_{\ell k}$, b_k, c, f are given functions, $A = [a_{\ell k}]_{\ell,k=1}^d$, $\mathbf{b} = [b_k]_{k=1}^d$, and Ω is a bounded open domain in $\mathbb{R}^d$ with Lipschitz boundary.

Isogeometric Galerkin approximation. The weak form of (7.147) reads as follows: find $u \in H_0^1(\Omega)$ such that

$$\mathrm{a}(u, v) = \mathrm{f}(v), \qquad \forall v \in H_0^1(\Omega),$$

where

$$\mathrm{a}(u, v) = \int_\Omega \left((\nabla v)^T A \nabla u + (\nabla u)^T \mathbf{b}\, v + cuv\right), \qquad \mathrm{f}(v) = \int_\Omega fv.$$

In the standard Galerkin method, we look for an approximation $u_{\mathcal{W}}$ of u by choosing a finite dimensional vector space $\mathcal{W} \subset H_0^1(\Omega)$, the so-called approximation space, and by solving the following (Galerkin) problem: find $u_{\mathcal{W}} \in \mathcal{W}$ such that

$$\mathrm{a}(u_{\mathcal{W}}, v) = \mathrm{f}(v), \qquad \forall v \in \mathcal{W}.$$

If $\{\varphi_1, \ldots, \varphi_N\}$ is a basis of $\mathcal{W}$, then we can write $u_{\mathcal{W}} = \sum_{j=1}^N u_j \varphi_j$ for a unique vector $\mathbf{u} = (u_1, \ldots, u_N)^T$, and, by linearity, the computation of $u_{\mathcal{W}}$ (i.e., of $\mathbf{u}$) reduces to solving the linear system

$$A^G \mathbf{u} = \mathbf{f},$$

where $\mathbf{f} = [\mathrm{f}(\varphi_i)]_{i=1}^N$ and

$$A^G = \left[a(\varphi_j, \varphi_i)\right]_{i,j=1}^N = \left[\int_\Omega \left((\nabla\varphi_i)^T A \nabla\varphi_j + (\nabla\varphi_j)^T \mathbf{b}\, \varphi_i + c\varphi_j\varphi_i\right)\right]_{i,j=1}^N \tag{7.148}$$

is the Galerkin stiffness matrix.

Now, suppose that the physical domain Ω can be described by a global geometry function $\mathbf{G} : [0, 1]^d \to \overline{\Omega}$, which is invertible and satisfies $\mathbf{G}(\partial([0, 1]^d)) = \partial\overline{\Omega}$. Let $\{\hat{\varphi}_1, \ldots, \hat{\varphi}_N\}$ be a set of basis functions defined on the parametric (or reference) domain $[0, 1]^d$ and vanishing on the boundary $\partial([0, 1]^d)$. In the isogeometric Galerkin approach, we find an approximation $u_{\mathscr{W}}$ of u by using the standard Galerkin method, in which the approximation space is chosen as $\mathscr{W} = \mathrm{span}(\varphi_1, \ldots, \varphi_N)$, where

$$\varphi_i(\mathbf{x}) = \hat{\varphi}_i(\mathbf{G}^{-1}(\mathbf{x})) = \hat{\varphi}_i(\hat{\mathbf{x}}), \qquad \mathbf{x} = \mathbf{G}(\hat{\mathbf{x}}). \tag{7.149}$$

The resulting stiffness matrix A^G is given by (7.148), with the basis functions φ_i defined as in (7.149). Assuming that $\mathbf{G}$ and $\hat{\varphi}_i$, $i = 1, \ldots, N$, are sufficiently regular, we can apply standard differential calculus to obtain the following expression for A^G in terms of $\mathbf{G}$ and $\hat{\varphi}_i$, $i = 1, \ldots, N$:

$$A^G = \Bigg[\int_{[0,1]^d} \Big((\nabla\hat{\varphi}_i)^T A_{\mathbf{G}} \nabla\hat{\varphi}_j + (\nabla\hat{\varphi}_j)^T (J_{\mathbf{G}})^{-1}\mathbf{b}(\mathbf{G})\, \hat{\varphi}_i + c(\mathbf{G})\hat{\varphi}_j\hat{\varphi}_i\Big)|\det(J_{\mathbf{G}})|\Bigg]_{i,j=1}^N, \tag{7.150}$$

where

$$A_{\mathbf{G}} = (J_{\mathbf{G}})^{-1} A(\mathbf{G}) (J_{\mathbf{G}})^{-T} = \left[a_{\mathbf{G},\ell k}\right]_{\ell,k=1}^d, \tag{7.151}$$

and $J_{\mathbf{G}}$ is the Jacobian matrix of $\mathbf{G}$, i.e.,

$$J_{\mathbf{G}} = \left[\frac{\partial G_i}{\partial \hat{x}_j}\right]_{i,j=1}^d = \left[\frac{\partial x_i}{\partial \hat{x}_j}\right]_{i,j=1}^d.$$

In the IgA framework, $\mathbf{G}$ is expressed in terms of the functions $\hat{\varphi}_i$, which are usually tensor-product B-splines or NURBS. Here, we do not require that $\mathbf{G}$ be expressed in terms of the $\hat{\varphi}_i$, and the $\hat{\varphi}_i$ are chosen as tensor-product B-splines over uniform knot sequences; for the case of NURBS, see [18].

Galerkin B-spline IgA discretization matrices. As in the IgA collocation framework considered in Sect. 7.5, in the Galerkin B-spline IgA based on (uniform) tensor-product B-splines, the functions $\hat{\varphi}_1, \ldots, \hat{\varphi}_N$ are chosen as the tensor-product B-splines $N_{\mathbf{2},[\boldsymbol{p}]}, \ldots, N_{\boldsymbol{n}+\boldsymbol{p}-\mathbf{1},[\boldsymbol{p}]}$ in (7.105). The resulting stiffness matrix (7.150) will be denoted by $A_{\mathbf{G},\boldsymbol{n}}^{[\boldsymbol{p}]}$ so as to emphasize its iependence on $\mathbf{G}$, $\boldsymbol{n}$, $\boldsymbol{p}$:

$$
\begin{aligned}
A^{[p]}_{\mathbf{G},n} &= \Bigg[\int_{[0,1]^d} \big((\nabla N_{i+\mathbf{1},[p]})^T A_{\mathbf{G}} \nabla N_{j+\mathbf{1},[p]} \\
&\qquad + (\nabla N_{j+\mathbf{1},[p]})^T (J_{\mathbf{G}})^{-1} \mathbf{b}(\mathbf{G})\, N_{i+\mathbf{1},[p]} \\
&\qquad + c(\mathbf{G}) N_{j+\mathbf{1},[p]} N_{i+\mathbf{1},[p]}\big) |\det(J_{\mathbf{G}})| \Bigg]_{\boldsymbol{i},\boldsymbol{j}=\mathbf{1}}^{\boldsymbol{n}+\boldsymbol{p}-\mathbf{2}} \\
&= \sum_{\ell,k=1}^{d} K^{[p]}_{\mathbf{G},n,\ell k} + \sum_{k=1}^{d} H^{[p]}_{\mathbf{G},n,k} + M^{[p]}_{\mathbf{G},n},
\end{aligned}
\tag{7.152}
$$

where, for $\ell, k = 1, \ldots, d$,

$$
\begin{aligned}
K^{[p]}_{\mathbf{G},n,\ell k} &= K^{[p]}_{n,\ell k}(|\det(J_{\mathbf{G}})| a_{\mathbf{G},\ell k}) \\
&= \left[\int_{[0,1]^d} |\det(J_{\mathbf{G}})| a_{\mathbf{G},\ell k} \frac{\partial N_{i+\mathbf{1},[p]}}{\partial \hat{x}_\ell} \frac{\partial N_{j+\mathbf{1},[p]}}{\partial \hat{x}_k}\right]_{\boldsymbol{i},\boldsymbol{j}=\mathbf{1}}^{\boldsymbol{n}+\boldsymbol{p}-\mathbf{2}}, && (7.153) \\
H^{[p]}_{\mathbf{G},n,k} &= H^{[p]}_{n,k}(|\det(J_{\mathbf{G}})| ((J_{\mathbf{G}})^{-1}\mathbf{b}(\mathbf{G}))_k) \\
&= \left[\int_{[0,1]^d} |\det(J_{\mathbf{G}})| ((J_{\mathbf{G}})^{-1}\mathbf{b}(\mathbf{G}))_k \frac{\partial N_{j+\mathbf{1},[p]}}{\partial \hat{x}_k} N_{i+\mathbf{1},[p]}\right]_{\boldsymbol{i},\boldsymbol{j}=\mathbf{1}}^{\boldsymbol{n}+\boldsymbol{p}-\mathbf{2}}, && (7.154) \\
M^{[p]}_{\mathbf{G},n} &= M^{[p]}_{n}(|\det(J_{\mathbf{G}})| c(\mathbf{G})) \\
&= \left[\int_{[0,1]^d} |\det(J_{\mathbf{G}})| c(\mathbf{G}) N_{j+\mathbf{1},[p]} N_{i+\mathbf{1},[p]}\right]_{\boldsymbol{i},\boldsymbol{j}=\mathbf{1}}^{\boldsymbol{n}+\boldsymbol{p}-\mathbf{2}}. && (7.155)
\end{aligned}
$$

Note that $A^{[p]}_{\mathbf{G},n}$ can be decomposed as follows:

$$
A^{[p]}_{\mathbf{G},n} = K^{[p]}_{\mathbf{G},n} + Z^{[p]}_{\mathbf{G},n}, \tag{7.156}
$$

where

$$
K^{[p]}_{\mathbf{G},n} = \sum_{\ell,k=1}^{d} K^{[p]}_{\mathbf{G},n,\ell k} = \left[\int_{[0,1]^d} (\nabla N_{i+\mathbf{1},[p]})^T |\det(J_{\mathbf{G}})| A_{\mathbf{G}}\, \nabla N_{j+\mathbf{1},[p]}\right]_{\boldsymbol{i},\boldsymbol{j}=\mathbf{1}}^{\boldsymbol{n}+\boldsymbol{p}-\mathbf{2}} \tag{7.157}
$$

is the matrix resulting from the discretization of the higher-order (diffusion) term in (7.147), and

$$\begin{aligned} Z_{\mathbf{G},\boldsymbol{n}}^{[\boldsymbol{p}]} &= \sum_{k=1}^{d} H_{\mathbf{G},\boldsymbol{n},k}^{[\boldsymbol{p}]} + M_{\mathbf{G},\boldsymbol{n}}^{[\boldsymbol{p}]} \\ &= \Bigg[\int_{[0,1]^d} |\det(J_{\mathbf{G}})| (\nabla N_{\boldsymbol{j}+\mathbf{1},[\boldsymbol{p}]})^T (J_{\mathbf{G}})^{-1} \mathbf{b}(\mathbf{G})\, N_{\boldsymbol{i}+\mathbf{1},[\boldsymbol{p}]} \\ &\qquad + |\det(J_{\mathbf{G}})| c(\mathbf{G}) N_{\boldsymbol{j}+\mathbf{1},[\boldsymbol{p}]} N_{\boldsymbol{i}+\mathbf{1},[\boldsymbol{p}]} \Bigg]_{\boldsymbol{i},\boldsymbol{j}=\mathbf{1}}^{\boldsymbol{n}+\boldsymbol{p}-\mathbf{2}} \end{aligned} \tag{7.158}$$

is the matrix resulting from the discretization of the terms in (7.147) with lower-order derivatives (the convection and reaction terms).

For $\ell, k = 1, \ldots, d$, define the matrices

$$K_{\boldsymbol{n},\ell k}^{[\boldsymbol{p}]} = K_{\boldsymbol{n},\ell k}^{[\boldsymbol{p}]}(1) = \left[\int_{[0,1]^d} \frac{\partial N_{\boldsymbol{i}+\mathbf{1},[\boldsymbol{p}]}}{\partial \hat{x}_\ell} \frac{\partial N_{\boldsymbol{j}+\mathbf{1},[\boldsymbol{p}]}}{\partial \hat{x}_k} \right]_{\boldsymbol{i},\boldsymbol{j}=\mathbf{1}}^{\boldsymbol{n}+\boldsymbol{p}-\mathbf{2}}, \tag{7.159}$$

$$H_{\boldsymbol{n},k}^{[\boldsymbol{p}]} = H_{\boldsymbol{n},k}^{[\boldsymbol{p}]}(1) = \left[\int_{[0,1]^d} \frac{\partial N_{\boldsymbol{j}+\mathbf{1},[\boldsymbol{p}]}}{\partial \hat{x}_k} N_{\boldsymbol{i}+\mathbf{1},[\boldsymbol{p}]} \right]_{\boldsymbol{i},\boldsymbol{j}=\mathbf{1}}^{\boldsymbol{n}+\boldsymbol{p}-\mathbf{2}}, \tag{7.160}$$

$$M_{\boldsymbol{n}}^{[\boldsymbol{p}]} = M_{\boldsymbol{n}}^{[\boldsymbol{p}]}(1) = \left[\int_{[0,1]^d} N_{\boldsymbol{j}+\mathbf{1},[\boldsymbol{p}]} N_{\boldsymbol{i}+\mathbf{1},[\boldsymbol{p}]} \right]_{\boldsymbol{i},\boldsymbol{j}=\mathbf{1}}^{\boldsymbol{n}+\boldsymbol{p}-\mathbf{2}}. \tag{7.161}$$

Lemma 7.4 *For $\boldsymbol{p}, \boldsymbol{n} \in \mathbb{N}^d$, we have*

$$K_{\boldsymbol{n},kk}^{[\boldsymbol{p}]} = \left(\bigotimes_{r=1}^{k-1} M_{n_r}^{[p_r]} \right) \otimes K_{n_k}^{[p_k]} \otimes \left(\bigotimes_{r=k+1}^{d} M_{n_r}^{[p_r]} \right) \tag{7.162}$$

for $k = 1, \ldots, d$,

$$\begin{aligned} K_{\boldsymbol{n},k\ell}^{[\boldsymbol{p}]} &= K_{\boldsymbol{n},\ell k}^{[\boldsymbol{p}]} \\ &= -\left(\bigotimes_{r=1}^{\ell-1} M_{n_r}^{[p_r]} \right) \otimes H_{n_\ell}^{[p_\ell]} \otimes \left(\bigotimes_{r=\ell+1}^{k-1} M_{n_r}^{[p_r]} \right) \otimes H_{n_k}^{[p_k]} \otimes \left(\bigotimes_{r=k+1}^{d} M_{n_r}^{[p_r]} \right) \end{aligned} \tag{7.163}$$

for $1 \le \ell < k \le d$,

$$H_{\boldsymbol{n},k}^{[\boldsymbol{p}]} = \left(\bigotimes_{r=1}^{k-1} M_{n_r}^{[p_r]} \right) \otimes H_{n_k}^{[p_k]} \otimes \left(\bigotimes_{r=k+1}^{d} M_{n_r}^{[p_r]} \right) \tag{7.164}$$

for $k = 1, \ldots, d$, and

$$M_{\boldsymbol{n}}^{[\boldsymbol{p}]} = \bigotimes_{r=1}^{d} M_{n_r}^{[p_r]}, \tag{7.165}$$

where the matrices $K_n^{[p]}$, $H_n^{[p]}$, $M_n^{[p]}$ *are defined for all* $p, n \in \mathbb{N}$ *as follows:*

$$K_n^{[p]} = \left[\int_0^1 N'_{j+1,[p]} N'_{i+1,[p]}\right]_{i,j=1}^{n+p-2}, \tag{7.166}$$

$$H_n^{[p]} = \left[\int_0^1 N'_{j+1,[p]} N_{i+1,[p]}\right]_{i,j=1}^{n+p-2} = \left[-\int_0^1 N_{j+1,[p]} N'_{i+1,[p]}\right]_{i,j=1}^{n+p-2}, \tag{7.167}$$

$$M_n^{[p]} = \left[\int_0^1 N_{j+1,[p]} N_{i+1,[p]}\right]_{i,j=1}^{n+p-2}, \tag{7.168}$$

with $N_{i,[p]}$, $i = 1, \dots, n+p$, *being the B-splines of degree p on the knot sequence*

$$\Big\{\underbrace{0, \dots, 0}_{p+1}, \frac{1}{n}, \frac{2}{n}, \dots, \frac{n-1}{n}, \underbrace{1, \dots, 1}_{p+1}\Big\};$$

see [22, pp. 232–235] *for the corresponding definitions and properties.*

Proof We only prove (7.162) as the proof of the other equations is the same. By the crucial property **P7**, for all $\boldsymbol{i}, \boldsymbol{j} = \mathbf{1}, \dots, \boldsymbol{n} + \boldsymbol{p} - \mathbf{2}$ we have

$$\left[\left(\bigotimes_{r=1}^{k-1} M_{n_r}^{[p_r]}\right) \otimes K_{n_k}^{[p_k]} \otimes \left(\bigotimes_{r=k+1}^{d} M_{n_r}^{[p_r]}\right)\right]_{\boldsymbol{i}\boldsymbol{j}} = (K_{n_k}^{[p_k]})_{i_k j_k} \prod_{\substack{r=1 \\ r\neq k}}^{d} (M_{n_r}^{[p_r]})_{i_r j_r}$$

$$= \int_0^1 N'_{i_k+1,[p_k]} N'_{j_k+1,[p_k]} \prod_{\substack{r=1 \\ r\neq k}}^{d} \int_0^1 N_{i_r+1,[p_r]} N_{j_r+1,[p_r]} = \int_{[0,1]^d} \frac{\partial N_{\boldsymbol{i}+\mathbf{1},[\boldsymbol{p}]}}{\partial \hat{x}_k} \frac{\partial N_{\boldsymbol{j}+\mathbf{1},[\boldsymbol{p}]}}{\partial \hat{x}_k}$$

$$= (K_{\boldsymbol{n},kk}^{[\boldsymbol{p}]})_{\boldsymbol{i}\boldsymbol{j}}.$$

□

Remark 7.11 As in the case of FDs, FEs, IgA collocation (see Remarks 7.3, 7.5, 7.8), $K_n^{[p]}$, $H_n^{[p]}$, $M_n^{[p]}$ are the diffusion, convection, reaction matrices resulting from the Galerkin B-spline IgA discretization of the univariate problem

$$\begin{cases} -u''(x) + u'(x) + u(x) = f(x), & x \in (0,1), \\ u(0) = u(1) = 0. \end{cases}$$

In other words, $K_n^{[p]}$, $H_n^{[p]}$, $M_n^{[p]}$ are the matrices resulting from the Galerkin B-spline IgA discretization of, respectively, the negative second derivative $-u''(x)$, the first derivative $u'(x)$, the identity operator $u(x)$. To see this, follow the above derivation of the Galerkin IgA matrices in the univariate case $d = 1$ or take a look at [22, Sect. 10.7.2]. In view of what follows, we recall from [22, Eqs. (10.185)–(10.187)] that

$$n^{-1} K_n^{[p]} = T_{n+p-2}(f_p) + R_n^{[p]}, \qquad \mathrm{rank}(R_n^{[p]}) \le 4(p-1), \tag{7.169}$$

$$-\mathrm{i}\, H_n^{[p]} = T_{n+p-2}(g_p) + S_n^{[p]}, \qquad \mathrm{rank}(S_n^{[p]}) \le 4(p-1), \tag{7.170}$$

$$n\, M_n^{[p]} = T_{n+p-2}(h_p) + V_n^{[p]}, \qquad \mathrm{rank}(V_n^{[p]}) \le 4(p-1), \tag{7.171}$$

where f_p, g_p, h_p are real trigonometric polynomials whose definitions are given in [22, Eqs. (10.182)–(10.184)].

Remark 7.12 It follows from Lemma 7.4, Eqs. (7.169)–(7.171), **T8** and **P8** that

$$K_{\boldsymbol{n},\ell k}^{[\boldsymbol{p}]} = \frac{n_\ell n_k}{N(\boldsymbol{n})} T_{\boldsymbol{n}+\boldsymbol{p}-\mathbf{2}}((H_{\boldsymbol{p}})_{\ell k}) + R_{\boldsymbol{n},\ell k}^{[\boldsymbol{p}]}, \qquad \ell, k = 1, \ldots, d, \tag{7.172}$$

where

$$\mathrm{rank}(R_{\boldsymbol{n},\ell k}^{[\boldsymbol{p}]}) \le N(\boldsymbol{n}+\boldsymbol{p}-\mathbf{2}) \sum_{i=1}^{d} \frac{4(p_i-1)}{n_i+p_i-2}, \qquad \ell, k = 1, \ldots, d, \tag{7.173}$$

and $H_{\boldsymbol{p}}(\boldsymbol{\theta})$ is the $d \times d$ symmetric matrix defined as follows:

$$(H_{\boldsymbol{p}})_{\ell k}(\boldsymbol{\theta}) = \begin{cases} f_{p_k}(\theta_k) \prod_{\substack{r=1\\ r\ne k}}^{d} h_{p_r}(\theta_r), & \text{if } \ell = k, \\ g_{p_\ell}(\theta_\ell) g_{p_k}(\theta_k) \prod_{\substack{r=1\\ r\ne \ell,k}}^{d} h_{p_r}(\theta_r), & \text{if } \ell \ne k. \end{cases} \tag{7.174}$$

Since $K_{\boldsymbol{n},\ell k}^{[\boldsymbol{p}]}(1) = K_{\boldsymbol{n},\ell k}^{[\boldsymbol{p}]}$, according to (7.172) and the discussion in Sect. 7.2 (see in particular (7.14)), we may predict that $H_{\boldsymbol{p}}(\boldsymbol{\theta})$ is, up to some normalization, the symbol of the negative Hessian operator. More precisely, we should call it the symbol of the *negative Hessian operator in the parametric variables* $\hat{x}_i$, because $K_{\boldsymbol{n},\ell k}^{[\boldsymbol{p}]}(1)$ is the matrix resulting from the Galerkin B-spline IgA discretization of the d-variate problem

$$\begin{cases} -\dfrac{\partial^2 \hat{u}}{\partial \hat{x}_\ell \partial \hat{x}_k} = f, & \text{in } (0,1)^d, \\ \hat{u} = 0, & \text{on } \partial((0,1)^d), \end{cases}$$

in which the physical domain Ω and the physical variables x_i are replaced by the parametric domain $(0,1)^d$ and the parametric variables $\hat{x}_i$. For instance, assuming $\boldsymbol{n} = \boldsymbol{\nu} n$ for some fixed vector $\boldsymbol{\nu} \in \mathbb{Q}^d$ with positive components, from (7.172), (7.173), **Z 1**, **GLT 3** and **GLT 4** we infer that

$$\left\{ n^{d-2} K_{\boldsymbol{n},\ell k}^{[\boldsymbol{p}]}(1) = \frac{\nu_\ell \nu_k}{N(\boldsymbol{\nu})} T_{\boldsymbol{n}+\boldsymbol{p}-\mathbf{2}}((H_{\boldsymbol{p}})_{\ell k}) + n^{d-2} R_{\boldsymbol{n},\ell k}^{[\boldsymbol{p}]} \right\}_n$$

$$\sim_{\mathrm{GLT}} \frac{\nu_\ell \nu_k}{N(\boldsymbol{\nu})} (H_{\boldsymbol{p}})_{\ell k}(\boldsymbol{\theta}) = (H_{\boldsymbol{p}}^{(\boldsymbol{\nu})})_{\ell k}(\boldsymbol{\theta}), \tag{7.175}$$

where $H_p^{(\nu)}(\theta) = \mathrm{diag}(\nu)H_p(\theta)\mathrm{diag}(\nu)/N(\nu)$. This means that, assuming $\boldsymbol{n} = \nu n$ and after normalization by n^{d-2}, the symbol of the negative Hessian operator (in the parametric variables) is $H_p^{(\nu)}(\theta)$, which coincides with $H_p(\theta)$ up to a trivial transformation. It was proved in [17, Theorem 2.2] that, for all $\boldsymbol{p} \geq \boldsymbol{1}$, the matrix $H_p(\theta)$ is SPSD for all $\theta \in [-\pi, \pi]^d$ and SPD for all $\theta \in [-\pi, \pi]^d$ such that $\theta_1 \cdots \theta_d \neq 0$.

GLT analysis of the Galerkin B-spline IgA discretization matrices. Using the theory of multilevel GLT sequences, we now derive the spectral and singular value distribution of the sequence of normalized IgA Galerkin matrices $\{n^{d-2}A_{\mathbf{G},\boldsymbol{n}}^{[\boldsymbol{p}]}\}_n$ under the assumptions that $\boldsymbol{n} = \nu n$ for some fixed vector ν, that the geometry map $\mathbf{G}$ is sufficiently smooth, and that the matrix of diffusion coefficients $A(\mathbf{x})$ is symmetric.

Theorem 7.7 *Let Ω be a bounded open domain in $\mathbb{R}^d$ with Lipschitz boundary and suppose that the following conditions on the PDE coefficients and the geometry map are satisfied:*

- $a_{\ell k} \in L^\infty(\Omega)$ *for every* $\ell, k = 1, \ldots, d$;
- $b_k \in L^\infty(\Omega)$ *for every* $k = 1, \ldots, d$;
- $c \in L^\infty(\Omega)$;
- $A(\mathbf{x}) = [a_{\ell k}(\mathbf{x})]_{\ell,k=1}^d$ *is symmetric for every* $\mathbf{x} \in \Omega$;
- $\mathbf{G}$ *is regular, i.e.,* $\mathbf{G} \in C^1([0,1]^d)$ *and* $\det(J_{\mathbf{G}}) \neq 0$ *over* $[0,1]^d$.

Let $\boldsymbol{p} \geq \boldsymbol{1}$, let $\nu \in \mathbb{Q}^d$ be a vector with positive components, and assume that $\boldsymbol{n} = \nu n$ (it is understood that n varies in the infinite subset of $\mathbb{N}$ such that $\boldsymbol{n} = \nu n \in \mathbb{N}^d$). Then

$$\{n^{d-2}A_{\mathbf{G},\boldsymbol{n}}^{[\boldsymbol{p}]}\}_n \sim_{\mathrm{GLT}} f_{\mathbf{G},\boldsymbol{p}}^{(\nu)} \tag{7.176}$$

and

$$\{n^{d-2}A_{\mathbf{G},\boldsymbol{n}}^{[\boldsymbol{p}]}\}_n \sim_{\sigma,\lambda} f_{\mathbf{G},\boldsymbol{p}}^{(\nu)}, \tag{7.177}$$

where

$$\begin{aligned} f_{\mathbf{G},\boldsymbol{p}}^{(\nu)}(\hat{\mathbf{x}},\theta) &= \sum_{\ell,k=1}^{d} |\det(J_{\mathbf{G}}(\hat{\mathbf{x}}))| a_{\mathbf{G},\ell k}(\hat{\mathbf{x}}) (H_{\boldsymbol{p}}^{(\nu)})_{\ell k}(\theta) \\ &= \mathbf{1}(|\det(J_{\mathbf{G}}(\hat{\mathbf{x}}))| A_{\mathbf{G}}(\hat{\mathbf{x}}) \circ H_{\boldsymbol{p}}^{(\nu)}(\theta))\mathbf{1}^T \\ &= \frac{\nu\big(|\det(J_{\mathbf{G}}(\hat{\mathbf{x}}))| A_{\mathbf{G}}(\hat{\mathbf{x}}) \circ H_{\boldsymbol{p}}(\theta)\big)\nu^T}{N(\nu)}, \\ H_{\boldsymbol{p}}^{(\nu)}(\theta) &= \frac{\mathrm{diag}(\nu)H_{\boldsymbol{p}}(\theta)\mathrm{diag}(\nu)}{N(\nu)}, \end{aligned} \tag{7.178}$$

and $A_{\mathbf{G}}(\hat{\mathbf{x}})$ and $H_{\boldsymbol{p}}(\theta)$ are defined, respectively, in (7.151) *and* (7.174).

Proof We follow the same argument as in the proof of Theorem 7.4. Throughout the proof, the letter C denotes a generic constant independent of n. While reading the proof, the reader should keep in mind the relation $\boldsymbol{n} = \boldsymbol{\nu} n$ and the notation $E_{\ell k}$ for the $d \times d$ matrix having 1 in position (ℓ, k) and 0 elsewhere $(1 \le \ell, k \le d)$.

Step 1. We show that

$$\left\| n^{d-2} K_{\mathbf{G},\boldsymbol{n}}^{[\boldsymbol{p}]} \right\| \le C, \tag{7.179}$$

$$\left\| n^{d-2} Z_{\mathbf{G},\boldsymbol{n}}^{[\boldsymbol{p}]} \right\| \le C n^{-1}. \tag{7.180}$$

To prove (7.179), we note that, for all $\boldsymbol{i}, \boldsymbol{j} = \mathbf{1}, \ldots, \boldsymbol{n} + \boldsymbol{p} - \mathbf{2}$, we have the following.

- If $|\boldsymbol{i} - \boldsymbol{j}|_\infty > |\boldsymbol{p}|_\infty$ then $|i_k - j_k| > p_k$ for some $k \in \{1, \ldots, d\}$, which implies that the intersection of the supports of $N_{\boldsymbol{i}+\mathbf{1},[\boldsymbol{p}]}$ and $N_{\boldsymbol{j}+\mathbf{1},[\boldsymbol{p}]}$ has zero measure by the local support property (7.97). Thus,

$$(K_{\mathbf{G},\boldsymbol{n}}^{[\boldsymbol{p}]})_{\boldsymbol{ij}} = \int_{[0,1]^d} (\nabla N_{\boldsymbol{i}+\mathbf{1},[\boldsymbol{p}]})^T |\det(J_{\mathbf{G}})| A_{\mathbf{G}} \nabla N_{\boldsymbol{j}+\mathbf{1},[\boldsymbol{p}]} = 0.$$

- By (7.97), (7.101) and the assumptions on the geometry map $\mathbf{G}$ and the PDE coefficients $a_{\ell k}$, we have

$$\begin{aligned}
\left|(K_{\mathbf{G},\boldsymbol{n}}^{[\boldsymbol{p}]})_{\boldsymbol{ij}}\right| &\le \sum_{\ell,k=1}^{d} \left|(K_{\mathbf{G},\boldsymbol{n},\ell k}^{[\boldsymbol{p}]})_{\boldsymbol{ij}}\right| \\
&\le \sum_{\ell,k=1}^{d} \int_{[0,1]^d} |a_{\mathbf{G},\ell k} \det(J_{\mathbf{G}})| \left|\frac{\partial N_{\boldsymbol{i}+\mathbf{1},[\boldsymbol{p}]}}{\partial \hat{x}_\ell}\right| \left|\frac{\partial N_{\boldsymbol{j}+\mathbf{1},[\boldsymbol{p}]}}{\partial \hat{x}_k}\right| \\
&\le C \sum_{\ell,k=1}^{d} \int_{\mathrm{supp}(N_{\boldsymbol{i}+\mathbf{1},[\boldsymbol{p}]})} \left|\frac{\partial N_{\boldsymbol{i}+\mathbf{1},[\boldsymbol{p}]}}{\partial \hat{x}_\ell}\right| \left|\frac{\partial N_{\boldsymbol{j}+\mathbf{1},[\boldsymbol{p}]}}{\partial \hat{x}_k}\right| \\
&\le C \sum_{\ell,k=1}^{d} 4 p_\ell n_\ell p_k n_k \mu_d(\mathrm{supp}(N_{\boldsymbol{i}+\mathbf{1},[\boldsymbol{p}]})) \le C n^{2-d}.
\end{aligned} \tag{7.181}$$

Thus, each row and column of $K_{\mathbf{G},\boldsymbol{n}}^{[\boldsymbol{p}]}$ has at most $(2|\boldsymbol{p}|_\infty + 1)^d$ nonzero entries whose moduli are bounded by $C n^{2-d}$, which implies (7.179) by **N 1**. The proof of (7.180) is conceptually identical. For all $\boldsymbol{i}, \boldsymbol{j} = \mathbf{1}, \ldots, \boldsymbol{n} + \boldsymbol{p} - \mathbf{2}$, we have the following.

- If $|\boldsymbol{i} - \boldsymbol{j}|_\infty > |\boldsymbol{p}|_\infty$ then $(Z_{\mathbf{G},\boldsymbol{n}}^{[\boldsymbol{p}]})_{\boldsymbol{ij}} = 0$ for the same reason for which $(K_{\mathbf{G},\boldsymbol{n}}^{[\boldsymbol{p}]})_{\boldsymbol{ij}} = 0$.
- By (7.97), (7.99)–(7.101) and the assumptions on the geometry map $\mathbf{G}$ and the PDE coefficients b_k, c, we have

$$
\begin{aligned}
\left|(Z^{[p]}_{\mathbf{G},n})_{ij}\right| &\le \sum_{k=1}^{d}\left|(H^{[p]}_{\mathbf{G},n,k})_{ij}\right| + \left|(M^{[p]}_{\mathbf{G},n})_{ij}\right| \\
&\le \sum_{k=1}^{d}\int_{[0,1]^d}\left|((J_{\mathbf{G}})^{-1}\mathbf{b}(\mathbf{G}))_k\det(J_{\mathbf{G}})\right|\left|\frac{\partial N_{j+\mathbf{1},[p]}}{\partial\hat{x}_k}\right|\left|N_{i+\mathbf{1},[p]}\right| \\
&\quad + \int_{[0,1]^d}\left|c(\mathbf{G})\det(J_{\mathbf{G}})\right|\left|N_{j+\mathbf{1},[p]}\right|\left|N_{i+\mathbf{1},[p]}\right| \\
&\le C\left[\sum_{k=1}^{d}\int_{\mathrm{supp}(N_{i+\mathbf{1},[p]})}\left|\frac{\partial N_{j+\mathbf{1},[p]}}{\partial\hat{x}_k}\right|\left|N_{i+\mathbf{1},[p]}\right|\right. \\
&\quad \left. + \int_{\mathrm{supp}(N_{i+\mathbf{1},[p]})}\left|N_{j+\mathbf{1},[p]}\right|\left|N_{i+\mathbf{1},[p]}\right|\right] \\
&\le C\mu_d(\mathrm{supp}(N_{i+\mathbf{1},[p]}))\left[\sum_{k=1}^{d}2p_k n_k + 1\right] \le Cn^{1-d}.
\end{aligned}
$$

Thus, each row and column of $Z^{[p]}_{\mathbf{G},n}$ has at most $(2|\boldsymbol{p}|_\infty + 1)^d$ nonzero entries whose moduli are bounded by Cn^{1-d}, which implies (7.180) by **N 1**.

Step 2. Let $L^1([0,1]^d, \mathbb{R}^{d\times d})$ be the space of functions $L : [0,1]^d \to \mathbb{R}^{d\times d}$ such that $L_{ij} \in L^1([0,1]^d)$ for all $i, j = 1, \ldots, d$. Consider the linear operator $K^{[p]}_n(\cdot) : L^1([0,1]^d, \mathbb{R}^{d\times d}) \to \mathbb{R}^{N(n+p-2)\times N(n+p-2)}$,

$$
K^{[p]}_n(L) = \left[\int_{[0,1]^d}(\nabla N_{i+\mathbf{1},[p]})^T L\,\nabla N_{j+\mathbf{1},[p]}\right]_{i,j=\mathbf{1}}^{n+p-\mathbf{2}}. \tag{7.182}
$$

The next four steps are devoted to showing that

$$
\{n^{d-2}K^{[p]}_n(L)\}_n \sim_{\mathrm{GLT}} \mathbf{1}(L(\hat{\mathbf{x}}) \circ H^{(\nu)}_p(\boldsymbol{\theta}))\mathbf{1}^T, \qquad \forall\, L \in L^1([0,1]^d, \mathbb{R}^{d\times d}). \tag{7.183}
$$

Once this is done, the theorem is proved. Indeed, by applying (7.183) with $L = |\det(J_{\mathbf{G}})|A_{\mathbf{G}}$, we get $\{n^{d-2}K^{[p]}_{\mathbf{G},n}\}_n \sim_{\mathrm{GLT}} f^{(\nu)}_{\mathbf{G},p}$. Moreover, $\{n^{d-2}Z^{[p]}_{\mathbf{G},n}\}_n \sim_\sigma 0$ by Step 1 and **Z 1** (or **Z 2**). Hence, the GLT relation (7.176) follows from the decomposition $n^{d-2}A^{[p]}_{\mathbf{G},n} = n^{d-2}K^{[p]}_{\mathbf{G},n} + n^{d-2}Z^{[p]}_{\mathbf{G},n}$ (see (7.156)) and **GLT 3**–**GLT 4**; the singular value distribution in (7.177) follows from (7.176) and **GLT 1**; and the spectral distribution in (7.177) follows from (7.176) and **GLT 2** applied to the decomposition $n^{d-2}A^{[p]}_{\mathbf{G},n} = n^{d-2}K^{[p]}_{\mathbf{G},n} + n^{d-2}Z^{[p]}_{\mathbf{G},n}$, taking into account what we have seen in Step 1, the inequality **N 2**, and the fact that, due to the symmetry assumption on $A(\mathbf{x})$ for all $\mathbf{x} \in \Omega$, the matrix $n^{d-2}K^{[p]}_{\mathbf{G},n}$ is symmetric.

Step 3. We first prove (7.183) in the constant-coefficient case where $L(\hat{\mathbf{x}}) = E_{\ell k}$ identically. In this case, $K^{[p]}_n(E_{\ell k}) = K^{[p]}_{n,\ell k}(1) = K^{[p]}_{n,\ell k}$ and (7.183) is just (7.175).

Step 4. We now prove (7.183) in the case where $L(\hat{\mathbf{x}}) = a(\hat{\mathbf{x}})E_{\ell k}$ with $a \in C([0,1]^d)$. To this end, we show that

$$\|Y_{\boldsymbol{n}}^{[\boldsymbol{p}]}\| \to 0 \tag{7.184}$$

where $Y_{\boldsymbol{n}}^{[\boldsymbol{p}]} = n^{d-2}K_{\boldsymbol{n}}^{[\boldsymbol{p}]}(aE_{\ell k}) - n^{d-2}D_{\boldsymbol{n}+\boldsymbol{p}-\boldsymbol{2}}(a)K_{\boldsymbol{n}}^{[\boldsymbol{p}]}(E_{\ell k})$. Once this is done, from **Z 1** (or **Z 2**) we have $\{Y_{\boldsymbol{n}}\}_n \sim_\sigma 0$. Hence, form Step 3 and **GLT 3**−**GLT 4** we obtain

$$\{n^{d-2}K_{\boldsymbol{n}}^{[\boldsymbol{p}]}(aE_{\ell k})\}_n \sim_{\rm GLT} a(\hat{\mathbf{x}})(H_{\boldsymbol{p}}^{(\boldsymbol{\nu})})_{\ell k}(\boldsymbol{\theta}),$$

as required. For every $\boldsymbol{i}, \boldsymbol{j} = \boldsymbol{1}, \ldots, \boldsymbol{n}+\boldsymbol{p}-\boldsymbol{2}$, we have the following.

- If $|\boldsymbol{i}-\boldsymbol{j}|_\infty > |\boldsymbol{p}|_\infty$ then $(K_{\boldsymbol{n}}^{[\boldsymbol{p}]}(L))_{\boldsymbol{i}\boldsymbol{j}} = 0$ for all $L \in L^1([0,1]^d, \mathbb{R}^{d\times d})$, because of the local support property of tensor-product B-splines already exploited in Step 1. Thus, $(Y_{\boldsymbol{n}}^{[\boldsymbol{p}]})_{\boldsymbol{i}\boldsymbol{j}} = 0$.
- If $|\boldsymbol{i}-\boldsymbol{j}|_\infty \le |\boldsymbol{p}|_\infty$ then, using (7.97), (7.101) and the fact that $\text{supp}(N_{\boldsymbol{i}+\boldsymbol{1},[\boldsymbol{p}]})$ is located near the point $\boldsymbol{i}/(\boldsymbol{n}+\boldsymbol{p}-\boldsymbol{2})$ because

$$\max_{\hat{\mathbf{x}}\in\text{supp}(N_{\boldsymbol{i}+\boldsymbol{1},[\boldsymbol{p}]})}\left|\hat{\mathbf{x}} - \frac{\boldsymbol{i}}{\boldsymbol{n}+\boldsymbol{p}-\boldsymbol{2}}\right|_\infty \le \frac{C}{\min(\boldsymbol{n})} \le Cn^{-1},$$

 we obtain

$$\begin{aligned}
|(Y_{\boldsymbol{n}}^{[\boldsymbol{p}]})_{\boldsymbol{i}\boldsymbol{j}}| &= n^{d-2}\left|\int_{[0,1]^d}\left[a(\hat{\mathbf{x}}) - a\Big(\frac{\boldsymbol{i}}{\boldsymbol{n}+\boldsymbol{p}-\boldsymbol{2}}\Big)\right]\frac{\partial N_{\boldsymbol{i}+\boldsymbol{1},[\boldsymbol{p}]}}{\partial \hat{x}_\ell}(\hat{\mathbf{x}})\frac{\partial N_{\boldsymbol{j}+\boldsymbol{1},[\boldsymbol{p}]}}{\partial \hat{x}_k}(\hat{\mathbf{x}})\mathrm{d}\hat{\mathbf{x}}\right| \\
&\le n^{d-2}\int_{\text{supp}(N_{\boldsymbol{i}+\boldsymbol{1},[\boldsymbol{p}]})}\left|a(\hat{\mathbf{x}}) - a\Big(\frac{\boldsymbol{i}}{\boldsymbol{n}+\boldsymbol{p}-\boldsymbol{2}}\Big)\right| 4p_\ell p_k n_\ell n_k \mathrm{d}\hat{\mathbf{x}} \\
&\le n^{d-2}Cn^2\omega_a\Big(\max_{\hat{\mathbf{x}}\in\text{supp}(N_{\boldsymbol{i}+\boldsymbol{1},[\boldsymbol{p}]})}\left|\hat{\mathbf{x}} - \frac{\boldsymbol{i}}{\boldsymbol{n}+\boldsymbol{p}-\boldsymbol{2}}\right|_\infty\Big)\int_{\text{supp}(N_{\boldsymbol{i}+\boldsymbol{1},[\boldsymbol{p}]})}\mathrm{d}\hat{\mathbf{x}} \\
&\le Cn^d\omega_a(Cn^{-1})\mu_d(\text{supp}(N_{\boldsymbol{i}+\boldsymbol{1},[\boldsymbol{p}]})) \le C\omega_a(Cn^{-1}).
\end{aligned}$$

Thus, each row and column of $Y_{\boldsymbol{n}}^{[\boldsymbol{p}]}$ has at most $(2|\boldsymbol{p}|_\infty + 1)^d$ nonzero entries whose moduli are bounded by $C\omega_a(Cn^{-1})$, which implies (7.184) by **N 1**.

Step 5. We now prove (7.183) in the case where $L(\hat{\mathbf{x}}) = a(\hat{\mathbf{x}})E_{\ell k}$ with $a \in L^1([0,1]^d)$. Take $a_m \in C([0,1]^d)$ such that $a_m \to a$ in $L^1([0,1]^d)$. We have

$$\{n^{d-2}K_{\boldsymbol{n}}^{[\boldsymbol{p}]}(a_mE_{\ell k})\}_n \sim_{\rm GLT} a_m(\hat{\mathbf{x}})(H_{\boldsymbol{p}}^{(\boldsymbol{\nu})})_{\ell k}(\boldsymbol{\theta})$$

by Step 4, and

$$a_m(\hat{\mathbf{x}})(H_{\boldsymbol{p}}^{(\boldsymbol{\nu})})_{\ell k}(\boldsymbol{\theta}) \to a(\hat{\mathbf{x}})(H_{\boldsymbol{p}}^{(\boldsymbol{\nu})})_{\ell k}(\boldsymbol{\theta})$$

in measure as $m \to \infty$. We prove that

$$\{n^{d-2}K_{\boldsymbol{n}}^{[\boldsymbol{p}]}(a_mE_{\ell k})\}_m \xrightarrow{\text{a.c.s.}} \{n^{d-2}K_{\boldsymbol{n}}^{[\boldsymbol{p}]}(aE_{\ell k})\}_n, \tag{7.185}$$

after which the desired result follows from **GLT 7**. By **N 3** and the bounds for the sum of partial derivatives of tensor-product B-splines (7.101), we have

$$\begin{aligned}
&\|K_{\boldsymbol{n}}^{[\boldsymbol{p}]}(aE_{\ell k}) - K_{\boldsymbol{n}}^{[\boldsymbol{p}]}(a_m E_{\ell k})\|_1 = \|K_{\boldsymbol{n}}^{[\boldsymbol{p}]}((a-a_m)E_{\ell k})\|_1\\
&\le \sum_{\boldsymbol{i},\boldsymbol{j}=\boldsymbol{1}}^{\boldsymbol{n}} |(K_{\boldsymbol{n}}^{[\boldsymbol{p}]}((a-a_m)E_{\ell k}))_{\boldsymbol{i}\boldsymbol{j}}|\\
&\le \sum_{\boldsymbol{i},\boldsymbol{j}=\boldsymbol{1}}^{\boldsymbol{n}} \int_{[0,1]^d} |a(\hat{\mathbf{x}}) - a_m(\hat{\mathbf{x}})| \left|\frac{\partial N_{\boldsymbol{i}+\boldsymbol{1},[\boldsymbol{p}]}}{\partial \hat{x}_\ell}(\hat{\mathbf{x}})\right| \left|\frac{\partial N_{\boldsymbol{j}+\boldsymbol{1},[\boldsymbol{p}]}}{\partial \hat{x}_k}(\hat{\mathbf{x}})\right| \mathrm{d}\hat{\mathbf{x}}\\
&\le \int_{[0,1]^d} |a(\hat{\mathbf{x}}) - a_m(\hat{\mathbf{x}})| \sum_{\boldsymbol{i}=\boldsymbol{1}}^{\boldsymbol{n}} \left|\frac{\partial N_{\boldsymbol{i}+\boldsymbol{1},[\boldsymbol{p}]}}{\partial \hat{x}_\ell}(\hat{\mathbf{x}})\right| \sum_{\boldsymbol{j}=\boldsymbol{1}}^{\boldsymbol{n}} \left|\frac{\partial N_{\boldsymbol{j}+\boldsymbol{1},[\boldsymbol{p}]}}{\partial \hat{x}_k}(\hat{\mathbf{x}})\right| \mathrm{d}\hat{\mathbf{x}}\\
&\le 4p_k n_k p_\ell n_\ell \|a - a_m\|_{L^1} \le Cn^2 \|a - a_m\|_{L^1},
\end{aligned}$$

hence

$$\begin{aligned}
\|n^{d-2} K_{\boldsymbol{n}}^{[\boldsymbol{p}]}(aE_{\ell k}) - n^{d-2} K_{\boldsymbol{n}}^{[\boldsymbol{p}]}(a_m E_{\ell k})\|_1 &\le Cn^d \|a - a_m\|_{L^1}\\
&\le CN(\boldsymbol{n}+\boldsymbol{p}-\boldsymbol{2})\|a - a_m\|_{L^1},
\end{aligned}$$

and the a.c.s. convergence (7.185) follows from **ACS 6**.

Step 6. Finally, we prove (7.183) for an arbitrary $L \in L^1([0,1]^d, \mathbb{R}^{d\times d})$. Write

$$L(\hat{\mathbf{x}}) = \sum_{\ell,k=1}^{d} L_{\ell k}(\hat{\mathbf{x}}) E_{\ell k}$$

and note that, by linearity,

$$K_{\boldsymbol{n}}^{[\boldsymbol{p}]}(L) = \sum_{\ell,k=1}^{d} K_{\boldsymbol{n}}^{[\boldsymbol{p}]}(L_{\ell k} E_{\ell k}).$$

Hence, by Step 5 and **GLT 4**,

$$\{n^{d-2} K_{\boldsymbol{n}}^{[\boldsymbol{p}]}(L)\}_n \sim_{\rm GLT} \sum_{\ell,k=1}^{d} L_{\ell k}(\hat{\mathbf{x}}) (H_{\boldsymbol{p}}^{(\boldsymbol{\nu})})_{\ell k}(\boldsymbol{\theta}) = \mathbf{1}(L(\hat{\mathbf{x}}) \circ H_{\boldsymbol{p}}^{(\boldsymbol{\nu})}(\boldsymbol{\theta}))\mathbf{1}^T,$$

which concludes the proof. □

We remark that the proof of Theorem 7.7 is essentially the same(!) as the proof of its univariate version [22, Theorem 10.15].

We also remark that the GLT relation (7.176) and the singular value distribution in (7.177) remain true even without the hypothesis that the matrix of diffusion coefficients $A(\mathbf{x})$ is symmetric for all $\mathbf{x} \in \Omega$. Indeed, as it is clear from the proof of Theorem 7.7 and especially from Step 2, the symmetry of A is used only in the proof of the eigenvalue distribution in (7.177). Actually, also the eigenvalue distribution remains true without the symmetry assumption on A, as long as we assume that the entries of A are continuous. This result, which has never been observed before in the literature, is proved in the next theorem.

Theorem 7.8 *Let Ω be a bounded open domain in $\mathbb{R}^d$ with Lipschitz boundary and suppose that the following conditions on the PDE coefficients and the geometry map are satisfied:*

- $a_{\ell k} \in C(\overline{\Omega})$ *for every* $\ell, k = 1, \ldots, d$*;*
- $b_k \in L^\infty(\Omega)$ *for every* $k = 1, \ldots, d$*;*
- $c \in L^\infty(\Omega)$*;*
- $\mathbf{G}$ *is regular, i.e.,* $\mathbf{G} \in C^1([0,1]^d)$ *and* $\det(J_\mathbf{G}) \neq 0$ *over* $[0,1]^d$.

Let $\boldsymbol{p} \geq \mathbf{1}$, let $\boldsymbol{\nu} \in \mathbb{Q}^d$ be a vector with positive components, and assume that $\boldsymbol{n} = \boldsymbol{\nu} n$ (it is understood that n varies in the infinite subset of $\mathbb{N}$ such that $\boldsymbol{n} = \boldsymbol{\nu} n \in \mathbb{N}^d$). Then, both (7.176) *and* (7.177) *are satisfied.*

Proof As noted before the statement of the theorem, we only have to prove the eigenvalue distribution in (7.177). The proof is similar to that of Theorem 7.5. The underlying idea is that, even in the case where A is not symmetric, the matrix $K_{\mathbf{G},\boldsymbol{n}}^{[\boldsymbol{p}]}$ is "almost" symmetric (as long as the entries of A are continuous). We can then derive the eigenvalue distribution in (7.177) from **GLT 2** applied to a suitable decomposition of $A_{\mathbf{G},\boldsymbol{n}}^{[\boldsymbol{p}]}$ obtained from (7.156) by replacing $K_{\mathbf{G},\boldsymbol{n}}^{[\boldsymbol{p}]}$ with one of its symmetric approximations. Let us work out the details. Throughout the proof, the letter C denotes a generic constant independent of n. By (7.153) and (7.157),

$$K_{\mathbf{G},\boldsymbol{n}}^{[\boldsymbol{p}]} = \sum_{\ell,k=1}^{d} K_{\boldsymbol{n},\ell k}^{[\boldsymbol{p}]}(|\det(J_\mathbf{G})| a_{\mathbf{G},\ell k}),$$

$$K_{\boldsymbol{n},\ell k}^{[\boldsymbol{p}]}(a) = \left[\int_{[0,1]^d} a \frac{\partial N_{\boldsymbol{i}+\mathbf{1},[\boldsymbol{p}]}}{\partial \hat{x}_\ell} \frac{\partial N_{\boldsymbol{j}+\mathbf{1},[\boldsymbol{p}]}}{\partial \hat{x}_k}\right]_{\boldsymbol{i},\boldsymbol{j}=\mathbf{1}}^{\boldsymbol{n}}.$$

Let $\tilde{K}_{\mathbf{G},\boldsymbol{n}}^{[\boldsymbol{p}]}$ be the symmetric approximation of $K_{\mathbf{G},\boldsymbol{n}}^{[\boldsymbol{p}]}$ given by

$$\tilde{K}_{\mathbf{G},\boldsymbol{n}}^{[\boldsymbol{p}]} = \sum_{\ell,k=1}^{d} \tilde{K}_{\boldsymbol{n},\ell k}^{[\boldsymbol{p}]}(|\det(J_\mathbf{G})| a_{\mathbf{G},\ell k}),$$

$$\tilde{K}_{\boldsymbol{n},\ell k}^{[\boldsymbol{p}]}(a) = S_{\boldsymbol{n}+\boldsymbol{p}-\mathbf{2}}(a) \circ n^{2-d} T_{\boldsymbol{n}+\boldsymbol{p}-\mathbf{2}}((H_{\boldsymbol{p}}^{(\boldsymbol{\nu})})_{\ell k}).$$

Note that $\|n^{d-2}\tilde{K}_{\boldsymbol{n},\ell k}^{[\boldsymbol{p}]}(a)\| \leq C$ for all $a \in C([0,1]^d)$, by Theorem 7.2. We prove that

$$\|n^{d-2} K_{\mathbf{G},\boldsymbol{n}}^{[\boldsymbol{p}]} - n^{d-2}\tilde{K}_{\mathbf{G},\boldsymbol{n}}^{[\boldsymbol{p}]}\|_1 = o(n^d). \tag{7.186}$$

Once this is done, the eigenvalue distribution in (7.177) follows from (7.176) and **GLT 2** applied to the decomposition

$$n^{d-2}A_{\mathbf{G},\boldsymbol{n}}^{[\boldsymbol{p}]} = n^{d-2}\tilde{K}_{\mathbf{G},\boldsymbol{n}}^{[\boldsymbol{p}]} + (n^{d-2}K_{\mathbf{G},\boldsymbol{n}}^{[\boldsymbol{p}]} - n^{d-2}\tilde{K}_{\mathbf{G},\boldsymbol{n}}^{[\boldsymbol{p}]}) + n^{d-2}Z_{\mathbf{G},\boldsymbol{n}}^{[\boldsymbol{p}]},$$

taking into account what we have seen in Step 1 of the proof of Theorem 7.7, the inequality **N 2**, and the symmetry of $\tilde{K}_{\mathbf{G},\boldsymbol{n}}^{[\boldsymbol{p}]}$. To prove (7.186), it is enough to show that

$$\|n^{d-2}K_{\boldsymbol{n},\ell k}^{[\boldsymbol{p}]}(a) - n^{d-2}\tilde{K}_{\boldsymbol{n},\ell k}^{[\boldsymbol{p}]}(a)\|_1 = o(n^d)$$

for every $a \in C([0,1]^d)$ and every $\ell, k = 1, \ldots, d$. With the notation used in the proof of Theorem 7.7 (see in particular (7.182)), we have $K_{\boldsymbol{n},\ell k}^{[\boldsymbol{p}]}(a) = K_{\boldsymbol{n}}^{[\boldsymbol{p}]}(aE_{\ell k})$. Thus,

$$\begin{aligned}
&\|n^{d-2}K_{\boldsymbol{n},\ell k}^{[\boldsymbol{p}]}(a) - n^{d-2}\tilde{K}_{\boldsymbol{n},\ell k}^{[\boldsymbol{p}]}(a)\|_1 \\
&\le \|n^{d-2}K_{\boldsymbol{n}}^{[\boldsymbol{p}]}(aE_{\ell k}) - n^{d-2}D_{\boldsymbol{n}+\boldsymbol{p}-\mathbf{2}}(a)K_{\boldsymbol{n}}^{[\boldsymbol{p}]}(E_{\ell k})\|_1 &\quad (7.187)\\
&\quad + \|n^{d-2}D_{\boldsymbol{n}+\boldsymbol{p}-\mathbf{2}}(a)K_{\boldsymbol{n}}^{[\boldsymbol{p}]}(E_{\ell k}) - D_{\boldsymbol{n}+\boldsymbol{p}-\mathbf{2}}(a)T_{\boldsymbol{n}+\boldsymbol{p}-\mathbf{2}}((H_{\boldsymbol{p}}^{(\boldsymbol{\nu})})_{\ell k})\|_1 &\quad (7.188)\\
&\quad + \|D_{\boldsymbol{n}+\boldsymbol{p}-\mathbf{2}}(a)T_{\boldsymbol{n}+\boldsymbol{p}-\mathbf{2}}((H_{\boldsymbol{p}}^{(\boldsymbol{\nu})})_{\ell k}) - S_{\boldsymbol{n}+\boldsymbol{p}-\mathbf{2}}(a)\circ T_{\boldsymbol{n}+\boldsymbol{p}-\mathbf{2}}((H_{\boldsymbol{p}}^{(\boldsymbol{\nu})})_{\ell k})\|_1. &\quad (7.189)
\end{aligned}$$

We consider separately the three trace-norms in (7.187)–(7.189) and we show that they are $o(n^d)$. Once this is done, the proof of the theorem is complete.

- The trace-norm (7.187) is $o(n^d)$ by **N 2**, because the spectral norm of its argument tends to 0; see Step 4 in the proof of Theorem 7.7.
- The trace-norm (7.188) is $o(n^d)$ by **N 2**, because of the following two properties: first, the rank of its argument is bounded by Cn^{d-1} (see Remark 7.12 and take into account that $K_{\boldsymbol{n},\ell k}^{[\boldsymbol{p}]}(E_{\ell k}) = K_{\boldsymbol{n},\ell k}^{[\boldsymbol{p}]}(1) = K_{\boldsymbol{n},\ell k}^{[\boldsymbol{p}]}$); second, the spectral norm of its argument is bounded by C, because

$$\begin{aligned}
\|D_{\boldsymbol{n}+\boldsymbol{p}-\mathbf{2}}(a)\| &\le \|a\|_\infty,\\
\|T_{\boldsymbol{n}+\boldsymbol{p}-\mathbf{2}}((H_{\boldsymbol{p}}^{(\boldsymbol{\nu})})_{\ell k})\| &\le \|(H_{\boldsymbol{p}}^{(\boldsymbol{\nu})})_{\ell k}\|_\infty,\\
\|n^{d-2}K_{\boldsymbol{n}}^{[\boldsymbol{p}]}(E_{\ell k})\| &\le C,
\end{aligned}$$

  where the first inequality is obvious, the second inequality follows from **T 3**, and the third inequality is proved by the same argument used in Step 1 of the proof of Theorem 7.7 to prove the inequality $\|n^{d-2}K_{\mathbf{G},\boldsymbol{n}}^{[\boldsymbol{p}]}\| \le C$.[4]

[4]Actually, the argument used in Step 1 of the proof of Theorem 7.7 to prove the inequality $\|n^{d-2}K_{\mathbf{G},\boldsymbol{n}}^{[\boldsymbol{p}]}\| \le C$ can be used to prove that $\|n^{d-2}K_{\boldsymbol{n}}^{[\boldsymbol{p}]}(L)\| \le C$ for all functions $L : [0,1]^d \to \mathbb{R}^{d\times d}$ such that $L_{ij} \in L^\infty([0,1]^d)$ for all $i, j = 1, \ldots, d$. Recall also that $K_{\mathbf{G},\boldsymbol{n}}^{[\boldsymbol{p}]} = K_{\boldsymbol{n}}^{[\boldsymbol{p}]}(|\det(J_{\mathbf{G}})|A_{\mathbf{G}})$ and $(|\det(J_{\mathbf{G}})|A_{\mathbf{G}})_{ij} \in L^\infty([0,1]^d)$ for all $i, j = 1, \ldots, d$ under the assumptions of both Theorem 7.7 and the present theorem.

- The trace-norm (7.189) is $o(n^d)$ by **N 2**, because the spectral norm of its argument tends to 0 by Theorem 7.2. □

Remark 7.13 (formal structure of the symbol) As we know from Sect. 7.5, problem (7.147) can be formally rewritten as in (7.81). If, for any $u : \overline{\Omega} \to \mathbb{R}$, we define $\hat{u} : [0,1]^d \to \mathbb{R}$ as in (7.89), then u satisfies (7.81) if and only if $\hat{u}$ satisfies the corresponding transformed problem (7.90), in which the higher-orderoperator takes the form $-\mathbf{1}(A_{\mathbf{G}}(\hat{\mathbf{x}}) \circ H\hat{u}(\hat{\mathbf{x}}))\mathbf{1}^T$. It is then clear that, similarly to the collocation case (see Remark 7.10), even in the Galerkin case the symbol $f_{\mathbf{G},\boldsymbol{p}}^{(\boldsymbol{\nu})}(\hat{\mathbf{x}},\boldsymbol{\theta}) = \mathbf{1}(|\det(J_{\mathbf{G}}(\hat{\mathbf{x}}))|A_{\mathbf{G}}(\hat{\mathbf{x}}) \circ H_{\boldsymbol{p}}^{(\boldsymbol{\nu})}(\boldsymbol{\theta}))\mathbf{1}^T$ preserves the formal structure of the higher-order operator associated with the transformed problem (7.90). However, in this Galerkin context, we notice the appearance of the determinant factor $|\det(J_{\mathbf{G}}(\hat{\mathbf{x}}))|$, which is not present in the collocation setting; cf. (7.178) and (7.131).

Remark 7.14 The matrix $A_{\mathbf{G},\boldsymbol{n}}^{[\boldsymbol{p}]}$ in (7.152), which we decomposed as in (7.156), can also be decomposed as follows, according to the diffusion, convection and reaction terms:

$$A_{\mathbf{G},\boldsymbol{n}}^{[\boldsymbol{p}]} = K_{\mathbf{G},\boldsymbol{n}}^{[\boldsymbol{p}]} + H_{\mathbf{G},\boldsymbol{n}}^{[\boldsymbol{p}]} + M_{\mathbf{G},\boldsymbol{n}}^{[\boldsymbol{p}]},$$

where the diffusion, convection and reaction matrices are given by

$$K_{\mathbf{G},\boldsymbol{n}}^{[\boldsymbol{p}]} = \left[\int_{[0,1]^d} (\nabla N_{\boldsymbol{i}+\mathbf{1},[\boldsymbol{p}]})^T |\det(J_{\mathbf{G}})| A_{\mathbf{G}}\, \nabla N_{\boldsymbol{j}+\mathbf{1},[\boldsymbol{p}]}\right]_{\boldsymbol{i},\boldsymbol{j}=\mathbf{1}}^{\boldsymbol{n}+\boldsymbol{p}-\mathbf{2}}, \tag{7.190}$$

$$H_{\mathbf{G},\boldsymbol{n}}^{[\boldsymbol{p}]} = \left[\int_{[0,1]^d} |\det(J_{\mathbf{G}})| (\nabla N_{\boldsymbol{j}+\mathbf{1},[\boldsymbol{p}]})^T (J_{\mathbf{G}})^{-1} \mathbf{b}(\mathbf{G})\, N_{\boldsymbol{i}+\mathbf{1},[\boldsymbol{p}]}\right]_{\boldsymbol{i},\boldsymbol{j}=\mathbf{1}}^{\boldsymbol{n}+\boldsymbol{p}-\mathbf{2}}, \tag{7.191}$$

$$M_{\mathbf{G},\boldsymbol{n}}^{[\boldsymbol{p}]} = \left[\int_{[0,1]^d} |\det(J_{\mathbf{G}})| c(\mathbf{G}) N_{\boldsymbol{j}+\mathbf{1},[\boldsymbol{p}]} N_{\boldsymbol{i}+\mathbf{1},[\boldsymbol{p}]}\right]_{\boldsymbol{i},\boldsymbol{j}=\mathbf{1}}^{\boldsymbol{n}+\boldsymbol{p}-\mathbf{2}}. \tag{7.192}$$

Let $\boldsymbol{p} \geq \mathbf{1}$ and assume $\boldsymbol{n} = \boldsymbol{\nu} n$ as in Theorem 7.7.

- Suppose that $|\det(J_{\mathbf{G}})| A_{\mathbf{G}} \in L^1([0,1]^d, \mathbb{R}^{d\times d})$. Then

$$\begin{aligned}\{n^{d-2} K_{\mathbf{G},\boldsymbol{n}}^{[\boldsymbol{p}]}\}_n &\sim_{\mathrm{GLT}} \mathbf{1}(|\det(J_{\mathbf{G}}(\hat{\mathbf{x}}))| A_{\mathbf{G}}(\hat{\mathbf{x}}) \circ H_{\boldsymbol{p}}^{(\boldsymbol{\nu})}(\boldsymbol{\theta}))\mathbf{1}^T \\ &= \frac{\boldsymbol{\nu}(|\det(J_{\mathbf{G}}(\hat{\mathbf{x}}))| A_{\mathbf{G}}(\hat{\mathbf{x}}) \circ H_{\boldsymbol{p}}(\boldsymbol{\theta}))\boldsymbol{\nu}^T}{N(\boldsymbol{\nu})}\end{aligned} \tag{7.193}$$

by (7.183) applied with $L = |\det(J_{\mathbf{G}})| A_{\mathbf{G}}$. If we also assume that $A(\mathbf{x})$ is symmetric for all $\mathbf{x} \in \Omega$ (so that $K_{\mathbf{G},\boldsymbol{n}}^{[\boldsymbol{p}]}$ is symmetric), then, by **GLT 1**,

$$\begin{aligned}\{n^{d-2} K_{\mathbf{G},\boldsymbol{n}}^{[\boldsymbol{p}]}\}_n &\sim_{\sigma,\lambda} \mathbf{1}(|\det(J_{\mathbf{G}}(\hat{\mathbf{x}}))| A_{\mathbf{G}}(\hat{\mathbf{x}}) \circ H_{\boldsymbol{p}}^{(\boldsymbol{\nu})}(\boldsymbol{\theta}))\mathbf{1}^T \\ &= \frac{\boldsymbol{\nu}(|\det(J_{\mathbf{G}}(\hat{\mathbf{x}}))| A_{\mathbf{G}}(\hat{\mathbf{x}}) \circ H_{\boldsymbol{p}}(\boldsymbol{\theta}))\boldsymbol{\nu}^T}{N(\boldsymbol{\nu})}.\end{aligned}$$

- Suppose that $|\det(J_{\mathbf{G}})|c(\mathbf{G}) \in L^1([0,1]^d)$. Note that this is the same as assuming that $c \in L^1(\Omega)$, because

$$\int_\Omega c = \int_{[0,1]^d} c(\mathbf{G})|\det(J_{\mathbf{G}})|.$$

Then, with the same argument used for proving (7.193), one can show that

$$\{n^d M_{\mathbf{G},\boldsymbol{n}}^{[\boldsymbol{p}]}\}_n \sim_{\mathrm{GLT}} \frac{|\det(J_{\mathbf{G}}(\hat{\mathbf{x}}))|c(\mathbf{G}(\hat{\mathbf{x}}))h_{\boldsymbol{p}}(\boldsymbol{\theta})}{N(\boldsymbol{\nu})}, \tag{7.194}$$

where

$$h_{\boldsymbol{p}}(\boldsymbol{\theta}) = \prod_{r=1}^d h_{p_r}(\theta_r). \tag{7.195}$$

Hence, by **GLT 1** and the symmetry of $M_{\mathbf{G},\boldsymbol{n}}^{[\boldsymbol{p}]}$,

$$\{n^d M_{\mathbf{G},\boldsymbol{n}}^{[\boldsymbol{p}]}\}_n \sim_{\sigma,\lambda} \frac{|\det(J_{\mathbf{G}}(\hat{\mathbf{x}}))|c(\mathbf{G}(\hat{\mathbf{x}}))h_{\boldsymbol{p}}(\boldsymbol{\theta})}{N(\boldsymbol{\nu})}.$$

7.7 Galerkin B-Spline IgA Discretization of Second-Order Eigenvalue Problems

Let $\mathbb{R}^+$ be the set of positive real numbers. Consider the following second-order eigenvalue problem: find eigenvalues $\lambda_j \in \mathbb{R}^+$ and eigenfunctions u_j, for $j = 1, 2, \ldots, \infty$, such that

$$\begin{cases} -\nabla \cdot A\nabla u_j = \lambda_j c u_j, & \text{in } \Omega, \\ u_j = 0, & \text{on } \partial\Omega, \end{cases} \tag{7.196}$$

where $A = [a_{\ell k}]_{\ell,k=1}^d$ and Ω is a bounded open domain in $\mathbb{R}^d$ with Lipschitz boundary. We assume that $a_{\ell k} \in L^1(\Omega)$ for all $\ell, k = 1, \ldots, d$, that $A(\mathbf{x}) = [a_{\ell k}(\mathbf{x})]_{\ell,k=1}^d$ is SPD for almost every $\mathbf{x} \in \Omega$, and that $c \in L^1(\Omega)$ with $c > 0$ a.e. in Ω. It can be shown that the eigenvalues λ_j must necessarily be real and positive. This can be formally seen by multiplying (7.196) by u_j and integrating over Ω:

$$\lambda_j = \frac{-\int_\Omega (\nabla \cdot A\nabla u_j)u_j}{\int_\Omega c u_j^2} = \frac{\int_\Omega (\nabla u_j)^T A \nabla u_j}{\int_\Omega c u_j^2} > 0.$$

Isogeometric Galerkin approximation. The weak form of (7.196) reads as follows: find eigenvalues $\lambda_j \in \mathbb{R}^+$ and eigenfunctions $u_j \in H_0^1(\Omega)$, for $j = 1, 2, \ldots, \infty$, such that

$$a(u_j, w) = \lambda_j(cu_j, w), \qquad \forall w \in H_0^1(\Omega),$$

where

$$a(u_j, w) = \int_\Omega (\nabla w)^T A \nabla u_j, \qquad (cu_j, w) = \int_\Omega cu_j w.$$

In the standard Galerkin method, we choose a finite dimensional vector space $\mathcal{W} \subset H_0^1(\Omega)$, the so-called approximation space, we let $N = \dim \mathcal{W}$ and we look for approximations of the eigenpairs (λ_j, u_j), $j = 1, 2, \ldots, \infty$, by solving the following discrete (Galerkin) problem: find $\lambda_{j,\mathcal{W}} \in \mathbb{R}^+$ and $u_{j,\mathcal{W}} \in \mathcal{W}$, for $j = 1, \ldots, N$, such that

$$a(u_{j,\mathcal{W}}, w) = \lambda_{j,\mathcal{W}}(c\, u_{j,\mathcal{W}}, w), \qquad \forall w \in \mathcal{W}. \tag{7.197}$$

Assuming that both the exact and numerical eigenvalues are arranged in non-decreasing order, the pair $(\lambda_{j,\mathcal{W}}, u_{j,\mathcal{W}})$ is taken as an approximation to the pair (λ_j, u_j) for all $j = 1, 2, \ldots, N$. The numbers $\lambda_{j,\mathcal{W}}/\lambda_j - 1$, $j = 1, \ldots, N$, are referred to as the (relative) *eigenvalue errors*. If $\{\varphi_1, \ldots, \varphi_N\}$ is a basis of $\mathcal{W}$, we can identify each $w \in \mathcal{W}$ with its coefficient vector relative to this basis. With this identification in mind, solving the discrete problem (7.197) is equivalent to solving the generalized eigenvalue problem

$$K^G \mathbf{u}_{j,\mathcal{W}} = \lambda_{j,\mathcal{W}} M^G \mathbf{u}_{j,\mathcal{W}}, \tag{7.198}$$

where $\mathbf{u}_{j,\mathcal{W}}$ is the coefficient vector of $u_{j,\mathcal{W}}$ with respect to $\{\varphi_1, \ldots, \varphi_N\}$ and

$$K^G = \left[\int_\Omega (\nabla\varphi_i)^T A \nabla\varphi_j\right]_{i,j=1}^N, \tag{7.199}$$

$$M^G = \left[\int_\Omega c\varphi_j\varphi_i\right]_{i,j=1}^N. \tag{7.200}$$

The matrices K^G and M^G are referred to as the Galerkin stiffness and mass matrix, respectively. Due to our assumption that $A > O$ a.e. and $c > 0$ a.e., both K^G and M^G are SPD, regardless of the chosen basis functions $\varphi_1, \ldots, \varphi_N$. Moreover, it is clear from (7.198) that the numerical eigenvalues $\lambda_{j,\mathcal{W}}$, $j = 1, \ldots, N$, are just the eigenvalues of the matrix

$$L^G = (M^G)^{-1} K^G. \tag{7.201}$$

In the isogeometric Galerkin method, we assume that the physical domain Ω is described by a global geometry function $\mathbf{G} : [0, 1]^d \to \overline{\Omega}$, which is invertible and satisfies $\mathbf{G}(\partial([0, 1]^d)) = \partial\overline{\Omega}$. We fix a set of basis functions $\{\hat\varphi_1, \ldots, \hat\varphi_N\}$ defined on the reference (parametric) domain $[0, 1]^d$ and vanishing on the boundary $\partial([0, 1]^d)$, and we find approximations to the exact eigenpairs (λ_j, u_j), $j = 1, 2, \ldots, \infty$, by using the standard Galerkin method described above, in which the approximation space is chosen as $\mathcal{W} = \operatorname{span}(\varphi_1, \ldots, \varphi_N)$, where

$$\varphi_i(\mathbf{x}) = \hat{\varphi}_i(\mathbf{G}^{-1}(\mathbf{x})) = \hat{\varphi}_i(\hat{\mathbf{x}}), \qquad \mathbf{x} = \mathbf{G}(\hat{\mathbf{x}}). \tag{7.202}$$

The resulting stiffness and mass matrices K^G and M^G are given by (7.199) and (7.200), with the basis functions φ_i defined as in (7.202). If we assume that $\mathbf{G}$ and $\hat{\varphi}_i$, $i = 1, \dots, N$, are sufficiently regular, we can apply standard differential calculus to obtain for K^G and M^G the following expressions:

$$K^G = \left[\int_{[0,1]^d} |\det(J_\mathbf{G})|(\nabla\hat{\varphi}_i)^T A_\mathbf{G} \nabla\hat{\varphi}_j\right]_{i,j=1}^N, \tag{7.203}$$

$$M^G = \left[\int_{[0,1]^d} c(\mathbf{G})|\det(J_\mathbf{G})|\hat{\varphi}_j\hat{\varphi}_i\right]_{i,j=1}^N, \tag{7.204}$$

where

$$A_\mathbf{G} = (J_\mathbf{G})^{-1} A(\mathbf{G})(J_\mathbf{G})^{-T} \tag{7.205}$$

and

$$J_\mathbf{G} = \left[\frac{\partial G_i}{\partial \hat{x}_j}\right]_{i,j=1}^d = \left[\frac{\partial x_i}{\partial \hat{x}_j}\right]_{i,j=1}^d.$$

GLT analysis of the Galerkin B-spline IgA discretization matrices. Following the approach of Sects. 7.5 and 7.6, we choose the basis functions $\hat{\varphi}_i$, $i = 1, \dots, N$, as the tensor-product B-splines $N_{\boldsymbol{i}+\mathbf{1},[\boldsymbol{p}]}$, $\boldsymbol{i} = \mathbf{1}, \dots, \boldsymbol{n} + \boldsymbol{p} - \mathbf{2}$. The resulting stiffness and mass matrices (7.203) and (7.204) are given by

$$K_{\mathbf{G},\boldsymbol{n}}^{[\boldsymbol{p}]} = \left[\int_{[0,1]^d} |\det(J_\mathbf{G})|(\nabla N_{\boldsymbol{i}+\mathbf{1},[\boldsymbol{p}]})^T A_\mathbf{G} \nabla N_{\boldsymbol{j}+\mathbf{1},[\boldsymbol{p}]}\right]_{\boldsymbol{i},\boldsymbol{j}=\mathbf{1}}^{\boldsymbol{n}+\boldsymbol{p}-\mathbf{2}},$$

$$M_{\mathbf{G},\boldsymbol{n}}^{[\boldsymbol{p}]} = \left[\int_{[0,1]^d} c(\mathbf{G})|\det(J_\mathbf{G})| N_{\boldsymbol{j}+\mathbf{1},[\boldsymbol{p}]} N_{\boldsymbol{i}+\mathbf{1},[\boldsymbol{p}]}\right]_{\boldsymbol{i},\boldsymbol{j}=\mathbf{1}}^{\boldsymbol{n}+\boldsymbol{p}-\mathbf{2}},$$

and it is immediately seen that they are the same as the reaction and diffusion matrices in (7.190) and (7.192). The numerical eigenvalues are simply the eigenvalues of the matrix

$$L_{\mathbf{G},\boldsymbol{n}}^{[\boldsymbol{p}]} = (M_{\mathbf{G},\boldsymbol{n}}^{[\boldsymbol{p}]})^{-1} K_{\mathbf{G},\boldsymbol{n}}^{[\boldsymbol{p}]}.$$

Theorem 7.9 *Let Ω be a bounded open domain in $\mathbb{R}^d$ with Lipschitz boundary and suppose that the following conditions on the PDE coefficients and the geometry map are satisfied:*

- $a_{\ell k} \in L^1(\Omega)$ *for all* $\ell, k = 1, \dots, d$;
- $c \in L^1(\Omega)$ *and* $c > 0$ *a.e. in* Ω;
- $A(\mathbf{x}) = [a_{\ell k}(\mathbf{x})]_{\ell,k=1}^d$ *is SPD for a.e.* $\mathbf{x} \in \Omega$;
- $|\det(J_\mathbf{G})| > 0$ *a.e. in* $[0,1]^d$ *and* $(|\det(J_\mathbf{G})| A_\mathbf{G})_{\ell k} \in L^1([0,1]^d)$ *for all* $\ell, k = 1, \dots, d$, *where* $A_\mathbf{G}$ *is defined in* (7.205).

Let $\boldsymbol{p} \geq \mathbf{1}$, let $\boldsymbol{\nu} \in \mathbb{Q}^d$ be a vector with positive components, and assume that $\boldsymbol{n} = \boldsymbol{\nu} n$ (it is understood that n varies in the infinite subset of $\mathbb{N}$ such that $\boldsymbol{n} = \boldsymbol{\nu} n \in \mathbb{N}^d$). Then

$$\{n^{-2} L_{\mathbf{G},\boldsymbol{n}}^{[\boldsymbol{p}]}\}_n \sim_{\mathrm{GLT}} e_{\mathbf{G},\boldsymbol{p}}^{(\boldsymbol{\nu})}(\hat{\mathbf{x}}, \boldsymbol{\theta}) \tag{7.206}$$

and

$$\{n^{-2} L_{\mathbf{G},\boldsymbol{n}}^{[\boldsymbol{p}]}\}_n \sim_{\sigma,\lambda} e_{\mathbf{G},\boldsymbol{p}}^{(\boldsymbol{\nu})}(\hat{\mathbf{x}}, \boldsymbol{\theta}), \tag{7.207}$$

where

$$e_{\mathbf{G},\boldsymbol{p}}^{(\boldsymbol{\nu})}(\hat{\mathbf{x}}, \boldsymbol{\theta}) = \frac{\boldsymbol{\nu}(A_{\mathbf{G}}(\hat{\mathbf{x}}) \circ H_{\boldsymbol{p}}(\boldsymbol{\theta}))\boldsymbol{\nu}^T}{c(\mathbf{G}(\hat{\mathbf{x}}))h_{\boldsymbol{p}}(\boldsymbol{\theta})} \tag{7.208}$$

and $H_{\boldsymbol{p}}(\boldsymbol{\theta})$, $h_{\boldsymbol{p}}(\boldsymbol{\theta})$ are defined in (7.174), (7.195), *respectively.*

Proof The components of $|\det(J_{\mathbf{G}})|A_{\mathbf{G}}$ belong to $L^1([0,1]^d)$ by assumption, and also the function $c(\mathbf{G})|\det(J_{\mathbf{G}})|$ belongs to $L^1([0,1]^d)$ because $c \in L^1(\Omega)$ by assumption and

$$\int_{[0,1]^d} c(\mathbf{G})|\det(J_{\mathbf{G}})| = \int_{\Omega} c.$$

Hence, by Remark 7.14,

$$\{n^{d-2} K_{\mathbf{G},\boldsymbol{n}}^{[\boldsymbol{p}]}\}_n \sim_{\mathrm{GLT}} \frac{\boldsymbol{\nu}(|\det(J_{\mathbf{G}}(\hat{\mathbf{x}}))|A_{\mathbf{G}}(\hat{\mathbf{x}}) \circ H_{\boldsymbol{p}}(\boldsymbol{\theta}))\boldsymbol{\nu}^T}{N(\boldsymbol{\nu})},$$
$$\{n^{d} M_{\mathbf{G},\boldsymbol{n}}^{[\boldsymbol{p}]}\}_n \sim_{\mathrm{GLT}} \frac{|\det(J_{\mathbf{G}}(\hat{\mathbf{x}}))|c(\mathbf{G}(\hat{\mathbf{x}}))h_{\boldsymbol{p}}(\boldsymbol{\theta})}{N(\boldsymbol{\nu})},$$

and the relations (7.206)–(7.207) follow from Theorem 7.1, taking into account that $K_{\mathbf{G},\boldsymbol{n}}^{[\boldsymbol{p}]}$ and $M_{\mathbf{G},\boldsymbol{n}}^{[\boldsymbol{p}]}$ are SPD, that $|\det(J_{\mathbf{G}})|$, $c > 0$ a.e. by assumption, and that $h_{\boldsymbol{p}}(\boldsymbol{\theta}) = \prod_{r=1}^d h_{p_r}(\theta_r) > 0$ for all $\boldsymbol{\theta} \in [-\pi, \pi]^d$ because $h_p(\theta) \geq (4/\pi^2)^{p+1}$ for all $p \geq 1$ and $\theta \in [-\pi, \pi]$; see [22, Eq. (10.165) and Remark 10.13]. □

Chapter 8
Future Developments

In the present Volume II, we have developed the theory of *multilevel (or multivariate)* GLT sequences, which, as illustrated in the applications of Chap. 7, allows the computation of the asymptotic singular value and eigenvalue distribution of matrices arising from the discretization of PDEs by virtually any approximation technique. In this final chapter, we list a series of possible future developments.

1. Develop the theories of multilevel *block* and *reduced* GLT sequences, as explained in items 1 and 2 of [22, Chap. 11]. We highlight that, for the unilevel block case, much work has already been carried out in three very recent papers [19, 25, 26]. In particular, papers [25, 26] systematically developed the theory of unilevel block GLT sequences, while paper [19] presented some of its most emblematic applications. It is worth noting that papers [25, 26] also suggested a new interesting algebraic-topological definition of GLT sequences, which allows a considerable simplification of the theory; it will certainly appear in the second editions of both Volumes I and II.
2. Try to design an automatic procedure for computing the symbol of a sequence of PDE discretization matrices, assuming to know that it is a (multilevel) GLT sequence. This objective was already proposed in item 3 of [22, Chap. 11] and it was partially pursued by Ahmed Ratnani at the Max Planck Institute for Plasma Physics (Munich, Germany). Ratnani is also the author of a so-called "GLT library".
3. Develop a new edition of both Volumes I and II. GLT sequences are an expanding research field, both from the theoretical and applicative point of view. It will then be necessary to develop a new edition of Volumes I and II in order to include the new important achievements in this young research area. Actually, in very recent years, also due to the publication of Volume I, the interest in the theory of GLT sequences has grown considerably, thus leading to an impressive number of new recent findings. We here mention some of the most important ones, in addition to those already mentioned in item 4 of [22, Chap. 11].

C. Garoni and S. Serra-Capizzano, *Generalized Locally Toeplitz Sequences: Theory and Applications*, https://doi.org/10.1007/978-3-030-02233-4_8

3.1 Giovanni Barbarino from Scuola Normale Superiore (Pisa, Italy) proved in [4] the following interesting result for unilevel GLT sequences, whose extension to multilevel GLT sequences is supposed to be straightforward.

Suppose $\{A_n\}_n \sim_{\rm GLT} \kappa$, *where* $\{A_n\}_n \in \mathscr{E} = \{\{B_n\}_n : \ B_n \in \mathbb{C}^{n\times n}\}$ *and* $\kappa \in \mathfrak{M}_1 = \{\xi : [0,1]\times[-\pi,\pi] \to \mathbb{C} : \ \xi$ *is measurable*$\}$. *Then, there exists a sequence of unitary matrices* $\{Q_n\}_n \in \mathscr{E}$ *which is independent of* $\{A_n\}_n$ *and satisfies*

$$A_n = Q_n D_n Q_n^* + Z_n,$$

where $\{Z_n\}_n \in \mathscr{E}$ *is a zero-distributed sequence and* $\{D_n\}_n \in \mathscr{E}$ *is a sequence of diagonal matrices such that* $\{D_n\}_n \sim_\lambda \kappa$.

This means in particular that, up to a small perturbation Z_n, the nth matrix A_n of a GLT sequence is close to a normal matrix $\tilde{A}_n$ whose eigenvectors are given by the columns of a fixed $n \times n$ unitary matrix Q_n; and, moreover, if $\{A_n\}_n \sim_{\rm GLT} \kappa$ then $\{\tilde{A}_n\}_n \sim_{\rm GLT} \kappa$ and $\{\tilde{A}_n\}_n \sim_\lambda \kappa$. An analogous result was somehow anticipated in [36, Remark 0.1] with the construction of the class of *Generalized Locally Circulant (GLC) sequences*, in which the role of Toeplitz matrices is played by circulant matrices and the normal matrices Q_n admit an explicit expression.

3.2 Giovanni Barbarino proved in [6] the following remarkable result about sequences of perturbed Hermitian matrices.

Let $\{X_n\}_n$, $\{Y_n\}_n$ *be sequences of matrices, with* X_n, Y_n *of size* d_n, *and set* $A_n = X_n + Y_n$. *Assume that the following conditions are met.*

1. *Every* X_n *is Hermitian and* $\{X_n\}_n \sim_\lambda f$.
2. $\|Y_n\|_2 = o(\sqrt{d_n})$.

Then $\{A_n\}_n \sim_\lambda f$.

This is a noteworthy extension of Corollary 2.3, because if $\|Y_n\| \le C$ for some constant C independent of n and $\|Y_n\|_1 = o(d_n)$, as in the hypotheses of Corollary 2.3, then

$$\|Y_n\|_2 = \sqrt{\sum_{i=1}^{d_n} \sigma_i(Y_n)^2} \le \sqrt{\sigma_{\max}(Y_n)\sum_{i=1}^{d_n} \sigma_i(Y_n)} = \sqrt{\|Y_n\|\,\|Y_n\|_1} = o(\sqrt{d_n}).$$

This extension of Corollary 2.3 increases the potential of the theory of GLT sequences as it allows us to replace property **GLT 2** with the following more powerful version:

GLT 2. If $\{A_{\boldsymbol n}\}_n \sim_{\rm GLT} \kappa$ and $A_{\boldsymbol n} = X_{\boldsymbol n} + Y_{\boldsymbol n}$, where
- every $X_{\boldsymbol n}$ is Hermitian,
- $N(\boldsymbol n)^{-1/2}\|Y_{\boldsymbol n}\|_2 \to 0$,

then $\{A_{\boldsymbol n}\}_n \sim_\lambda \kappa$.

Such an improved **GLT 2** property is the key for computing the asymptotic spectral distribution of matrices arising from the discretization of PDEs with unbounded coefficients, as illustrated in [6]. It was also conjectured in [6], on the basis of numerical experiments, that the assumption "$\|Y_n\|_2 = o(\sqrt{d_n})$" in the above extension of Corollary 2.3 can be replaced by the weaker assumption "$\|Y_n\|_1 = o(d_n)$", thus yielding an even more powerful version of **GLT 2**. Yet, this conjecture is still unsolved and solving it is certainly an interesting topic for future research.

3.3 A "GLT program" for future research has been outlined in [23, Remark 15]. Roughly speaking, it suggests the idea that from the symbol of the GLT sequence arising from the discretization of a continuous eigenvalue problem one can obtain the distribution of the continuous eigenvalues through an appropriate limit process "from the discrete to the continuous". Although this may seem like science fiction, the idea deserves a serious consideration, because Davide Bianchi from University of Insubria (Como, Italy) was able to select a class of continuous eigenvalue problems for which the above GLT program applies(!)

3.4 On the applicative side, it was shown in [14] that the theory of GLT sequences finds applications also in the context of *fractional differential equations*. Considering the importance of this topic nowadays, the GLT analysis of at least one model fractional differential equation should be included in a future edition of Volumes I and/or II.

References

1. AURICCHIO F., BEIRÃO DA VEIGA L., HUGHES T. J. R., REALI A., SANGALLI G. *Isogeometric collocation methods*. Math. Models Methods Appl. Sci. 20 (2010) 2075–2107.
2. AVRAM F. *On bilinear forms in Gaussian random variables and Toeplitz matrices*. Probab. Theory Related Fields 79 (1988) 37–45.
3. BARBARINO G. *Equivalence between GLT sequences and measurable functions*. Linear Algebra Appl. 529 (2017) 397–412.
4. BARBARINO G. *Normal form for GLT sequences*. Full text available at: https://arxiv.org/abs/1805.08708v2.
5. BARBARINO G., GARONI C. *From convergence in measure to convergence of matrix-sequences through concave functions and singular values*. Electron. J. Linear Algebra 32 (2017) 500–513.
6. BARBARINO G., SERRA-CAPIZZANO S. *Non-Hermitian perturbations of Hermitian matrix-sequences and applications to the spectral analysis of approximated PDEs*. Technical Report 2018-004 (2018), Department of Information Technology, Uppsala University. Full text available at: http://www.it.uu.se/research/publications/reports/2018-004.
7. BÖTTCHER A., GUTIÉRREZ-GUTIÉRREZ J., CRESPO P. M. *Mass concentration in quasicommutators of Toeplitz matrices*. J. Comput. Appl. Math. 205 (2007) 129–148.
8. BÖTTCHER A., SILBERMANN B. *Introduction to Large Truncated Toeplitz Matrices*. Springer-Verlag, New York (1999).
9. BÖTTCHER A., SILBERMANN B. *Analysis of Toeplitz Operators*. Second Edition, Springer-Verlag, Berlin (2006).
10. BREZIS H. *Functional Analysis, Sobolev Spaces and Partial Differential Equations*. Springer, New York (2011).
11. CARAMELLO O. *Theories, Sites, Toposes: Relating and Studying Mathematical Theories through Topos-Theoretic 'Bridges'*. Oxford University Press, Oxford (2017).
12. COTTRELL J. A., HUGHES T. J. R., BAZILEVS Y. *Isogeometric Analysis: Toward Integration of CAD and FEA*. John Wiley & Sons, Chichester (2009).
13. DONATELLI M., GARONI C., MAZZA M., SERRA- CAPIZZANO S., SESANA D. *Preconditioned HSS method for large multilevel block Toeplitz linear systems via the notion of matrix-valued symbol*. Numer. Linear Algebra Appl. 23 (2016) 83–119.
14. DONATELLI M., MAZZA M., SERRA-CAPIZZANO S. *Spectral analysis and structure preserving preconditioners for fractional diffusion equations*. J. Comput. Phys. 307 (2016) 262–279.
15. DONATELLI M., NEYTCHEVA M., SERRA-CAPIZZANO S. *Canonical eigenvalue distribution of multilevel block Toeplitz sequences with non-Hermitian symbols*. Oper. Theory Adv. Appl. 221 (2012) 269–291.

C. Garoni and S. Serra-Capizzano, *Generalized Locally Toeplitz Sequences: Theory and Applications*, https://doi.org/10.1007/978-3-030-02233-4

16. GARONI C. *Topological foundations of an asymptotic approximation theory for sequences of matrices with increasing size.* Linear Algebra Appl. 513 (2017) 324–341.
17. GARONI C., MANNI C., SERRA-CAPIZZANO S., SESANA D., SPELEERS H. *Lusin theorem, GLT sequences and matrix computations: an application to the spectral analysis of PDE discretization matrices.* J. Math. Anal. Appl. 446 (2017) 365–382.
18. GARONI C., MANNI C., SERRA-CAPIZZANO S., SPELEERS H. *NURBS versus B-splines in isogeometric discretization methods: a spectral analysis.* Submitted.
19. GARONI C., MAZZA M., SERRA-CAPIZZANO S. *Block generalized locally Toeplitz sequences: from the theory to the applications.* Axioms 7 (2018) 49.
20. GARONI C., SERRA-CAPIZZANO S. *The theory of locally Toeplitz sequences: a review, an extension, and a few representative applications.* Bol. Soc. Mat. Mex. 22 (2016) 529–565.
21. GARONI C., SERRA-CAPIZZANO S. *The theory of generalized locally Toeplitz sequences: a review, an extension, and a few representative applications.* Oper. Theory Adv. Appl. 259 (2017) 353–394.
22. GARONI C., SERRA-CAPIZZANO S. *Generalized Locally Toeplitz Sequences: Theory and Applications* (*Volume I*). Springer, Cham (2017).
23. GARONI C., SERRA-CAPIZZANO S. *Generalized locally Toeplitz sequences: a spectral analysis tool for discretized differential equations.* Lecture Notes in Mathematics 2219 (2018) 161–236.
24. GARONI C., SERRA-CAPIZZANO S., SESANA D. *Spectral analysis and spectral symbol of d-variate $\mathbb{Q}_p$ Lagrangian FEM stiffness matrices.* SIAM J. Matrix Anal. Appl. 36 (2015) 1100–1128.
25. GARONI C., SERRA-CAPIZZANO S., SESANA D. *Block locally Toeplitz sequences: construction and properties.* Springer INdAM Series, Proceedings Volume of the INdAM Meeting "Structured Matrices in Numerical Linear Algebra: Analysis, Algorithms and Applications", Cortona (Arezzo), Italy, 04–08/09/2017 (in press).
26. GARONI C., SERRA-CAPIZZANO S., SESANA D. *Block generalized locally Toeplitz sequences: topological construction, spectral distribution results, and star-algebra structure.* Springer INdAM Series, Proceedings Volume of the INdAM Meeting "Structured Matrices in Numerical Linear Algebra: Analysis, Algorithms and Applications", Cortona (Arezzo), Italy, 04–08/09/2017 (in press).
27. GRENANDER U., SZEGŐ G. *Toeplitz Forms and Their Applications.* Second Edition, AMS Chelsea Publishing, New York (1984).
28. KATZNELSON Y. *An Introduction to Harmonic Analysis.* Third Edition, Cambridge University Press, Cambridge (2004).
29. PARTER S. V. *On the distribution of the singular values of Toeplitz matrices.* Linear Algebra Appl. 80 (1986) 115–130.
30. ROYDEN H. L., FITZPATRICK P. M. *Real Analysis.* Fourth Edition, Pearson Education Asia Limited and China Machine Press (2010).
31. RUDIN W. *Principles of Mathematical Analysis.* Third Edition, McGraw-Hill, New York (1976).
32. RUDIN W. *Real and Complex Analysis.* Third Edition, McGraw-Hill, Singapore (1987).
33. SERRA-CAPIZZANO S. *Distribution results on the algebra generated by Toeplitz sequences: a finite dimensional approach.* Linear Algebra Appl. 328 (2001) 121–130.
34. SERRA-CAPIZZANO S. *More inequalities and asymptotics for matrix valued linear positive operators: the noncommutative case.* Oper. Theory Adv. Appl. 135 (2002) 293–315.
35. SERRA-CAPIZZANO S. *Generalized locally Toeplitz sequences: spectral analysis and applications to discretized partial differential equations.* Linear Algebra Appl. 366 (2003) 371–402.
36. SERRA-CAPIZZANO S. *The GLT class as a generalized Fourier analysis and applications.* Linear Algebra Appl. 419 (2006) 180–233.
37. SERRA-CAPIZZANO S., TILLI P. *On unitarily invariant norms of matrix-valued linear positive operators.* J. Inequal. Appl. 7 (2002) 309–330.
38. TILLI P. *A note on the spectral distribution of Toeplitz matrices.* Linear and Multilinear Algebra 45 (1998) 147–159.

39. TILLI P. *Locally Toeplitz sequences: spectral properties and applications.* Linear Algebra Appl. 278 (1998) 91–120.
40. TILLI P. *Some results on complex Toeplitz eigenvalues.* J. Math. Anal. Appl. 239 (1999) 390–401.
41. TYRTYSHNIKOV E. E. *A unifying approach to some old and new theorems on distribution and clustering.* Linear Algebra Appl. 232 (1996) 1–43.
42. TYRTYSHNIKOV E. E., ZAMARASHKIN N. L. *Spectra of multilevel Toeplitz matrices: advanced theory via simple matrix relationships.* Linear Algebra Appl. 270 (1998) 15–27.
43. ZAMARASHKIN N. L., TYRTYSHNIKOV E. E. *Distribution of eigenvalues and singular values of Toeplitz matrices under weakened conditions on the generating function.* Sb. Math. 188 (1997) 1191–1201.

Index

Symbols

C. Garoni and S. Serra-Capizzano, *Generalized Locally Toeplitz Sequences: Theory and Applications*, https://doi.org/10.1007/978-3-030-02233-4

N

O

P